MW01502902

# Trace Elements in Coal and Coal Combustion Residues

*Edited by*
**Robert F. Keefer**
**Kenneth S. Sajwan**

**LEWIS PUBLISHERS**
Boca Raton    Ann Arbor    London    Tokyo

**Library of Congress Cataloging-in-Publication Data**

Trace elements in coal and coal combustion residues / edited by Robert F. Keefer and Kenneth S. Sajwan.
    p.  cm. — (Advances in trace substances research)
    Includes bibliographical references and index.
    ISBN 0-87371-890-9
    1. Coal—Analysis. 2. Coal ash—Analysis. 3. Trace elements—Analysis. I. Keefer, Robert F. II. Sajwan, Kenneth S. III. Series.
TP325.T73   1993
662.6′22—dc20                                          93-17908
                                                          CIP

PRINTED IN THE UNITED STATES OF AMERICA
1 2 3 4 5 6 7 8 9 0
Printed on acid-free paper

# ADVANCES IN TRACE SUBSTANCES RESEARCH

## Series Preface

The need to synthesize, critically analyze, and put into perspective the ever-mounting body of information on trace chemicals in the environment provided the impetus for the creation of this series. In addition to examining the fate, behavior and transport of these substances, the transfer into the food chain and risk assessment to the consumers, including humans, will also be taken into account. It is hoped then that this information will be user-friendly to students, researchers, regulators, and administrators.

The series will have "topical" volumes to address more specific issues as well as volumes with heterogeneous topics for a quicker dissemination. It will have international scope and will cover issues involving natural and anthropogenic sources in both the aquatic and terrestrial ecosystems. To ensure a high quality publication, volume editors and the editorial board will subject each article to peer review.

Thus, **Advances in Trace Substances Research** should provide a forum where experts can discuss contemporary environmental issues dealing with trace chemicals; hopefully, this forum can lead to solutions resulting in a cleaner and healthier environment.

**Domy C. Adriano**
**Editor-in-Chief**

# Preface and Acknowledgments

Accumulation of tremendous amounts of ash from fossil fuel combustion for electrical power generation is becoming a major environmental concern in the United States. Furthermore, stringent environmental requirements — including the Clean Air Act, Clean Water Act, RCRA (Resource Conservation and Recovery Act), and state and local environmental regulations — result in even more waste production with subsequent contact with the environment. The concentrations of trace metals in coal residues are extremely variable and depend on the composition of the original coal, conditions during combustion, efficiency of emission control devices, storage and handling of by-products, and climate.

The papers in this book were presented as part of the International Conference on Metals in Soils, Waters, Plants, and Animals held at Orlando, FL in April-May 1990. The purpose of the conference was to present current knowledge on the sources, pathways, behavior, fate, and effects of metals and metalloids in soils, waters, plants, and animals. The topics covered a wide range of metal-metalloid relationships and the major portion has been published.* The research papers presented herein have been subjected to peer review. The editors have arranged the articles systematically by topic, beginning with the keynote address by Dr. Keefer entitled "Coal Ashes — Industrial Wastes or Beneficial By-Products?", followed by "An Overview of Electric Power Research Institute (EPRI) Research Related to Effective Management of Coal Combustion Residues" and sections on Environmental Effects from Power Plants, Tests for and Monitoring of Fossil Fuel Dispersion and Ash Disposal, Transport and Leachability of Metals from Coal and Ash Piles, and Use of Coal Ash for Growth of Higher Plants.

This book is intended for a variety of readers, including public health and environmental professionals, industrial hygienists, consultants, and academics. It may prove valuable to scientists conducting research on coal and coal combustion by-products.

The editors wish to thank the authors for their diligence in providing the changes requested by the reviewers and for their patience in waiting so long for these to go into print.

**Robert F. Keefer**
**Kenneth S. Sajwan**

*Adriano, D. C. "Metals in Soils, Waters, Plants and Animals," *Water, Air, and Soil Pollution* 57-58: 1-930 (1991).

# The Editors

**Robert F. Keefer** is a Professor in the Division of Plant and Soil Sciences, College of Agriculture and Forestry, at West Virginia University. After receiving his B.S. in general agriculture from Cornell University and his M.S. and Ph.D. in soil science (agronomy) from Ohio State University, he worked two years as an organic chemist with Hercules Powder Company, and began his academic and research career in 1965 with West Virginia University. He has taught Soil Fertility, Soil Conservation and Management, Advanced Soil Fertility, the Chemistry of Soil Organic Matter, and part of a team-taught course on Plant Disorders. He devoted a year to teaching, conducting research, and developing a graduate program in soil science at Makerere University in Uganda and assisted one of his graduate students in Togo (West Africa).

Dr. Keefer's research has been broad — beginning with soil fertility and soil test correlation, then branching into plant nutrition/soil chemistry relationship, especially dealing with micronutrients. Interest over the years shifted to use of agricultural manures, municipal wastes such as sewage sludge, and industrial by-products such as coal ashes and sawdust, particularly with respect to environmental aspects of plant nutrition, toxicities, and heavy metal transport in soils, waters, plants, and animals.

The geographical positioning of West Virginia in the center of the eastern U.S. coal field and the recent concern with maintaining or improving the quality of our environment led Dr. Keefer to develop a research program emphasizing constructive use of coal combustion by-products in reclaiming surface-mined land. His proficiency in this area is evidenced by the many calls he receives on this topic. Dr. Keefer's recent sabbatical was devoted mainly to producing a digest of research on the chemical composition and leaching characteristics of coal combustion by-products by the Electric Power Research Institute (EPRI) that will be published soon. He was a major contributor to a book section on arsenic mobilization and bioavailability in soils forthcoming in *Advances in Environmental Science and Technology,* entitled "Arsenic in the Environment."

**Kenneth S. Sajwan** is an Associate Professor of Biology and the Coordinator of Environmental Studies in the School of Sciences and Technology at Savannah State College. He received his Ph.D. in Soil Science (Agronomy) from Colorado State University. In addition, he holds a Ph.D. (Post Harvest Technology), M.S. (Agronomy), and B.S. (Agriculture & Animal Husbandry) from India.

Prior to joining Savannah State College, Dr. Sajwan worked as an Assistant Research Ecologist in the Biogeochemical Ecology Division of the University of Georgia's Savannah River Ecology Laboratory. Earlier he worked as an ARS Scientist-Agronomy for the Indian Council of Agricultural Research; an Associate Professor-Water Use Management at the University of Roorkee, India; a World Bank Agricultural Consultant to Colombia, South America; and a Postdoctoral Fellow-Soil Chemistry at the University of Kentucky. Dr. Sajwan is also very active in teaching and has taught at the Indian Institute of Technology, University of Roorkee, Colorado State University, and the University of South Carolina.

Dr. Sajwan's primary research interests include biogeochemistry of trace metals, soil-plant environmental chemistry, and groundwater quality and chemical equilibria in soils. Currently, Dr. Sajwan is investigating the potential benefits and environmental impact of applications of coal ash and organic waste mixtures to agricultural lands for crop production.

IN GRATITUDE TO


*Domy C. Adriano*
Series Editor and Guest Editor of Metals in Soils, Waters, Plants, and Animals
Savannah River Ecology Laboratory, University of Georgia

for his major contribution in organizing the International Conference on Metals in Soils, Waters, Plants, and Animals, of which the present work is a part; and for the initial and final editing of the present work

*Willard L. Lindsay*
Centennial Professor, Colorado State University

for introducing us and strengthening our background in soil chemistry and plant nutrition

and

*Michael H. Smith*
Director, Savannah River Ecology Laboratory

for inspiring us to enhance environmental quality and conservation

# Contributors

*M. Agrawal*, Department of Botany, Banaras Hindu University, Varanasi, UP 221005, India.

*M. A. Anderson*, Department of Soil and Environmental Sciences, University of California, Riverside, CA 92521.

*M. A. Arthur*, Boyce Thompson Institute for Plant Research, Cornell University, Ithaca, NY 14853.

*D. E. Baker*, Environmental Resources Research Institute, 126 Land & Water Research Bldg., Pennsylvania State University, University Park, PA 16802.

*A. O. Beers*, New York State Electric and Gas Corp., Binghamton, NY 13903

*P. M. Bertsch*, Biogeochemical Ecology Division, Savannah River Ecology Laboratory, Aiken, SC 29801.

*D. K. Bhumbla*, Division of Plant and Soil Sciences, Box 6108, West Virginia University, Morgantown, WV 26506.

*K. Chandra*, School of Sciences and Technology, Savannah State College, Savannah, GA 31404.

*J. P. Coetzee*, Environmental Resources Research Institute, 126 Land & Water Research Bldg., Pennsylvania State University, University Park, PA 16802. (Visiting Scientist, University of Potchefstroom, Republic of South Africa.)

*T. E. Davis*, Department of Geological Sciences, California State University, Los Angeles, CA 90032.

*A. A. Elseewi*, Southern California Edison Co., P.O. Box 800, Rosemead, CA 91770.

*G. S. Ghuman*, School of Sciences and Technology, Savannah State College, Savannah, GA 31404.

*M. Heit*, Environmental Measurements Laboratory, U.S. Department of Energy, 376 Hudson St., New York, NY 10014.

*S. Hodgkiss*, Environmental Measurements Laboratory, U.S. Department of Energy, 376 Hudson St., New York, NY 10014.

*R. W. Hurst*, Chempet Research Corp., 330 N. Zachary Ave., Ste. 107, Moorpark, CA 93021.

*J. James*, School of Sciences and Technology, Savannah State College, Savannah GA 31404.

*A. K. Jha*, Department of Botany, Banaras Hindu University, Varanasi, UP 221005, India.

*R. F. Keefer*, Division of Plant and Soil Sciences, Box 6108, West Virginia University, Morgantown, WV 26506.

*C. S. Klusek*, Environmental Measurements Laboratory, U.S. Department of Energy, 376 Hudson St., New York, NY 10014.

*J. A. Laurence*, Boyce Thompson Institute for Plant Research, Cornell University, Ithaca, NY 14853.

*S. V. Mattigod*, Geosciences Department, Battelle Pacific Northwest Laboratories, Richland, WA 99352.

*M. P. Menon*, School of Sciences and Technology, Savannah State College, Savannah, GA 31404.

*G. L. Mills*, Biogeochemical Ecology Division, Savannah River Ecology Laboratory, Aiken, SC 29801.

*I. P. Murarka*, Environmental Assessment Division, Electric Power Research Institute, P.O. Box 10412, Palo Alto, CA 94303.

*A. L. Page*, Department of Soil and Environmental Sciences, University of California, Riverside, CA 92521.

*F. G. Pannebaker*, Environmental Resources Research Institute, 126 Land & Water Research Bldg., Pennsylvania State University, University Park, PA 16802.

*G. Rubin*, Biometrics Unit, Cornell University, Ithaca, NY 14853.

*K. S. Sajwan*, School of Sciences and Technology, Savannah State College, Savannah, GA 31404.

*S. S. Sandhu*, Division of Natural Sciences and Mathematics, Claflin College, Orangeburg, SC 29115.

*R. S. Schneider*, Boyce Thompson Institute for Plant Research, Cornell University, Ithaca, NY 14843.

*A. P. Schwab*, Department of Agronomy, Kansas State University, Manhattan, KS 66506.

*J. P. Senft*, Environmental Resources Research Institute, 126 Land & Water Research Bldg., Pennsylvania State University, University Park, PA 16802.

*J. Singh*, Department of Botany, Banaras Hindu University, Varanasi, UP 221005, India.

*J. S. Singh*, Department of Botany, Banaras Hindu University, Varanasi, UP 221005, India.

*R. N. Singh*, Division of Plant and Soil Sciences, Box 6108, West Virginia University, Morgantown, WV 26506.

*L. H. Weinstein*, Boyce Thompson Institute for Plant Research, Cornell University, Ithaca, NY 14853.

*P. B. Woodbury*, Boyce Thompson Institute for Plant Research, Cornell University, Ithaca, NY 14853.

*L. W. Zelazny*, Department of Crop and Soil Environmental Sciences, Virginia Polytechnic Institute and State University, Blacksburg, VA 24061.

# Contents

## INTRODUCTION

## ENVIRONMENTAL EFFECTS FROM POWER PLANTS

## TESTS FOR AND MONITORING OF FOSSIL FUEL
## DISPERSION AND ASH DISPOSAL

## USE OF COAL ASH FOR PLANT GROWTH

# Trace Elements in Coal and Coal Combustion Residues

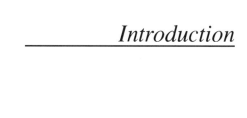

*Introduction*

# 1

# Coal Ashes — Industrial Wastes or Beneficial By-Products?

R. F. Keefer

West Virginia University, Morgantown, WV

## ABSTRACT

Coal ashes can be defined as chemical products similar to calcined or fired clay. The three types of ash produced are fly ash, bottom ash, and boiler slag; the amount produced depends on boiler type, characteristics of the coal burned, and/or manner of handling the ash. The abundant amounts produced and the versatility of coal ash make it an ideal substance to take the place of some natural resources; furthermore, its use is supported by the federal government. Coal ashes have many potential uses which include backfill, cement and concrete manufacture, adhesives, wallboard, and soil amelioration/reclamation. When fly ash is applied to soils, concentrations of some of the many elements present in the ash increase. This increase in enrichment ratios requires careful monitoring. Only B becomes toxic to plants, but As, Mo, Se, and others may be potentially toxic to animals consuming the plants grown on the ash-amended soil. Limited research has not shown any detrimental effect on water quality from ash applied to soil. Use of fly ash as a soil amendment has some disadvantages (which may be overcome using proper management), but offers many advantages such as increased plant yield through improved physical and chemical conditions. As a consequence, research needs to be directed not only toward pollution aspects but also toward beneficial uses of ash.

## INTRODUCTION

Coal, our abundant natural resource, is being used extensively by power-generating plants to produce electricity. In the process of burning coal, three general types of ash are produced — fly ash, bottom ash, and boiler slag. The proportions of each produced depend on the type of boiler used, the characteristics of the coal burned, and the manner of handling the ash. Chemically, coal ash is similar to a calcined or fired clay.

0-87371-890-9/93/$0.00+$.50
© 1993 by Lewis Publishers

## RESULTS AND DISCUSSION

Coal ash is used because it is abundant, economical, and versatile; also, it preserves natural resources, lowers utility rates, and is supported by U. S. governmental agencies.[1] There are many uses, depending on the type of ash considered.

### Bottom Ash and Boiler Slag Applications

Bottom ash is defined as a granular material similar to concrete sand; boiler slag is defined as a shiny black granular material that is very hard and abrasive. Uses for these two materials include snow and ice control, structural embankments, backfill drainage, raw material for cement manufacture, a partial replacement of natural aggregates in concrete blocks, a stabilized road base for road construction, flexible asphalt pavement, roofing shingles, blasting abrasives, golf course bunkers, amber glass manufacture, and a resistant filler in fiberglass pipes (plastics).[1]

### Fly Ash Applications

Fly ash is a powdery material consisting of tiny separate glass cenospheres which are chemically oxides of silicon, aluminum, iron, and calcium. Although this material has many uses, little is actually used; therefore, disposal becomes a problem unless or until it is used more effectively. These uses include cement replacement in concrete, engineered structural fills; paints and coatings by replacing talc and $TiO_2$ as a pigment, and a cheap filler for plastics in which it can act as ball bearings to lubricate the mix when extruding.[1] Additionally, fly ash can be used as a grout in sealing off air to stop mine fires, and in preventing mine subsidence by filling mine rooms; as raw material in lightweight aggregate for concrete or concrete products; as insulation in panels; as gypsum wallboard; as adhesives, as sanitary landfill cover, and in sludge fixation and stabilization.[1] Researchers[2-8] have found that fly ash can also be used in soil amelioration/reclamation. When fly ash is mixed with very acid soils, especially for reclaiming surface mine land, it improves certain soil characteristics, e.g., available water-holding ability.

### Elements Found in Fly Ash, Soils, Plants, and Animals

A total of 25 potentially toxic elements have been found in measurable amounts in fly ash[9] (Table 1). When fly ash is applied to soils, there are four elements — Be, Cd, F, and Ni — that are not found at elevated levels in ash-amended soils.[10] On the other hand, eight elements — Ba, B, Ca, Mo, Pb, S, Se, and Sr — generally increase in concentration in the soil to which fly ash is applied.[11] Six of these elements, i.e., Ba, B, Mo, S, Se, and Sr, along with Al and As, have been shown to increase in concentration in plants grown on

**Table 1**
**Elements Found in Fly Ash, Soils, Plants, and Animals**

| In Fly Ash | Not Found in Soils with Fly Ash Application | Increased in Soils with Fly Ash Application | Increased in Plants with Fly Ash Application | Toxic to Plants | Potentially Toxic to Animals |
|---|---|---|---|---|---|
| Ag | | | | | |
| Al | | | Al | | |
| As | | | As | | *As* |
| Ba | | Ba | Ba | | |
| Be | Be | | | | |
| Bi | | | | | |
| B | | B | B | *B* | |
| Ca | | Ca | | | |
| Cd | Cd | | | | Cd |
| Co | | | | | |
| Cr | | | | | |
| Cu | | | | | Cu |
| F | F | | | | F |
| Mn | | | | | |
| Mo | | Mo | Mo | | *Mo* |
| Ni | Ni | | | | |
| Pb | | Pb | | | Pb |
| S | | S | S | | |
| Sb | | | | | |
| Se | | Se | Se | | *Se* |
| Sr | | Sr | Sr | | |
| Ti | | | | | |
| Tl | | | | | |
| V | | | | | |
| W | | | | | |
| Zn | | | | | |

fly ash-amended soils.[10] This does not pose a phytotoxicity problem for the plants, except B in certain boron-sensitive plants. However, from an animal health standpoint, some of these elements may be potentially toxic to animals.[12,13] Those elements that may be potentially dangerous to animals include As, Cd, Cu, F, Mo, Pb, and Se. Since Cd, Cu, F, and Pb do not accumulate to any extent in plants grown on fly ash amended soils, there does not appear to be a problem with these elements.[10] On the other hand, Mo and Se are required in small amounts by animals, but become toxic when present in large amounts.[14] Arsenic is the one element that needs careful monitoring, as it is not only toxic to animals but also accumulates in plants grown on fly ash amended soils. The elements As, Mo, and Se are present in soils in the form of oxides as arsenites, arsenates, molybdates, selenites, and selenates (all of which are anions). The main electrical charges associated with soil colloids (clays and organic matter) are negative. Therefore, most of the elements which are adsorbed by soil colloids have positive charge. Since the elements of concern all have negative charges, we are now looking at chemical reactions which have been less extensively studied and are not so clear. This calls for a different type of research to deal with these elements.

## Use of Fly Ash for Plant Growth

### Disadvantages

*Change in equilibrium.* Plant nutrient deficiencies may develop when pH increases due to fly ash application. This assumes that the ash is alkaline initially, although some kinds of fly ash are acidic. The typical deficiencies are N and P, but others may also become deficient if conditions change the nutrient availability. Consequently, supplemental fertilizers are often needed on soils amended with fly ash. Salinity of the soil often increases because fly ash often has many salts present. This changes the electrical conductivity of the solution and sometimes retards plant growth. Since fly ash is the remains of burned coal, all of the organic matter has been oxidized as $CO_2$ and $H_2O$. Therefore, no organic matter is added to soils when fly ash is used. This then dilutes any organic matter present, and on surface-mined land where there is no organic matter it makes conditions worse. This lack of organic matter also considerably reduces microbial activity, which is vital to a healthy soil. The large amount of Ca, Mg, and K provided in fly ash may cause antagonistic interaction among these elements, to the detriment of growing plants.

*Toxicity from B.* Fly ash is often high in B, resulting in an increased concentration of B in soils to which fly ash is applied. Increased uptake of B by plants creates a phytotoxic condition in many plants, although some, such as sweet clover and alfalfa, require large amounts of B and are relatively tolerant to high levels of it in the growth medium.[15] Toxicity from B in fly ash-amended soils becomes less of a problem with time, since B is present in soils as neutral complexes or negatively charged anions which are not adsorbed onto soil colloids and readily leach.

*Animal toxicity due to As, Mo, Se, and some other metals.* Small amounts of Mo and Se are required by animals, but when either of these elements is present in high concentrations in the diet, it becomes toxic. Any amount of As present in animal feed is also toxic. Because As accumulates in plants grown on fly ash-amended soils (Table 1), to avoid problems animal producers must be sure that the feed contains no As. Nickel and Pb are also toxic to animals, but these do not seem to accumulate in plants grown on fly ash-amended soils.

*Accumulation in soils — B, Ba, Ca, Mo, Pb, S, Se, and Sr.* Some elements do accumulate in soils amended with fly ash; however, except for B, Mo, and Se, these elements do not seem to pose a problem for either plants or animals.

### Advantages

*Increased dry-matter yields.* Fly ash in the growth medium may lead to increases in dry-matter yields. This has been ascribed to a correction of a plant nutrient deficiency, prevention of toxic effects from high concentrations of Al, Mn, or some metals, and/or improved pH, especially in very acid soils.

*Improved physical properties of the growth medium.* Application of fly ash to soils seems to improve certain physical properties, such as aeration and water-holding capacity. This is especially important in reclamation of very acid surface-mined land which, in many cases, has been compacted during the last stages of its reclamation.

*Substitution for Topsoil.* To a limited extent, fly ash can act as a topsoil substitute. The improved physical conditions mentioned above, along with supply of certain plant nutrients such as Ca, Mg, K, B, Mn, Mo, Ni, S, and Zn, provide at least a partial substitute for topsoil. However, since fly ash has no organic matter, N, or P, these must be provided in the form of organic residues or fertilizers. Thus, fly ash has limitations for this use.

*Retarding pyrite oxidation.* One of the most serious problems facing the coal mining industry in the eastern U. S. is overcoming pyrite ($Fe_2S$) oxidation, which is often the source of acid mine drainage. If some means could be devised to prevent $Fe_2S$ from oxidizing, most of this problem could be solved. Research by Keefer, Singh, and associates at West Virginia University[6] has indicated that fly ash can act to suppress pyrite oxidation. The fly ash was applied in massive amounts to the surface of very acid surface-mined land with the premise of acting as a blanket to retard oxygen diffusion to pyrite particles in the upper layers of the mined land. Furthermore, it was postulated that by encouraging plant growth on this fly ash, additional oxygen would be used by plant roots, with less reaching the pyrite. Also, oxidation products of pyrite — i.e., $Fe^{+2}$, $Fe^{+3}$, and $H^+$ — will be neutralized by alkalinity from calcium oxides in the fly ash, preliminary data indicate.

*Maintaining desirable water quality.* Many people fear that applying fly ash to land may contaminate our natural waters. Experiments are under way by researchers at West Virginia University[6,16] to address this issue. Preliminary data indicate that the water in the fly ash layer is actually of better quality than in the very acid surface-mined land. Other experiments are being devised to see if further management techniques could be beneficial in improving water quality where acid mine drainage is a problem.

## Enrichment Ratios

Plants grown on fly ash or fly ash-amended soils become enriched with certain elements. Those that have relatively high enrichment ratios and are of immediate concern are As, B, Pb, Mo, Ni, Se, and Sr (Figure 1). Data concerning this was shown by Tolle et al.[9] and Adriano et al.[10] The data of Adriano et al. gave no specific type of plant or conditions, but data of Tolle et al. was specifically for oats and alfalfa in both greenhouse and field conditions. These general data of Adriano et al. and specific data on oats and alfalfa grown in the field were plotted (Figure 1). There appears to be some agreement between these researchers, but there are notable exceptions. For example, As was greatly enriched in alfalfa but in lesser amount by oats; B, Pb, and Ni showed considerably greater enrichment in oats than in alfalfa and general data of Adriano et al. Enrichment of Sr was high (value of about 6) for the Adriano et al. group but was low (about 1) for both oats and alfalfa by Tolle et al. Selenium was enriched to values of 7 to 8 for both groups. From this limited data, it appears than enrichment ratios bear further consideration. The range and mean of

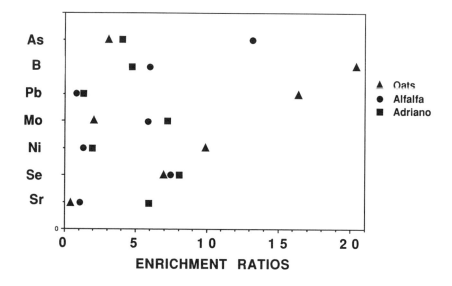

**FIGURE 1.** Enrichment ratios of elements in plants grown on fly ash or fly ash-amended soils.

Table 2
**Range and Mean of Elemental Values (mg/kg) for Coal, Fly Ash, Bottom Ash, and Soil**

| | Coal | | Fly Ash | | Bottom Ash | | Soil | |
|---|---|---|---|---|---|---|---|---|
| Element | Range | Mean | Range | Mean | Range | Mean | Range | Mean |
| As | 0.5–106 | 15 | 2–6,300 | 156 | 0.5–168 | 4 | 1–50 | 5 |
| B | 1–356 | 50 | 10–5,000 | 370 | 1–513 | 161 | 2–150 | 30 |
| Mo | 0–73 | 3 | 1–236 | 44 | 1–443 | 14 | 0.2–5 | 2 |
| Se | 0.4–8 | 4 | 0.2–134 | 14 | 0.08–14 | 4 | 0.1–4 | 0.3 |

the four elements — As, B, Mo, and Se — have been summarized[17] (Table 2). All four of these elements are concentrated from the original coal when it is burned, and they reside in the fly ash and bottom ash. Native soil concentrations of these elements are quite low, so enrichment undoubtedly does occur.

## Emphasis in Present Work and Direction of Future Research

Of the 11 chapters to follow in this book, seven consider coal ashes as some form or other of pollutants, and only four explore their beneficial uses. It appears that researchers need to change their emphasis from the negative aspect of pollution (even though this is important) to a more positive approach to beneficial uses of coal ashes or, at least, how to overcome the pollution aspects. There seem to be so many potential beneficial uses of these by-products that we need to be heading our research in that direction.

## REFERENCES

1. Golden, D. M., EPRI coal ash utilization research, *Proc. 8th Int. Ash Util. Symp.*, 1, 1-1, 1987.
2. Buck, J. K., Direct (soil-less) revegetation of anthracite waste using coal fly ash as a major amendment, *Proc. 8th Int. Ash Util. Symp.*, 1, 28-1 1987.
3. Ghazi, H. E., Bhumbla, D. K., Keefer, R. F., and Singh, R. N., Utilization of fly ash for reclaiming abandoned mine spoils for the production of agricultural crops, *Proc. 8th Int. Ash Util. Symp.*, 1, 27-1 1987.
4. Erickson, A. E., Jacobs, L. W., and Sierzega, P. E., Improving crop yield potentials of coarse textured soils with coal ash amendments, *Proc. 8th Int. Ash Util. Symp.*, 1, 26-1 1987.
5. Bhumbla, D. K., Keefer, R. F., and Singh, R. N., Selenium uptake by alfalfa and wheat grown on a mine spoil reclaimed with fly ash, *Proc. Conf. Mine Drainage and Surface Mine Reclamation,* U.S. Department of Interior Inf. Circ. 9184, Vol. 2(1), 15, 1988.
6. Bhumbla, D. K., Singh, R. N., and Keefer, R. F., Retardation of pyrite oxidation by massive fly ash application to mine spoils, in *Proc. 1990 Mining and Reclamation Conf. and Exhib.,* Vol. 2, Skousen, J., Sencindiver, J., and Samuel, D., Eds., West Virginia University, Morgantown, WV, 1990, 469.
7. Keefer, R. F., Bhumbla, D. K., and Singh, R. N., Chapter 11 in *Trace Elements in Coal and Coal Combustion Residues,* Lewis Publishers, Chelsea, MI, 1993.
8. Singh, R. N., Keefer, R. F., and Bhumbla, D. K., Arsenic in fly ash-amended acidic mine spoils, *Proc. Int. Conf. Metals in Soils, Waters, Plants, and Animals,* Abstr. 236, 1990.
9. Tolle, D. A., Arthur, M. F., and Van Voris, P., Microcosm/field comparison of trace element uptake in crops grown in fly ash-amended mine soil, *Sci. Total Env.,* 31, 243, 1983.
10. Adriano, D. C., Page, A. L., Elseewi, A. A., Chang, A. C., and Straughan, I., Utilization and disposal of fly ash and other coal residues in terrestrial ecosystems: a review, *J. Environ. Qual.,* 9, 333, 1980.
11. National Research Council, Trace-Element Geochemistry of Coal Resource Development Related to Environmental Quality and Health, National Academy of Sciences, Washington, D.C., 1980.
12. Gutenmann, W. H., Bache, C. A., Youngs, W. D. and Lisk, D. J., Selenium in fly ash, *Science,* 191, 966, 1976.
13. Furr, A. K., Parkinson, T. F., Hinrichs, R. A., Van Campen, D. R., Bache, C. A., Gutenmann, W. H., St. John, L. E., Jr., Pakkala, I. S., and Lisk, D. J., National survey of elements and radioactivity in fly ashes. Absorption of elements by cabbage grown in fly ashes, *Environ. Sci. Technol.,* 11, 1104, 1977.
14. Allaway, W. H., Agronomic controls over the environmental cycling of trace elements, *Adv. Agron.,* 20, 235, 1968.
15. Eton, F. M., Deficiency, toxicity, and accumulation of boron in plants, *J. Agric. Res.,* 69, 237, 1944.
16. Bhumbla, D. K., Singh, R. N., and Keefer, R. F., Leachability of metals from fly ashes and its effect on water quality of amended acidic mine spoils, *Proc. Int. Conf. Metals in Soils, Waters, Plants, and Animals,* Abstr. 241, 1990.
17. Rai, D. , Ainsworth, C. C., Eary, L. E., Mattigod, S. V., and Jackson, D. R., *Inorganic and Organic Constituents in Fossil Fuel Combustion Residues,* Vol. 1, Report EA-5176, Electric Power Research Institute, Palo Alto, CA, 1987.

# An Overview of Electric Power Research Institute (EPRI) Research Related to Effective Management of Coal Combustion Residues

**I. P. Murarka,[1] S. V. Mattigod,[2] and R. F. Keefer[3]**

[1]Electric Power Research Institute, Palo Alto, CA
[2]Battelle, Pacific Northwest Laboratories, Richland, WA
[3]West Virginia University, Morgantown, WV

## ABSTRACT

The Electric Power Research Institute (EPRI) has sponsored research in solid waste environmental studies designed to develop knowledge for efficient, environmentally sound, cost-effective management of solid residues from coal combustion. These residues have been characterized physically and chemically. Water-soluble components have been extracted, identified, and recovered in leachates from the residues in laboratory and field studies. Geochemical parameters of dissolution/precipitation, adsorption/desorption, and oxidation/reduction have been used to evaluate solubility-controlling factors involved in leachates from the residues. Computer models have been developed to predict geochemical transformations, leaching and hydrologic transport of the elements in the natural environment. The data collected have been used as inputs into the models to verify solubility-controlling factors. The models can also be used for specific cases to evaluate coal combustion residue disposal types and methods in specific hydrological settings.

## INTRODUCTION

Coal combustion for electric power generation results in solid residues that need environmentally safe disposal. In the U.S., combustion of more than 800 million metric tons of coal annually results in approximately 75 million metric tons of solid residues.[1] The increasing use of coal coupled with air-quality control requirements is expected to generate over 120 million metric tons of solid residues by the end of this century. Even though these solid residues represent only about

2% of all the solid by-products now being produced in the U.S., these combustion residues are the sixth most abundant material for resource recovery.[2]

Two classes of by-products are generated from coal-fired power plants. The low-volume liquid effluents are generated from boiler cleaning, cooling tower blowdown, demineralizer regeneration, and floor and yard drains. The high-volume solid by-products are the noncombustible residues from coal combustion, and the residues generated in the process of scrubbing $SO_\chi$ and $NO_\chi$ components in flue gases. Fly ash is the suspended solids in the flue gas that is collected by mechanical filters or electrostatic precipitators to reduce the emission of particulates into air. Bottom ash and boiler slag are the heavy solids that remain in the boiler fire boxes. The flue gas desulfurization (FGD) sludge is generated as a result of the removal of gaseous sulfur from flue gases to meet the air emission standards. The total solid residue generated comprises about 70% fly ash, 25% bottom ash and slag, and 5% FGD sludge.[1] In 1988, about 25% of the solid residues (fly ash and bottom ash) was utilized. Some of these uses include cement manufacture and partial substitution in cement manufacture, in lightweight aggregate, in road fills and bases, and in asphalt mixes. Remaining residues are disposed of in landfills and ponds. Frequently, the low-volume residues are codisposed with the high-volume solid residues.

The primary concern in disposing of any solid residue on land is the possibility of contaminating surface and underlying waters because these residues contain several inorganic constituents that may impact plants, animals, and human health. Therefore, the electric utilities are interested in safely managing these solid residues. The most important aspect of effectively managing any solid residue is to acquire thorough knowledge of the physical and chemical properties of the residues, how the inorganic constituents of concern may be released (i.e., the rates and quantities of leaching), and how these mobilized constituents will migrate into the ambient environment. Therefore, EPRI, an independent research and development organization, is sponsoring research by several contractors under the Solid Waste Environmental Studies (SWES) project to develop such knowledge for efficient, environmentally sound, and cost-effective management of solid residues from coal combustion. This ongoing project has developed pertinent information that can be used to ensure that the solid residues are disposed of in an environmentally sound manner. These research efforts are focused on local issues, such as developing laboratory and field information on leaching of inorganic components from the solid residues generated by the utility industries. Additionally, predictive tools are being developed to incorporate the new information about the geochemical behavior of leached elements and their attenuation or immobilization by the geological materials underlying the disposal facilities. Further, optimally cost-effective and environmentally sound solid residue management strategies will be developed and implemented. Finally, a network will be devised for monitoring the environmental effectiveness of chosen management strategies. This paper summarizes selected results of research supported by the SWES project of EPRI. Greater details of the information presented here can be found in a digest of EPRI research results on coal combustion by-products composition and leaching.[3]

## CHARACTERIZATION OF COAL COMBUSTION SOLID RESIDUES

### Physical Properties

Physical properties of the solid residues have been measured because these parameters are helpful in evaluating the proper use of these materials, determining the appropriate disposal methods, and monitoring the postdisposal conditions. While the physical properties of fly ash and bottom ash have been measured frequently, similar measurements on FGD sludge are rather sparse.

Many physical properties of fly ash and bottom ash are similar.[4] The specific gravities of fly ash particles have a wider range (1.59 to 3.10) than particles of bottom ash (2.17 to 2.78). However, fly ash bulk densities have narrower ranges (1.01 to 1.43 metric ton/m$^3$) than bottom ash densities (0.74 to 1.58 metric ton/m$^3$). Mean particle diameters of fly ash particles are considerably smaller (20 to 80 µ/m) than those of bottom ash particles (500 to 7000 µm). Fly ash particles occur mainly in silt and sand sizes, whereas bottom ash particles occur in sand and gravel sizes. The significant difference in particle sizes is reflected in greater surface area and lower hydraulic conductivity of fly ash. Because of the surface area and the contact with flue gases, fly ash contains relatively higher concentrations of volatile elements — As, B, and Se.

The morphological properties of fly ash particles have been studied extensively by light and scanning electron microscopy.[4-6] Among the several morphological classes of fly ash particles, hollow spheres (cenospheres) (Figure 1A) and microsphere-filled spheres (plerospheres) (Figure 1B) constitute the bulk (67 to 95%) of the fine particles.[4] These spheres consist of an aluminosilicate matrix (Figure 2). A small fraction of iron-rich solid spherical particles has also been observed (Figure 3). The formation of such spheres results from many physical and chemical reactions that occur during coal combustion.[5,6] Similar morphological data for bottom ash and FGD sludge are not currently available.

### Chemical Properties

Because fossil fuels are geological materials, the elemental compositions of their solid residues following combustion are qualitatively similar to that of other lithospheric materials such as soils.[4] About 90% by weight of these solid residues consists of Al, Si, Fe, Ca, and O (Table 1). Except in FGD sludge, Mg, Na, K, and S occur as minor components. All other elements occur in trace quantities, i.e., 1% or less.

Even though the elemental compositions of solid residues are well known, these bulk properties do not correlate with the elemental leachability. This is because leachability depends on the ionic forms of these elements and their distribution within the solid matrix. The dominant ionic forms of the most abundant elements — Al, Si, Fe, Ca, and O — in the solid residues have definitely been identified as mullite ($Al_6Si_2O_{13}$), glass and quartz ($SiO_2$), lime (CaO), periclase (MgO), anhydrite ($CaSO_4$), spinels ($Fe_3O_4$, $Fe_2O_3$), hannebachite ($CaSO_3 \cdot 1/2H_2O$), gypsum ($CaSO_4 \cdot 2H_2O$), and calcite ($CaCO_3$), whereas comparable data on trace

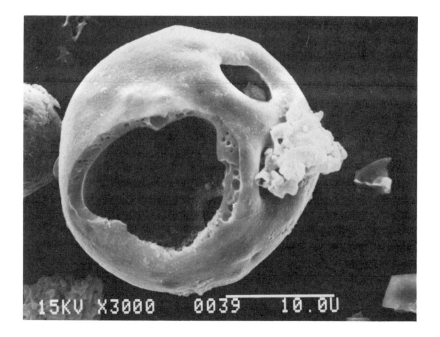

A

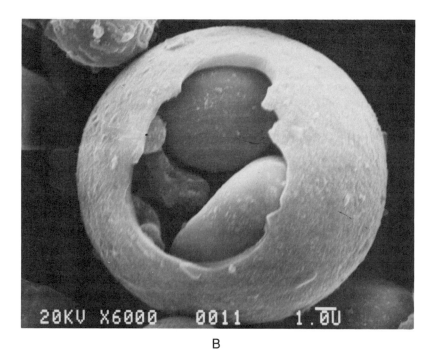

B

**FIGURE 1.**   Typical morphology of a cenosphere (A) and a plerosphere (B) found in fly ash.

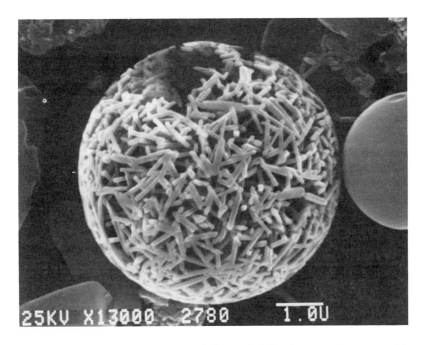

**FIGURE 2.**  Mullite (aluminosilicate) crystals in a typical fly ash cenosphere exposed from etching with hydrofluoric acid.

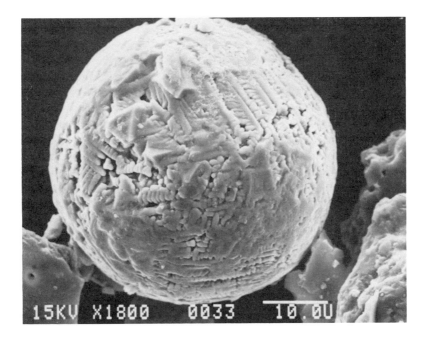

**FIGURE 3.**  Typical magnetic sphere isolated from a fly ash sample.

**Table 1**
**Concentration Ranges of Major and Trace Elements in Water Extracts from Solid Coal Combustion Residues and Soils**

| Element (wt%) | Fly Ash | Bottom Ash | FGD[a] Sludge | Soils |
|---|---|---|---|---|
| Al | 0.10 20.86 | 3.05–18.50 | 0.64–9.69 | 1.00–30.00 |
| Ca | 0.11–22.30 | 0.22–24.10 | BD[b]–34.50 | 0.70–50.00 |
| Fe | 1.00–27.56 | 0.40–20.10 | 0.13–13.80 | 0.70–55.00 |
| Mg | 0.04–7.72 | 0.20–4.80 | BD–1.99 | 0.06–0.60 |
| K | 0.17–6.72 | 0.26–3.30 | 0.001–1.70 | 0.04–3.50 |
| Si | 1.02–31.78 | 5.10–31.20 | 0.27–17.70 | 23.00–25.00 |
| Na | 0.01–7.10 | 0.08–4.13 | 0.02–5.53 | 0.075–1.00 |
| S | 0.04–6.44 | <0.04–7.40 | 0.08–22.80 | 0.003–1.00 |
| | | | | |
| Element (μg/g) | | | | |
| As | 2–440 | 0.02–168 | 0.8–53 | 1–50 |
| Ba | 1–13800 | 110–9360 | 25–2280 | 100–3000 |
| B | 10–5000 | 2–513 | 42–530 | 2–100 |
| Cd | 0.1–130 | 0.1–4 | 0.06–25 | 0.1–0.7 |
| Cr | 4–900 | 0.2–5820 | 2–180 | 1–1000 |
| Cu | 33–2200 | 4–930 | 6–340 | 2–100 |
| Pb | 3–2100 | 0.4–1100 | 0.3–300 | 2–200 |
| Mn | 25–3000 | 60–1900 | 40–310 | 20–3000 |
| Hg | 0.01–12 | 0.01–4 | 0.01–6 | 0.01–0.3 |
| Mo | 1–140 | 1–440 | <4–50 | 0.2–5 |
| Ni | 2–4300 | <10–2900 | <5–145 | 5–500 |
| Se | 0.2–130 | 0.1–10 | <2–160 | 0.1–2 |
| Sr | 30–7600 | 170–6400 | 70–3000 | 50–1000 |
| V | 12–1180 | 12–540 | <50–260 | 20–500 |
| Zn | 14–3500 | 4–1800 | 8–610 | 10–300 |

*Sources:* Mattigod et al.,[3] Eary et al..[6]

[a]  FGD = flue gas desulfurization.
[b]  BD = below detection limit.

elements is not readily available.[4] Some fractions of the trace elements — Cr, Co, Ni, and V — are known to exist in spinels. Major fractions of volatile trace elements (As, Mo, Se, Pb, and Zn) may exist on particle surfaces as either oxides or sulfates, and therefore may be rapidly mobilized after disposal.[7]

## COMPOSITION OF EXTRACTS AND LEACHATES

### Laboratory Experiments

Generally, studies related to environmentally sound disposal of solid residues rely on various tests to assess the leachability of inorganic constituents. These extraction tests differ in the type and concentrations of extractants used, duration, temperature, solid/solution ratio, and the degree of agitation used during the extraction process. Therefore, these tests cannot reliably simulate leaching under field conditions. Because of the number of variables involved, the results of

**Table 2**
**Ranges of Major and Trace Element Concentrations (mg/L) in Water Extracts from Solid Residues**

| Element | Fly Ash | Bottom Ash | FGD[a] Sludge |
|---|---|---|---|
| Al | 0.12–62 | <0.05–43 | 0.2–6.8 |
| Ca | 67–634 | 2.1–774 | 319–553 |
| Fe | <0.005–3 | <0.005–1.84 | 0.02–0.03 |
| Mg | <0.05–118 | <0.05–22 | 0.1–143 |
| K | 0.72–191 | 0.8–42 | <0.3–56 |
| Si | <0.05–46 | 0.48–30 | 0.3–17 |
| Na | 1.87–2008 | 0.3–862 | 2.6–693 |
| S | 33–3583 | 3–1857 | 397–2165 |
| As | <0.08–14 | <.08–1.6 | <.08–0.6 |
| Ba | 0.05–2.2 | 0.005–0.4 | 0.03–0.8 |
| B | 0.1–109 | <0.01–9.3 | 0.9–24 |
| Cd | <0.01–1.8 | <0.002–0.003 | 0.1–8 |
| Cr | 0.02–44 | <0.01–0.07 | <0.01–1.5 |
| Cu | <0.01–24 | <0.01–0.03 | <0.01–0.03 |
| Pb | <0.05–3.8 | <0.06–0.3 | <0.04–0.2 |
| Mn | <0.001–290 | <0.001–0.5 | 0.003–0.06 |
| Mo | 0.01–6.8 | <0.01–0.09 | <0.01–4 |
| Ni | <0.01–8.5 | <0.01–0.1 | <0.01–0.3 |
| Se | <0.05–0.4 | NA[b] | 0.004–0.008 |
| V | <0.003–1.1 | NA | 0.02–0.25 |
| Zn | <0.01–121 | <0.01–0.1 | <0.01–0.4 |

*Sources:* Mattigod et al.,[3] Eary et al.[6]

[a]  FGD = flue gas desulfurization.
[b]  NA = data not available.

various extraction methods show that the extract concentrations can vary widely[4] as a function of the laboratory procedure.

Nevertheless, a few qualitative trends can be observed from these results (Table 2). Among the major elements, Ca, Na, and S solubilize easily, K and Mg dissolve more slowly, and Al, Si, and Fe are only very slowly released. Trace elements, such as As, B, Cr, Cu, Mo, Ni, and Zn, that are reported to be enriched on particle surfaces dissolve more readily than other elements. The pH of these extracts can range from highly acidic (3.3) to highly alkaline (12.3).[4,7]

## EPA Regulatory Tests Applied to Coal Combustion Residues

The U.S. Environmental Protection Agency has adopted specific laboratory extraction tests for regulating hazardous materials.[8–10] The first test used was the Extraction Procedure (EP), a 24-hour extraction of the material with water and adjustment of pH with acetic acid. This was replaced in 1990 by a new test, the Toxicity Characterization Leaching Procedure (TCLP), with extraction of the material using a sodium acetate buffer at pH < 5 or acetic acid at a pH > 5. Both the EP and the TCLP tests have been applied to the solid residues from coal combustion.[11] The results obtained with the TCLP on a limited number of fly ash and bottom ash samples showed that the range of concentrations of specified

**Table 3**
**Concentration of Inorganic Constituents (mg/L) Extracted from**
**Fly Ash and Bottom Ash by the TCLP Test**

| Constituent | Fly Ash | Bottom Ash |
|---|---|---|
| As | 0.0039–2.68 | <0.0006–0.0328 |
| Ba | 0.113–0.71 | 0.13–1.61 |
| Cd | <0.01–0.564 | <0.01 |
| Cr | <0.02–4.64 | <0.02 |
| Pb | <0.01–2.94 | <0.001 |
| Hg | <0.0001 | <0.0001–<0.0003 |
| Se | <0.003–0.041 | <0.001–0.0038 |
| Ag | <0.01 | <0.01 |

*Source:*  Ainsworth and Rai.[10]

*Note:*  Data obtained from testing 7 fly ash and 9 bottom ash samples.

**Table 4**
**Elements Considered by EPA as Being Toxic in Drinking Water**
**with the Limit of Acceptability[a]**

| Element | National Primary Drinking Water Standards × 100 (mg/L) |
|---|---|
| Ag | 5.0 |
| As | 5.0 |
| Ba | 100.0 |
| Cd | 1.0 |
| Cr | 5.0 |
| Pb | 5.0 |
| Hg | 0.2 |
| Se | 1.0 |

[a]  **Anon.,** *Fed. Register.,* 45, May 19, 1980, p. 33127.

constituents was well below the regulatory levels (Table 3). These residues are therefore classified as non- hazardous wastes and can be disposed of on land without the risk of contaminating groundwaters to the extent of exceeding the drinking water standards (Table 4).[12] There has been no reported contamination from fly ash or bottom ash.

## PREDICTIONS OF LEACHATE COMPOSITIONS

### Chemistry and Geochemistry of Residue Disposal

A quantitative assessment of the chemical composition of leachates is crucial to plan environmentally sound disposal of the solid residues. Because laboratory extraction procedures yield imprecise estimates of field leachate compositions, and field studies alone do not provide the cause-and-effect relationships for the observed behavior, the SWES research has focused on developing an improved understanding

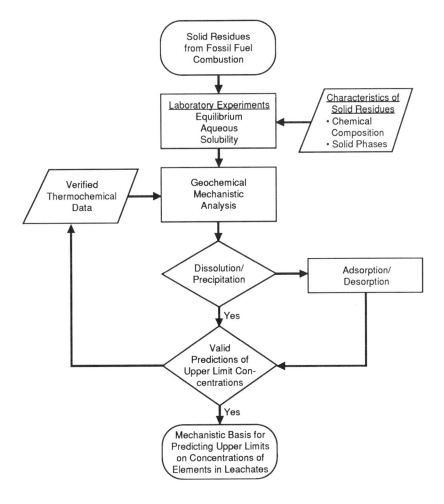

**FIGURE 4.**  Geochemical scheme for predicting porewater and leachate compositions from disposed of solid residues resulting from coal combustion.

of the chemistry of the solid residues and leachates, both in laboratory and field studies.

The geochemical reactions of dissolution/precipitation, adsorption/desorption, and oxidation/reduction are recognized as controlling the mobilization of various constituents from solid residues.[11,12,14,15] Therefore, an integrated scheme has been developed[4] to use these fundamental geochemical reactions as the basis to interpret and predict the chemistry of leachates and their interactions with various geological materials (Figure 4). This approach provides two distinct advantages over specific field studies: (1) laboratory studies can be conducted for a wide range of conditions, such as pH, redox, complexations, and ionic strength of leachates and porewaters that would be encountered at different field sites, and (2) laboratory studies are less costly than field studies, and provide useful and widely applicable data.

## Laboratory Studies on Geochemical Parameters

Therefore, based on the scheme of fundamental geochemical reactions, a number of laboratory studies have been conducted to determine the maximum concentrations for major elements (Al, Ca, Fe, S, and Si) expected in leachates from fly ash and FGD sludge. These studies have established that the leachate concentrations of these elements can be predicted from the dissolution/precipitation reactions of secondary phases, such as aluminum hydroxy sulfate ($AlOHSO_4$), aluminum hydroxide ($AlOH_3$), gypsum, calcite, hannebachite, and iron hydroxide ($FeOH_3$).[16–18]

Maximum concentrations of several of the trace elements — As, Ba, Cd, Cu, Cr, Mn, and Mo — have also been obtained from laboratory studies.[17–19] These laboratory-based predictions can be extrapolated to field disposal sites for fly ash and FGD sludge even though some chemical conditions, such as pH or redox, differ significantly between these two materials.

## Porewater and Leachate Composition from Field Experiments

Field measurements of porewater and leachate compositions have been conducted at a fly ash test cell, an FGD sludge disposal site, and at four fly ash/bottom ash disposal sites in Pennsylvania.

### Fly Ash Test Cell

A fly ash test cell was constructed to monitor the characteristics of the porewater and the leachates generated under field conditions.[20] Data were collected over a period of years and analyzed for geochemical parameters. The dissolution/precipitation of certain solid phases identified by laboratory studies as the solubility-controlling factors in the leaching of fly ash was confirmed by the field observations of aqueous concentrations of Al, Ba, Ca, Cr, Cu, Fe, S, and Sr.

### Disposal of FGD Wastes

Field studies conducted at an FGD sludge disposal site[21] showed that several redox couples of S were at equilibrium in the pore waters. The concentrations of Ca, Ba, and Sr in these porewaters appeared to be regulated by sulfate, sulfite, and carbonate solid phases. The presence of these solid phases in FGD sludges had been previously identified from equilibrium solubility studies in the laboratory. Concentrations of Cu, Fe, Ni, Pb, and Zn in porewaters were at or below the detection limits. To verify results of the laboratory studies, FGD sludge suspensions were spiked with these metal ions. The results indicated that the metal ion concentrations in the suspensions decreased to the levels observed in the field samples and that these concentrations matched those predicted from the solubility phenomena. Direct identification of the solid phases confirmed that these metal ions had precipitated as sulfides.

*Laboratory and Field Studies at Fly Ash Disposal Sites*

Similar supplemental field and laboratory studies were conducted on samples from four fly ash disposal sites in Pennsylvania.[22] Observed concentrations of Al, Ba, Ca, Mo, Si, Sulfate, and Sr were shown to be regulated by the laboratory-predicted solubility-controlling solid phases. These field studies demonstrate that reliable predictions can be made of concentrations of various major and trace elements in the porewaters and leachates from disposal sites by well-designed laboratory studies.

## Development of Computer Models

Several computer models have been developed to predict geochemical trans-formations, leaching, and hydrologic transport of solutes in the natural subsurface environment. One computer model, the FOssil fuel combustion Waste Leaching code (FOWL™), has been developed to predict the concentrations and quantities of 18 major and trace constituents (Al, Ba, Ca, Cr, Mo, Si, Sr, S, As, B, Cd, Cu, Fe, Mg, Na, Ni, Se, and Zn) in leachates generated from fly ash, bottom ash, oil ash, and FGD sludges.[23,24] A revised, improved version of this model (FOWL™ 2.0) will be released soon by EPRI. These FOWL™ computer models can be obtained from the Electric Power Research Institute (P. O. Box 10412, Palo Alto, CA 94304). The predictions can be made for both landfills and pond sites ranging in size from a small test plot to a full-size disposal facility. This computer code can also be used to predict the leaching duration of constituents that could readily be used at a disposal facility.

Element-specific codes are also being developed to predict the geochemical fate of a specific trace element. The first example of such a code is the Chromium Attenuation (CHROMAT™) model.[24] This code predicts the concentrations of Cr in the leachates and porewaters using the appropriate geochemical reactions, including redox transformations, adsorption/desorption, and dissolution/precipi-tation. Results from such models can be used as inputs for analyzing the fate of this element during subsurface transport.

## APPLICATIONS TO SOLID RESIDUE DISPOSAL SYSTEMS

Ponding and landfilling are currently used as the principal methods of land disposal for solid residues from coal combustion. Ponding or sluicing the fly ash and bottom ash with water generates high volumes of waste water. This produces rapid dissolution of soluble components that leach quickly. Subsequent leachates from ponds contained very low elemental concentrations. Ponding was the preferred method of disposal during the 1950s and 1960s. Recently, landfilling is being used with increasing fre-quency[1] because this method of disposal is dry and therefore generates only low volumes of leachates. On the other hand, elemental concentrations of these leachates are concentrated from the relatively small volume of leachates that emanate over time.

The laboratory and the field measurements coupled with the predictive codes serve two useful goals. In the case of existing solid residue disposal sites, these tools help monitor any mobilized constituents and to determine whether the concentrations of these constituents are less than the specified regulatory levels. For new disposal sites being developed, these studies and computer models help to determine the suitability of sites, appropriate methods for disposal, and predictions on the future environmental performance of these sites. For example, the FOWL™ computer model was used in evaluating disposal designs involving four residue types, two disposal methods, and hydrological settings.[25]

Continuing research supported by EPRI in the laboratory and field is focusing on completing the development of a comprehensive data base used for confidently predicting the leaching and attenuation of inorganic soluble constituents during coal combustion residue disposal.

## CONCLUSIONS

Research supported by EPRI has evaluated physical and chemical properties of coal combustion residues to complete a data base on these materials. A computer model (FOWL™) has been devised to predict geochemical transformation, leaching, and hydrological factors involved in disposal of coal combustion residues in the natural environment. The data collected from laboratory and field experiments have been used as inputs into the computer model verifying assumptions about solubility-controlling factors in elements present in the coal combustion residues. The model can also be used by utility companies and others involved to evaluate specific disposal sites for these residues.

## ACKNOWLEDGMENTS

We thank Dr. Dhanpat Rai for his valuable comments and Laurel Grove for her editorial assistance.

## REFERENCES

1. Murarka, I. P., Boyd, R. H., and Harbert, H. P., Electric utility industry solid wastes, in *Solid Waste Disposal and Reuse in the United States,* Murarka, I. P., Ed., CRC Press, Inc., Boca Raton, FL, 1987, 95.
2. Torrey, S., *Coal Ash Utilization: Fly Ash, Bottom Ash and Slag,* Noyes Publishing, Park Ridge, NJ, 1978, 23.
3. Keefer, R. F., and Murarka, I. P., *EPRI Research Results on Coal Combustion By-Products Composition and Leaching: A Digest,* Electric Power Research Institute, Palo Alto, CA, in press.

4. Mattigod, S. V., Rai, D., Eary, L. E., and Ainsworth, C. C., Geochemical factors controlling the mobilization of inorganic constituents from fossil fuel combustion residues. I. Review of the major elements, *J. Environ. Qual.,* 19, 188, 1990.

5. Raask, E., Cenospheres in pulverized-fuel ash, *J. Inst. Fuel,* 41, 339, 1968.

6. Raask, E., Fusion of silicate particles in coal flames, *Fuel,* 48, 366, 1969.

7. Eary, L. E., Rai, D., Mattigod, S. V., and Ainsworth, C. C., Geochemical factors controlling the mobilization of inorganic constituents from fossil fuel combustion residues. II. Review of the minor elements, *J. Environ. Qual.,* 19, 202, 1990.

8. Test Methods for Evaluating Solid Wastes: Physical/Chemical Methods, Report SW-846, U.S. Environmental Protection Agency, Washington, D.C., 1982.

9. U.S. Environmental Protection Agency, Hazardous waste management. I. Toxicity characteristic leaching procedure, Fed. Regist. 51, January 14, 1986.

10. U.S. Environmental Protection Agency, Hazardous waste management system; identification and listing of hazardous waste; toxicity characteristics revisions; final rule, Fed. Regist. 55, March 29, 1990.

11. Ainsworth, C. C., and Rai, D., *Chemical Characterization of Fossil Fuel Wastes,* Report EA-5321, Electric Power Research Institute, Palo Alto, CA, 1987.

12. Roy, W. R. and Griffin, R. A., A proposed classification system for coal fly ash in multidisciplinary research, *J. Environ. Qual.,* 11, 563, 1982.

13. **Anon.**, EPA Extraction Procedure, Fed. Regist., 45, May 19, 1980, p. 33127.

14. Talbot, R. W., Anderson, M. A., and Andren, A. W., Qualitative model of heterogeneous equilibria in a fly ash pond, *Environ. Sci. Technol.,* 12, 1056, 1978.

15. Mattigod, S. V., Chemical composition of aqueous extracts of fly ash: ionic speciation as a controlling factor, *Environ. Technol. Lett.,* 4, 485, 1983.

16. Rai, D., Eary, L. E., Mattigod, S. V., Ainsworth, C. C., and Zachara, J. M., Leaching behavior of fossil fuel wastes: mineralogy and geochemistry of calcium, in *Fly Ash and Coal Conversion By-Products Characterization, Utilization, and Disposal,* Vol. 86, Proc. Mater. Res. Soc. Symp., McCarthy, G. J., Glasser, F. P., Roy, D. M., and Diamond, S., Eds., Materials Research Society, Pittsburgh, 1987, 3.

17. Rai, D., Mattigod, S. V., Eary, L. E., and Ainsworth, C. C., Fundamental approach for predicting pore-water composition in fossil fuel combustion wastes, in *Fly Ash and Coal Conversion By-Products Characterization, Utilization, and Disposal,* Vol. 113, Proc. Mater. Res. Soc. Symp., McCarthy, G. J. and Glasser, F. P., Eds., Materials Research Society, Pittsburgh, 1988, 317.

18. Mattigod, S. V., Rai, D., Zachara, J. M., and Amonette, J. E., Mineralogy of weathered flue gas desulfurization sludge, in *Fly Ash and Coal Conversion By-products Characterization, Utilization, and Disposal,* Vol. 136, Proc. Mater. Res. Soc. Symp., Berry, E. E., Ed. Materials Research Society, Pittsburgh, 1989, 3.

19. Rai, D. and Szelmeczka, R. W., Aqueous behavior of chromium in coal fly ash, *J. Environ. Qual.,* 19, 378, 1990.

20. Fruchter, J. S., Rai, D., Zachara, J. M., and Schmidt, R. L., *Leachate Chemistry at the Montour Fly Ash Test Cell,* Report EA- 5922, Electric Power Research Institute, Palo Alto, CA, 1988.

21. Rai, D., Zachara, J. M., Moore, D. A., McFadden, K. M., and Resch, C. T., Field Investigation of a Flue Gas Desulfurization (FGD) Sludge Disposal Pit, Report EA-5923, Electric Power Research Institute, Palo Alto, CA, 1989.

22. Rai, D., personal communication, 1990.

23. Hostetler, C. J., Erikson, R. L., and Rai, D., *The Fossil Fuel Combustion Waste Leaching (FOWL$^{TM}$) Code: Version 1, User's Manual,* Report EA-5472-CCM, Electric Power Research Institute, Palo Alto, CA, 1988.

24. Felmy, A. R., personal communication, 1991.
25. Aschenbrenner, R. P., Gumtz, G. D., and Freitag, J. A., Coal combustion waste disposal design factors — their influence on ground water impacts, in *Proc. 83rd Annu. Meet. Exhib. Air Waste Manage. Assoc.*, 1990.

*Environmental Effects from Power Plants*

# 3

# Coal-Based Environmental Problems in a Low-Rainfall Tropical Region

**M. Agrawal, J. Singh, A. K. Jha, and J. S. Singh**

Banaras Hindu University, Varanasi, India

## ABSTRACT

Several thermal power projects and open-cast coal mines have ensued in northern India's Singrauli belt, which has massive coal reserves. Thermal power plant (TPP) emissions have remarkably affected the air quality and vegetation, particularly in the vicinity of emission sources and at sites affected by emissions from both power plants and adjacent coal mines. Vegetation at sites receiving higher pollution loads have reduced total chlorophyll content, leaf area, and specific leaf area as well as increased contents of sulfate-S and trace elements. Coal mine spoils exhibited changes with time in physiochemical and biological properties, species composition, and biomass levels. Total soil N, mineral N, $NaHCO_3$-extractable P, and exchangeable K increased with age of mine spoils; these parameters were lower than native forest soil even after 20 years of succession. Only a few species dominated in the vegetation succession. Impact of microsites, even after 20 years of soil and vegetation development, remained important. Undulating surface and flat surface microsites were better habitats than slope and coalpatch margin microsites for plant growth. On the 12-year-old flat surface microsites, 24 out of 30 plant species seeded showed satisfactory growth performance.

## INTRODUCTION

Availability of energy in terms of electricity has been a powerful motivator of economic development and social change throughout the world. In India also, a high priority has been given to programs related to electricity generation and distribution for achieving economic self-reliance. The national energy policy, accepted by the Government of India in May 1981, recognizes coal as the principal source of energy. Most of the coal consumed in India is burned at TTPs, which contribute 85% of the total installed electric generating capacity in the country. The present installed capacity is around 65,000 MW and is expected to exceed 100,000 MW by the end of 1995. In other Third World countries, the utilization of coal has also increased tremendously during the past decade. For example, in China the contribution of coal to total energy has increased from 71.81% in 1980 to 76.03% in 1986.[1]

Coal combustion inevitably produces large amounts of gaseous and particulate pollutants. The level of $SO_2$ emission depends upon the sulfur content in the coal and the physical devices used for $SO_2$ control. The conversion of S constituents to $SO_x$ is 90 to 100%. More than 98% of $SO_x$ emitted is in the form of $SO_2$. A TPP with daily consumption of 10,000 metric tons of coal may emit 50 to 90 metric tons of $SO_2$ daily depending upon the S content of coal.[2] Sulfur dioxide emission in India was 6.76 million metric tons in 1979 and is expected to reach around 13.19 million metric tons by the year 1995.[3] In China total emissions of $SO_2$ have risen from 10.0 million metric tons in 1980 to 14.2 million metric tons in 1984.[4] Coal has contributed to more than 80% of the $SO_2$ emission in China by power stations, boilers, and heating systems.

The burning of coal releases vast quantities of particulate matter, the major component being fly ash. Heavier particles of ash are removed as bottom ash. The lighter particles are trapped in electrostatic precipitators or cyclone separators, and the excess is discharged into the atmosphere through stacks which fall out in the environment surrounding power generating plants. The emission rate depends upon ash content, burning rate, combustion efficiency, particle size and density, and flue-gas temperature. As the Indian coal has very high ash content (30 to 50%), the annual production of fly ash was estimated at more than 40 million metric tons, and this is expected to increase to about 120 million metric tons by 1999-2000.[5] Coal was suggested to contribute more than 69% of the dust levels in China.[4] If the emission values are considered at 100% efficiency, the total fly ash emission at a 285 MW capacity TPP would be 180 metric tons/day. Because of anticipated increasing dependence on coal as a source of energy, the problem of fly ash disposal is becoming a massive environmental concern. Trace elements present in fly ash may be potentially hazardous. Wet ash dumps near water reservoirs may leach potentially harmful elements from ash into the water body. Dumping of dry fly ash may produce particulate pollution. Fly ash may affect plants directly through deposition on plant surfaces or indirectly through accumulation in the soil.

Coal mining in India started in the year 1774 in the Raniganj coalfield, and by the year 1830 several additional mines had developed in that area. By the end of the 19th century, coal mining had started in Assam, Singreni, Warda Valley, Jharia, and central India. Systematic work in coal mines received a boost following nationalization of mines in 1971 and 1973. The total coal production for India during 1989–1990 was 226 million metric tons. Open-cast coal mining contributed about 56% of the current coal production and will contribute about 60% by 2000 A.D. Coal India Ltd. has projected a total land requirement for coal mining of 116,691.60 ha by 1994-1995, of which 31.18% is presently forest land. In China, the area of land disturbed by mining is about 2 million ha and the rate of land degradation due to mining is 2000 ha/year; by the end of the century this rate will exceed 33,000 ha/year.[6]

Surface coal mining results in deforestation of large areas, with loss of flora and fauna and disturbance in the hydrological and soil biological systems. Huge amounts of infertile overburden materials are dumped on lands adjacent to the pit-scarred landscape. The major adverse effects of coal mining are the disruption of the geology-soil-plant stability system, increased nutrient export from this system, and depletion of the soil organic pool. Furthermore, environmental monitoring studies have shown high levels of particulate pollution, particularly of fine suspended particulate matter (<5 μm), and high dustfall rates around open-cast coal mines.[7] The large quantities of overburden material also contribute to the dust problem, especially in the areas experiencing low rainfall.

The presence of massive coal reserves and availability of water nearby have encouraged open-cast coal mines and high-capacity superthermal power plants. One such energy center is the Singrauli belt in northern India (along Uttar Pradesh and Madhya Pradesh border), where several power projects and open-cast coal mines are under construction or in operation. This region has become one of the most rapidly growing industrial centers in India, and a whole range of potential environmental problems have arisen.[8,9] Vast stretches of the region were once covered with natural, dry tropical forests. The construction and operation of TPP and coal mines have led to serious land degradation, deforestation, and socioeconomic problems. Undesirable changes in the air, water, and soil quality have occurred due to gaseous and particulate emission, liquid effluents, and solid wastes.

This report presents a case study of environmental problems due to surface coal mining and coal combustion in TPPs in a low-rainfall region of India experiencing rapid industrialization. Plant responses to air quality and fly ash, and revegetation of mine spoils are examined. Information from other low- to moderate-rainfall areas is included where appropriate.

## STUDY AREA

The Singrauli coalfields cover about 2200 km² in the Sidhi and Shahdol districts of Madhya Pradesh and 80 km² in the Sonbhadra district of Uttar Pradesh. About 10,850 million metric tons of coal reserves are present in a 300 km² area of Northern Coalfields Limited (NCL), Singrauli. It has acquired 19,352 ha of land, out of which 9508, 2716, and 6150 ha are forest, government, and tenancy lands, respectively. About 90,000 metric tons of coal are excavated daily in NCL; there are 11 coal mines. Coal reserves[10] in different mines range from 14.67 to 348.93 million metric tons with stripping ratio (coal reserves to overburden) from 1:1.15 to 1:4.30 (Table 1). The overburden in these coal mines ranges from 35.9 to 1,110 million m³.

There are six TPPs in the area, with total installed capacity of 4265 MW.[11] Construction of units for additional capacity of 4470 MW is under way, and a

**Table 1**
**Stripping Ratio for Certain Selected Coal Mine Areas in Singrauli Coal Field**

| Coal Mine | Coal Reserves (million metric tons) | Overburden (million m³) | Stripping Ratio[a] |
|---|---|---|---|
| Jhingurda | 121.00 | 139.0 | 1:1.15 |
| Gorbi | 14.67 | 35.9 | 1:1.47 |
| Jayant | 348.93 | 831.9 | 1:2.60 |
| Bina | 128.70 | 225.4 | 1:1.75 |
| Kakri | 71.93 | 177.0 | 1:2.25 |
| Khadia | 299.20 | 1110.0 | 1:4.25 |
| Dudhichua | 344.96 | 1054.0 | 1:3.29 |
| Almori | 319.25 | 1500.0 | 1:4.30 |

*Source:*   Sinha and Pradham.[10]

[a]  Stripping ratio = coal reserves to overburden.

further addition of 4500 MW is planned by the end of 1995. The projected coal consumption in the area at maximum capacity of 13,277 MW is estimated at 1,097,000 metric tons per day. The coal used is bituminous[9] with a high ash content (30 to 35%) and a low calorific value (3000 to 4205 Cal/kg). The coal has 7 to 13% moisture, 23 to 26% volatile matter, and concentrations of S of 0.4 to 2%, N of 0.45 to 0.9%, and fixed C at 23 to 25%. A realistic assessment of the fly ash emission rate at 35% ash content and 3000 to 4000 K cal/kg calorific value would be 1.38 metric tons/MW.

## Location of Study Sites

The study sites lie between the latitudes of 24° and 24°12′31″ N and the longitudes of 82°40′ and 82°44′30″ E in Sonbhadra district of Uttar Pradesh and Singrauli region of Sidhi district in Madhya Pradesh (Figure 1). The elevation ranges from 200 to 450 m above mean sea level.

## Climate

The climate is tropical monsoonal and there are three distinct seasons: a mild winter (December-February), a hot summer (May-June) and a warm rainy season (July-September). Mean monthly minimum temperature within the annual cycle ranges from 6.4 to 28°C and mean monthly maximum temperature from 20 to 42°C. The annual rainfall averages 1092 mm, of which about 85% occurs between late June and September (Table 2).

## Geology and Soil

The Singrauli series consists of a triangular patch of land lying between the Rihand and Deohar rivers. The parent material is carboniferous in nature. The rocks belong to Raniganj stage and are fine- to coarse-grained feldspathic sandstones,

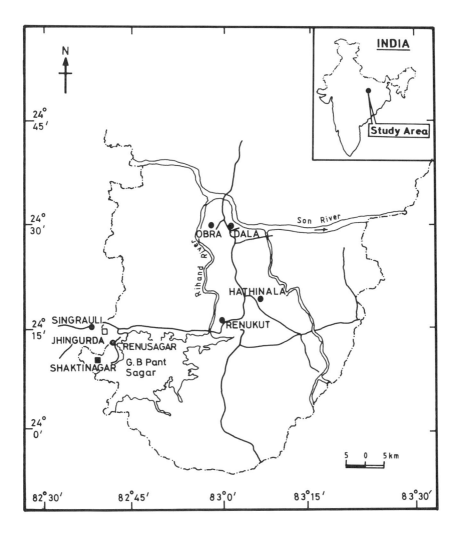

**FIGURE 1.**   Location of the study area: TPPs are Shaktinagar and Renusagar, and the coal-mine spoil is Jhingurda.

white and grey clays with ferruginous bands, carbonaceous shales, and coal seams. The Raniganj stage contains the thickest coal seam recorded for India. Coal seam formation belongs to the Paleozoic Epoch and falls in the Damudas series of the lower Gondwana field, being 230 million years old. Soils are shallow, leached, residual, sandy loam, reddish to reddish brown Ultisols.

## Vegetation

The undulating, nearly flat area is cultivated, with adjacent forest confined mostly to the hilly portions. The native vegetation is a typical mixed dry deciduous forest dominated by *Boswellia serrata* Roxb. ex Colebr., *Lagerstroemia parviflora*

**Table 2**
**Climatological Data of Singrauli Area (average of 4 years, 1986-1989)**

| Month | Average Temperature (°C) | | Rainfall (mm) | Relative Humidity (%) | Wind Velocity (km/hr) |
|-------|------|------|---------------|----------------------|----------------------|
|       | Min. | Max. |               |                      |                      |
| January | 14.3 | 20.4 | 9.1 | 38.6 | 6.2 |
| February | 16.0 | 24.6 | 8.3 | 36.5 | 6.3 |
| March | 23.0 | 31.5 | 12.5 | 43.5 | 7.2 |
| April | 29.3 | 38.0 | 9.0 | 34.8 | 7.9 |
| May | 33.2 | 41.1 | 5.1 | 30.1 | 11.2 |
| June | 32.6 | 39.4 | 166.0 | 60.0 | 12.2 |
| July | 29.7 | 33.1 | 272.5 | 71.3 | 8.9 |
| August | 27.6 | 36.2 | 266.6 | 74.8 | 9.1 |
| September | 29.1 | 36.3 | 280.2 | 67.5 | 7.5 |
| October | 24.6 | 29.3 | 52.6 | 52.5 | 5.9 |
| November | 20.2 | 24.5 | 5.2 | 42.1 | 5.1 |
| December | 16.1 | 20.8 | 4.7 | 38.0 | 5.9 |

Roxb., *Wrightia tomentosa* R and S, *Anogeissus latifolia* Wall, and *Butea monosperma* Lamk. Rapid industrialization and increasing human activities recently have resulted in conversion of substantial stretches of natural forests into savanna, grasslands, and marginal croplands. Surface mining for coal has also caused extensive damage to the natural ecosystems with growing dumps of overburden.

## THERMAL POWER PLANTS

The present three-year study (1987 to 1989) was conducted in areas affected mainly by Renusagar and Shaktinagar power plant emissions. The Renusagar thermal power plant (RTPP) belongs to Hindustan Aluminium Corporation (HINDALCO) at Renukoot with installed capacity of 285 MW. Its stack height is 100 m, equipped with an electrostatic precipitator (ESP) since 1989. The superthermal power plant at Shaktinagar (STPP) is operated and managed by National Thermal Power Corporation of India. This consists of five units of 200 MW capacity each and two units of 500 MW capacity each. The stack heights are 220 m, and all are equipped with ESP facilities. The coal required for the boiler is available from the nearby coalfields, and water for steam generation is taken from the G.B. Pant Sagar (lake). The daily consumption of coal is about 3,300 and 22,950 metric tons at RTPP and STPP, respectively.

Several study sites were selected to the northeast of STPP on an 18-km transect, through RTPP, considering prevailing wind direction, physical relief, and undulations.

### Ambient Air Quality

Relatively large amounts of suspended particulate matter (SPM) and settled dust were evident (Tables 3 and 4). The SPM and $SO_2$ concentrations in ambient

**Table 3**
**Average Seasonal Suspended Particulate Matter (SPM) and SO$_2$ Concentrations at Selected Study Sites During 1989–1990**

| Study Site | Distance and Direction from TPP | SPM ($\mu$g/m$^3$) | | | SO$_2$ ($\mu$g/m$^3$) | | |
|---|---|---|---|---|---|---|---|
| | | Summer | Winter | Rainy | Summer | Winter | Rainy |
| | Shaktinagar | | | | | | |
| S$_1$ | 1.0 km N | 221.5[a] | 196.1 | 201.8 | 92.7 | 66.8 | 118.3 |
| | | ±10.6 | ±9.9 | ±11.5 | ±9.3 | ±5.5 | ±8.9 |
| S$_2$ | 3.0 km NE | 664.0 | 229.6 | 418.6 | 85.5 | 56.6 | 96.6 |
| | | ±16.6 | ±10.5 | ±16.9 | ±8.1 | ±4.8 | ±8.6 |
| S$_3$ | 4.5 km NE | 962.9 | 210.5 | 263.2 | 84.2 | 50.9 | 109.0 |
| | | ±20.8 | ±12.6 | ±14.5 | ±8.6 | ±5.9 | ±9.1 |
| S$_4$ | 6.0 km NE | 795.5 | 401.5 | 554.9 | 187.5 | 77.9 | 233.3 |
| | | ±11.8 | ±14.7 | ±21.0 | ±13.3 | ±6.3 | ±11.1 |
| | Renusagar | | | | | | |
| R$_1$ | 0.5 km N | 266.1 | 239.6 | 412.6 | 100.0 | 86.4 | 149.6 |
| | | ±10.3 | ±11.5 | ±17.3 | ±8.5 | ±4.8 | ±9.5 |
| R$_2$ | 1.0 km N | 439.3 | 211.5 | 228.8 | 106.7 | 62.6 | 123.3 |
| | | ±13.1 | ±9.7 | ±12.8 | ±9.3 | ±5.1 | ±10.8 |
| R$_3$ | 4.0 km N | 201.6 | 153.8 | 183.4 | 88.3 | 52.1 | 99.1 |
| | | ±12.1 | ±10.1 | ±10.0 | ±8.2 | ±4.4 | ±8.5 |
| R$_4$ | 5.0 km N | 153.3 | 142.8 | 169.8 | 52.8 | 48.8 | 93.8 |
| | | ±9.1 | ±9.3 | ±10.7 | ±6.7 | ±3.8 | ±8.1 |

*Note:* Analysis of variance indicated significant differences at $p<0.01$ in SO$_2$ and SPM concentrations among the study sites and among the seasons.

[a] Mean ± 1 SE.

**Table 4**
**Average Seasonal Dustfall at Different Study Sites During 1989–1990**

| Study Site | Distance and Direction from TPP | Dustfall (g/m$^3$/day) | | |
|---|---|---|---|---|
| | | Summer | Rainy | Winter |
| | Shaktinagar | | | |
| S$_1$ | 1.0 km N | 2.10 ± 0.11[a] | 1.10 ± 0.06 | 2.00 ± 0.07 |
| S$_2$ | 3.0 km NE | 1.99 ± 0.10 | 0.89 ± 0.05 | 1.06 ± 0.06 |
| S$_3$ | 4.5 km NE | 1.63 ± 0.07 | 0.86 ± 0.09 | 1.12 ± 0.08 |
| S$_4$ | 6.0 km NE | 3.06 ± 0.12 | 1.58 ± 0.08 | 2.64 ± 0.10 |
| | Renusagar | | | |
| R$_1$ | 0.5 km N | 2.63 ± 0.11 | 1.43 ± 0.06 | 1.89 ± 0.09 |
| R$_2$ | 1.0 km N | 2.48 ± 0.15 | 1.51 ± 0.03 | 1.67 ± 0.09 |
| R$_3$ | 4.0 km N | 1.64 ± 0.08 | 0.99 ± 0.04 | 1.11 ± 0.06 |
| R$_4$ | 5.0 km N | 1.03 ± 0.05 | 0.71 ± 0.04 | 0.92 ± 0.05 |

*Note:* Analysis of variance indicated significant differences at $p<0.01$ in dustfall among the study sites and among the seasons.

[a] Mean ± 1 SE.

air did not always show a gradient of decreasing levels at increasing distances from the source. For example, SPM and $SO_2$ values were usually maximized at 6 km from STPP and 0.5 km from RTPP. The high levels of SPM and settled dust at these sites may be partly due to their proximity to Bina, Jayant, and Kakri coal mines, which produce huge clouds of dust following blasting operations. The loose overburden adds to the quantity of dust, especially during summer due to high wind velocity and negligible rainfall. Studies have revealed that the pollution load in the coal mine areas with respect to SPM level is quite high.[12] Generally, during the rainy season the rains and other meteorological situations reduced the dust levels by 60%. Precipitation accelerates the deposition of particulate matter and gaseous pollutants on the ground. The wind variations were positively related, while the humidity changes were inversely related with SPM concentrations in the air. A highly significant positive correlation ($r = 0.972$, $p<0.01$) occurred between temperature and SPM concentration during summer.

The ambient $SO_2$ concentration also varied in different seasons; it remained highest during winter, followed by summer and rainy seasons (Table 3). Higher winter $SO_2$ concentration may be due to frequent late-night and early-morning thermal inversions, which inhibit vertical mixing and dispersion of pollutants. A similar trend was reported in an industrial area in Dewas district of Madhya Pradesh.[13] Over a three-year study period, there was a marked increase in the ambient $SO_2$ concentration. The mean winter $SO_2$ levels, at 0.5 km from RTPP were 79.55, 88.45, and 124.82 $\mu g/m^3$ during 1987, 1988, and 1989, respectively. The mean annual SPM concentration 6 km from STPP also increased from 468.35 $\mu g/m^3$ during 1987 to 675.18 $\mu g/m^3$ during 1989.

The data on annual variations in concentrations of $SO_2$ and SPM during 1981–1985 in China have indicated increased $SO_2$ and SPM levels in south China. In north China the SPM level has decreased during this period. These variations were related with the differences in meteorological conditions of south and north China.[4] In a horticultural area near power plants in Western Australia, the 24-hr average concentration of $SO_2$ exceeded 100 and 150 $\mu g/m^3$ on 49 and 19 occasions, respectively, during 1979–1980.[14]

The amounts of settled dust were often higher in the immediate vicinity of the power plants and at a site affected by the emissions of both power plants and adjacent coal mines (Table 4). The settled dust levels were 2.1, 1.99, 1.63, 3.06, 2.63, 2.48, and 1.64 $g/m^2/day$ at 1, 3, 4.5, and 6 km from STPP and 0.5, 1, and 4 km from RTPP, respectively. Studies conducted around a 540 MW thermal power plant located at Korba (Madhya Pradesh, India) having a coal consumption of 1500 to 1700 metric tons/day and installed with mechanical dust collectors showed particulate fallout rates (metric tons/$km^2$/month) of 291.0 to 848.2 at 1 km, 134.1 to 548.2 at 2.5 km, and 64.3 to 478.2 at a distance of 4 km from the source.[15] The dustfall rate was higher during the dry season.

Gaseous $SO_2$ and particulates were usually higher around RTPP (capacity 285 MW) as compared to STPP (capacity 2000 MW) (Table 3). These variations are mainly due to the tall stack (220 m) and the efficient control devices installed at STPP.

**Table 5**
**Foliar Deposition (mg/cm³ leaf surface) on Different Plant Species Growing Around Shaktinagar and Renusagar TPPs During Summer Season of 1989.**

| Species | Distance from STPP | | | | Distance from RTPP | | |
|---|---|---|---|---|---|---|---|
| | 1.0 km | 3.0 km | 4.5 km | 6.0 km | 0.5 km | 1.0 km | 5.0 km |
| Mangifera | 7.63[a] | 6.19 | 4.22 | 9.56 | 9.13 | 10.10 | 6.21 |
| indica | ±0.91 | ±0.89 | ±0.96 | ±0.88 | ±1.0 | ±1.1 | ±0.71 |
| Psidium | 4.60 | 4.90 | 3.56 | 5.85 | 4.93 | 5.01 | 3.75 |
| guajava | ±0.62 | ±0.58 | ±0.56 | ±0.49 | ±0.68 | ±0.67 | ±0.71 |
| Cassia | 3.20 | 3.80 | 3.13 | 4.63 | 3.45 | 3.91 | 3.11 |
| siamea | ±0.59 | ±0.21 | ±0.32 | ±0.60 | ±0.49 | ±0.52 | ±0.56 |
| Delonix | 6.88 | 6.73 | 5.29 | 8.61 | 8.99 | 7.26 | 6.55 |
| regia | ±0.78 | ±0.48 | ±0.61 | ±0.78 | ±0.77 | ±0.61 | ±0.59 |
| Eucalyptus | 1.29 | 1.36 | 1.11 | 1.31 | 1.53 | 1.59 | 1.11 |
| hybrid | ±0.09 | ±0.04 | ±0.05 | ±0.11 | ±0.09 | ±0.13 | ±0.08 |
| Bougain- | 3.89 | 4.01 | 3.61 | 4.69 | 4.95 | 3.89 | 3.51 |
| villea | ±0.69 | ±0.47 | ±0.39 | ±0.36 | ±0.53 | ±0.29 | ±0.44 |
| spectabilis | | | | | | | |

*Note:* Analysis of variance test indicated significant differences at $p<0.01$ in foliar dust deposition among species and among seasons.

[a] Mean ± 1 SE.

## Plant Performance in Relation to Air Quality

The leaf surfaces of plants were covered with particulate matter, mainly fly ash and black coal particles. The size of particles settling at sites close to emission sources was large (85±6 to 115±8 μm). A sizable amount of dust remained entrapped at the base and between the leaves of unfolding buds, especially the apical ones, and on branches facing the TPP emission directly. These buds often died and the branches receiving more pollution load were partly defoliated from the tip downward. The flowers were found coated with a thick deposit of dust. Fruit setting was considerably affected at sites 1, 3, and 6 km from STPP and 0.5, 1, and 4.5 km from RTPP. Dustfall, especially at the time of flowering and fruiting, hampered the fruit setting, reducing the economic yield of plants. Fruit setting in *Mangifera indica* was especially affected.

The foliar dust load was maximum during summer, followed by winter, and least in the rainy season. A significant correlation ($p<0.05$) occurred between settled dust and foliar dust depositions during summer for all the plants, except *Eucalyptus hybrid*. The heavily polluted site (6 km from STPP) showed foliar dust deposition of 9.56, 8.61, 4.69, 5.85, 4.63, and 1.31 mg/cm² on foliar surfaces of *Mangifera indica, Delonix regia, Bougainvillea spectabilis, Psidium guajava, Cassia siamea,* and *Eucalyptus hybrid,* respectively (Table 5). The deposited dust may have physically, chemically and physiologically affected the leaves. Dusts have been reported to interfere with stomatal function,[16] increase leaf temperature[17] and transpiration,[18] and reduce photosynthesis.[19]

**Table 6**
**Total Chlorophyll Content (mg/g dry leaf) of Selected Plant Species Growing at Different Distances from Shaktinagar and Renusagar TPPs During Different Seasons of 1989–1990**

| Species | Season[a] | Distance from STPP | | Distance from RTPP | |
|---------|---------|---------|---------|---------|---------|
| | | 1.0 km | 6.0 km | 0.5 km | 8.0 km |
| Mangifera | S | 3.79 ± 0.55[b] | 3.55 ± 0.47 | 3.63 ± 0.29 | 4.9 ± 0.36 |
| indica | R | 4.85 ± 0.69 | 4.65 ± 0.58 | 4.72 ± 0.27 | 5.29 ± 0.31 |
| | W | 4.38 ± 0.61 | 4.01 ± 0.55 | 4.21 ± 0.34 | 4.86 ± 0.44 |
| Psidium | S | 3.34 ± 0.48 | 3.21 ± 0.29 | 3.27 ± 0.21 | 3.96 ± 0.31 |
| guajava | R | 4.09 ± 0.59 | 3.96 ± 0.35 | 3.98 ± 0.30 | 4.47 ± 0.42 |
| | W | 3.05 ± 0.36 | 2.98 ± 0.22 | 3.05 ± 0.29 | 3.72 ± 0.39 |
| Delonix | S | 5.40 ± 0.81 | 5.20 ± 0.46 | 5.18 ± 0.44 | 6.30 ± 0.56 |
| regia | R | 5.78 ± 0.76 | 5.66 ± 0.57 | 5.70 ± 0.41 | 6.50 ± 0.48 |
| | W | 4.71 ± 0.69 | 4.62 ± 0.51 | 4.69 ± 0.39 | 6.01 ± 0.61 |

Note: Differences in chlorophyll content due to plant species, study sites, and seasons were significant at $p < 0.01$. Species × sites and species × seasons interactions were not significant.

[a] S = summer; R = rainy; W = winter.
[b] Mean ± 1 SE.

The individual plant species differed greatly in their dust filtering capacity. On the basis of the dust load per unit leaf area, *M. indica* and *D. regia* were the most efficient dust collectors, whereas *E. hybrid* was a poor dust collector. It is known that the dust-capturing capacity of the plants depends on the orientation and morphology of leaf, trichome density, presence of cuticle, and moisture content on the foliar surface.

The average leaf area of the plants growing in the vicinity of the power plants and other heavily polluted sites was smaller than the leaf area of plants growing at less polluted sites. The reduction in leaf area as a result of pollutant exposure has been shown in several laboratory and field experiments with cultivated plants[20,21] and in field research with native plants.[9,22]

Variations in concentrations of total chlorophyll (Table 6), specific leaf area (SLA) (Table 7) and $SO_4^{2-}$ S (Table 8) were evident in *M. indica, P. guajava,* and *D. regia* at different distances from source in different seasons. In all plants, total chlorophyll and SLA were reduced considerably at sites receiving higher pollution loads. Reductions in chlorophyll content were maximum during summer in *M. indica,* the values being 3.97, 3.55, 3.63, and 4.91 mg/g dry leaf at 1 and 6 km from STPP and 0.5 and 8 km from RTPP, respectively. However, in *P. guajava* and *D. regia,* maximum chlorophyll reductions at all the sites occurred during winter, the values being 3.05, 2.98, 3.05, and 3.72 mg/g dry leaf for *P. guajava* and 4.1, 4.62, 4.69, and 6.01 mg/g dry leaf for *D. regia* at 1 and 6 km from STPP and 0.5 and 8 km from RTPP, respectively. Minimum chlorophyll reductions at all sites were observed during the rainy season.

The chlorophyll content of the leaves is a useful diagnostic tool for subtle pollutant effects.[23,24] Significant reductions in chlorophyll content of western

**Table 7**
**Specific Leaf Area (cm²) of Selected Plant Species Growing at Different Distances from Shaktinagar and Renusagar TPPs During Different Seasons of 1989–1990**

| Species | Season[a] | Distance from STPP | | Distance from RTPP | |
|---|---|---|---|---|---|
| | | 1.0 km | 6.0 km | 0.5 km | 8.0 km |
| Mangifera | S | 14.23 ± 0.91[b] | 13.00 ± 0.88 | 13.05 ± 0.69 | 18.33 ± 0.83 |
| indica | R | 18.45 ± 0.96 | 18.06 ± 0.91 | 17.34 ± 0.87 | 19.75 ±0.66 |
| | W | 14.85 ± 0.85 | 14.51 ± 0.76 | 14.44 ± 0.71 | 19.49 ± 0.63 |
| Psidium | S | 15.05 ± 0.77 | 14.95 ± 0.82 | 14.97 ± 0.68 | 16.34 ± 0.71 |
| guajava | R | 15.88 ± 0.93 | 15.93 ± 0.90 | 16.01 ± 0.83 | 17.46 ± 0.84 |
| | W | 11.06 ± 0.69 | 10.88 ± 0.66 | 10.56 ± 0.79 | 12.14 ± 0.51 |
| Delonix | S | 7.40 ± 0.58 | 7.00 ± 0.60 | 7.12 ± 0.71 | 8.50 ± 0.63 |
| regia | R | 7.66 ± 0.61 | 7.56 ± 0.51 | 7.51 ± 0.51 | 8.70 ± 0.58 |
| | W | 6.10 ± 0.51 | 5.70 ± 0.59 | 5.80 ± 0.59 | 7.80 ± 0.47 |

Note: Differences in specific leaf area due to plant species, study sites, and seasons were significant at $p<0.01$. Species × sites and species × seasons interactions were significant at $p<0.01$.

[a]  S = summer; R = rainy; W = winter.
[b]  Mean ± 1 SE.

**Table 8**
**Sulfate Sulfur Content (%) in Foliage of Selected Plant Species Growing at Different Distances from Shaktinager and Renusager TPPs During Different Seasons of 1989–1990**

| Species | Season[a] | Distance from STPP | | Distance from RTPP | |
|---|---|---|---|---|---|
| | | 1.0 km | 6.0 km | 0.5 km | 8.0 km |
| Mangifera | S | 0.15 ± 0.03[b] | 0.156 ± 0.01 | 0.154 ± 0.02 | 0.130 ± 0.009 |
| indica | R | 0.153 ± 0.02 | 0.162 ± 0.01 | 0.160 ± 0.01 | 0.139 ± 0.01 |
| | W | 0.177 ± 0.03 | 0.189 ± 0.03 | 0.185 ± 0.01 | 0.145 ± 0.02 |
| Psidium | S | 0.141 ± 0.01 | 0.149 ± 0.008 | 0.150 ± 0.009 | 0.124 ± 0.008 |
| guajava | R | 0.155 ± 0.02 | 0.157 ± 0.02 | 0.159 ± 0.01 | 0.132 ± 0.01 |
| | W | 0.137 ± 0.02 | 0.175 ± 0.03 | 0.174 ± 0.02 | 0.155 ± 0.007 |
| Delonix | S | 0.188 ± 0.01 | 0.196 ± 0.01 | 0.205 ± 0.03 | 0.169 ± 0.02 |
| regia | R | 0.198 ± 0.02 | 0.218 ± 0.03 | 0.225 ± 0.003 | 0.188 ± 0.03 |
| | W | 0.221 ± 0.09 | 0.241 ± 0.03 | 0.235 ± 0.02 | 0.192 ± 0.02 |

Note: Differences in $SO_4{}^{2-}$–S content due to plant species, study site, and seasons were significant at $p<0.01$. Species × sites and species × seasons interactions were not significant.

[a]  S = summer; R = rainy; W = winter.
[b]  Mean ± 1 SE.

wheatgrass exposed to different levels of $SO_2$ under field conditions were reported.[25] A decrease in chlorophyll content in leaves of different tree species growing around TPPs has been reported in various studies in India.[24,26,27] Since the $SO_2$ concentrations around power plants are not very high, perhaps $SO_2$ produced $SO_3{}^{2-}$ and $HSO_3{}^-$, involving free radical reactions.[28] The formation of these free radicals may lead to oxidation of chlorophyll, development of chlorotic symptoms, and early senescence in leaves. Furthermore, the deposition of fly ash particles on leaf surfaces may decrease photosynthetic activity of plants by reducing the quantum of available solar radiation as well as by interfering with gaseous exchange.

The growth performance and the physiological status of a plant are reflected by dry weight accumulation per unit foliar surface. The SLA of plants declined considerably at sites receiving higher dust load and experiencing higher $SO_2$ concentrations (Table 7). Reductions in foliar dry weights of plants growing in the vicinity of TPPs have been reported.[9,13] The SLA is also influenced by a number of factors, including light, temperature, and nutrient status.[29] All these factors are considerably changed due to particulate and gaseous emission in the area.

Foliar S analysis indicated that tree species present at sites experiencing higher annual $SO_2$ concentrations, accumulated more foliar $SO_4^{2-}$ S than those at sites with low $SO_2$ level (Table 8). The foliar S content in all the plants increased from summer to winter season. During winter the $SO_4^{2-}$ S content in *M. indica* was 0.177, 0.189, 0.185, and 0.145% at 1 and 6 km from STPP and 0.5 and 8 km from RTPP, respectively. Increase of more than 0.3% S in 1-year-old conifer needles under field conditions after 9 months' exposure to ambient $SO_2$ has been reported.[30] A good correlation between ambient air concentrations of $SO_2$ and foliar S contents around a TPP in Korea was observed; the average correlation coefficient was 0.831, which was significant at the 1% level.[31] Foliar S analysis can be reliably associated with $SO_2$ concentrations in the vicinity of TPPs, but it does not correlate well with visible injury due to low ambient concentrations and other environmental factors. A significant negative correlation between foliar $SO_4^{2-}$ S and total chlorophyll was observed in *M. indica* (r= –0.884, p<0.05). A linear relationship between the total $SO_2$ absorbed and net photosynthetic inhibition has been reported.[30]

The $SO_2$ gas enters the leaf interior through stomata, where it is oxidized first to sulfite and then to sulfate, resulting in elevated sulfur levels.[32] The sulfate fraction is usually considered to represent a pool of accumulated S, when uptake of $SO_2$ exceeds the metabolic demand for synthesis of organic S compounds.[33] The interaction between particulate matter and $SO_2$ has been reported in the atmosphere.[34] Sulfur dioxide gets adsorbed on the dust particles, remains there in hydrated form, and is introduced into the living organisms.[34] Thus, $SO_2$ enters the plant both in free gas phase as well as gradually in the adsorbed phase.

## Fly Ash and Trace Elements

The concentrations of metal elements in the leaves of *M. indica* were higher in the plants growing in proximity to the source than in plants growing at a distance, except for Mn, which was maximum at 2 km from RTPP (Table 9). The height of the stack plays a decisive role in the distribution of trace elements emitted from the source. Prevailing wind conditions strongly affect the distribution and deposition of airborne trace metals on vegetation.[35] A diverse trace element mixture is discharged into the atmosphere as a vapor or as an aerosol. The trace elements can be removed from the atmosphere by wet and/or dry deposition,

**Table 9**
**Trace Element Concentrations ($\mu$g/g) in Leaves of *M. indica* at Different Study Sites Around Renusagar TPP During Winter Season of 1989–1990**

| Study Site Location | Mn | Fe | Cd | Cu | Pb | Ni |
|---|---|---|---|---|---|---|
| 0.5 km N | 106.0[a] | 26.85 | 1.55 | 30.0 | 12.89 | 2.38 |
| | ±18.8 | ±3.89 | ±0.06 | ±2.69 | ±2.22 | ±0.35 |
| 1.0 km N | 130.1 | 22.35 | 1.49 | 28.0 | 12.67 | 2.04 |
| | ±26.6 | ±2.63 | ±0.08 | ±3.02 | ±2.51 | ±0.40 |
| 2.0 km N | 138.5 | 20.0 | 1.34 | 25.3 | 9.50 | 1.75 |
| | ±20.5 | ±3.11 | ±0.05 | ±2.22 | ±1.88 | ±0.11 |
| 3.0 km N | 96.8 | 21.5 | 1.28 | 28.1 | 8.96 | 1.63 |
| | ±16.9 | ±2.53 | ±0.08 | ±2.18 | ±1.10 | ±0.28 |
| 5.0 km NE | 89.0 | 18.8 | 1.29 | 21.9 | 8.16 | 1.92 |
| | ±18.2 | ±2.19 | ±0.07 | ±1.95 | ±1.71 | ±0.31 |
| 7.0 km NE | 74.9 | 17.6 | 1.21 | 19.9 | 8.01 | 1.61 |
| | ±15.0 | ±2.38 | ±0.09 | ±2.38 | ±1.02 | ±0.22 |
| 8.0 km NE | 69.1 | 17.1 | 1.11 | 18.1 | 7.48 | 1.59 |
| | ±16.2 | ±1.99 | ±0.06 | ±2.53 | ±1.11 | ±0.26 |

*Note:* Analysis of variance indicated significant differences ($p<0.01$) in the foliar trace element concentrations due to distance from the power plant. Element × distance interaction was also significant ($p<0.05$).

[a] Mean ± 1 SE.

or sedimentation, diffusion, and impaction on solid objects. During transport and deposition, the metal may be dissolved partly or totally in liquid droplets. Dissolved and undissolved metals may be adsorbed, bound chemically, or taken up by leaves, bark of stems, or branches. Only part of the substances deposited on the canopy reach the soil by precipitation. The particulates rich in Ca and trace elements act as a sink for the dry deposited $SO_2$. Butler[36] showed that up to 20% of the $CaCO_3$ particulate mass, in 3 to 10 $\mu$m particle size deposited on a test surface, was converted to $CaSO_4$ per day, when exposed to $SO_2$-rich environment. Liberti[34] found that the alkalinity of various atmospheric particulate samples is a determining factor in favor of the conversion of $SO_2$. Williams and Stensland[37] showed that road dust, a large source of calcium comparable on a charge-equivalent basis to the amount of anthropogenic $SO_2$, is capable of $SO_2$ adsorption which becomes a significant regional sink of $SO_2$.

The present data indicate that the alkaline material emitted is sufficient to almost neutralize the total $SO_2$ emission. The elemental composition of fly ash could vary widely depending upon coal characteristics. Usually, fly ash contains adequate amounts of essential plant nutrients, except for N, P, and K. The elemental composition of fly ash collected from stacks at RTPP[9] was determined (Table 10). The chemical composition revealed large percentages of Si, Al, and Fe, with lesser quantities of Ca, K, Na, B, and Mg. The fly ash had negligible amounts of N (<0.01%) and a low amount of P (0.2%), mostly in unavailable form. The size of the particles[9] varied greatly — from >3.4 $\mu$m to 400 $\mu$m. However, most fly ash particles ranged from >3.4 to <14.5 $\mu$m size. The pH was 7.8 and electrical conductivity was 0.85 mmho/cm$^2$, respectively.

**Table 10**
**Chemical Composition of Fly Ash[9] as**
**Determined Through Atomic Absorption**
**Spectrophotometry**

| Element (extractable) | Concentration |
|---|---|
| | Values in % |
| Si | 51.94 |
| Na | 0.49 |
| K | 0.60 |
| Mg | 0.29 |
| Ca | 1.38 |
| Fe | 6.13 |
| Al | 23.94 |
| P | 0.22 |
| | Values in mg/kg |
| B | 400 |
| Co | 21 |
| Ni | 86 |
| Cu | 68 |
| Mo | 50 |
| Zn | 72 |
| Cd | 5 |
| As | 36 |

The distribution of Mn particles depends upon the stack height with maximum pollutant concentrations at a distance of 10 to 14 times the height of the stack[38] from the source of emission. In the present study, the maximum Mn concentration occurred at a distance of 2 km instead of the expected 1 km, probably due to the undulating topography of the area. The availability of Mn to the plants corresponds to the solubility of manganese oxides, which depends on oxidizing conditions and pH. Manganese and Fe oxides are important in fixing trace elements in the soil, rendering them unavailable to the plants.[39] Iron is another essential plant nutrient whose continuous supply is required for proper plant growth. However, free Fe ions may cause oxidation of cellular reductants and production of organic and oxygen radicals, which may eventually lead to the death of the cell.

Contamination of vegetation by Pb, Cd, Cu, and Ni in industrial regions mainly through aerial deposition has been reported.[40] Lead particles deposited on leaf surfaces could affect the intracellular physiological processes of plants if Pb is solubilized and moves into the cells of the leaf. However, much of the Pb present on the leaf surface is removable. Numerous studies have shown considerable reduction of growth with increasing lead concentrations.[41,42]

Airborne Cd deposited on leaves may enter the leaf interior through foliar absorption and adherence, and/or through roots.[43] The soil Cd concentration and pH are two important factors that influence the plant accumulation of Cd. Concentration of Cd at toxic levels results in reduced plant growth and chlorosis[44] and decreased rates of photosynthesis and transpiration.[45]

Copper is another indispensible element for the growth of plants. Many enzymes have been shown to require Cu as a co-factor. However, Cu at toxic levels causes general chlorosis and growth retardation. Mukherji and Das Gupta[46]

Table 11
Extractable Trace Element Concentrations ($\mu$g/g) in Soil at Different Study Sites Around Renusagar TPP During Winter of 1989–1990

| Study Site Location | Mn | Fe | Cd | Cu | Pb | Ni |
|---|---|---|---|---|---|---|
| 0.5 km N | 120.0[a] | 85.0 | 0.023 | 6.0 | 8.6 | 2.2 |
| | ±31.0 | ±8.76 | ±0.006 | ±0.21 | ±0.89 | ±0.08 |
| 1.0 km N | 156.1 | 67.6 | 0.021 | 5.6 | 5.3 | 2.0 |
| | ±29.2 | ±9.0 | ±0.005 | ±0.36 | ±0.55 | ±0.06 |
| 3.0 km N | 110.1 | 66.6 | 0.017 | 5.0 | 5.0 | 2.0 |
| | ±16.5 | ±6.81 | ±0.008 | ±0.20 | ±0.38 | ±0.09 |
| 5.0 km NE | 81.5 | 54.8 | 0.016 | 4.3 | 1.6 | 1.6 |
| | ±18.0 | ±8.0 | ±0.004 | ±0.41 | ±0.40 | ±0.05 |
| 8.0 km NE | 79.5 | 50.1 | 0.013 | 4.1 | 4.2 | 1.3 |
| | ±15.4 | ±4.9 | ±0.003 | ±0.29 | ±0.62 | ±0.08 |

Note: Analysis of variance indicated significant differences at $p<0.05$ in the soil trace element concentrations due to distance from the power plant. Element $\times$ distance interaction was not significant.

[a] Mean ± 1 SE.

observed that elevated Cu levels increased activity of catalase, IAA-oxidase, and peroxidase activity in lettuce seedlings. Excess Cu may also reduce metabolic activity in the soil.

Ni is reported to be necessary for the functioning of certain enzyme systems.[47] It also is important in microbiological fixation of $N_2$ in soil.[48] The symptoms of Ni toxicity includes chlorosis or necrosis of leaves, stunted growth of the roots and shoots, reduced dry matter production, and deformation of various plant parts.

The concentrations of Mn, Fe, Cd, Cu, Pb, and Ni in soil samples usually showed a significant decline as distance from the TPP increased (Table 11). The trace metals in the soil accumulated through atmospheric particulate matter as well as through bulk precipitation. The soil was a good sink of $SO_2$; efficient soil uptake of $SO_2$ occurs, especially when moist and alkaline. Several soil factors are reported to influence the absorption of trace elements by terrestrial plants. Roots act to limit uptake of elements from the soil solution by secreting $H^+$, chelating agents, reducing compounds, or more likely by specific ion absorption through the cell plasma membrane.

The present study indicated deterioration of the ambient air quality near the TPP area. The data on responses of *Mangifera indica, Psidium guajava,* and *Delonix regia* to TPP pollutants provide concrete evidence of the significant plant growth reduction up to a distance of 8 km from the emission sources.

## COAL MINE SPOIL

The drastically disturbed mine spoil ecosystems are usually physically, chemically, biologically, and nutritionally recalcitrant media for plant growth. Mine spoils present undesirable conditions for both plant and microbial growth because

**Table 12**
**Certain Physicochemical and Biological Properties of Mine Spoils and Forest Soils**

| | Mine Spoil Age (years) | | | Forest Soil |
|---|---|---|---|---|
| | 5 | 10 | 20 | |
| | Values in % | | | |
| Soil texture | | | | |
| >2.0 mm | 25 ± 2.0[a] | 23 ± 1.7 | 18 ± 1.0 | 8 ± 1.2 |
| 2 mm–0.2 mm | 61 ± 1.5 | 63 ± 1.0 | 60 ± 1.5 | 64 ± 0.8 |
| 0.2 mm–0.1 mm | 4 ± 0.3 | 6 ± 0.6 | 10 ± 1.4 | 14 ± 0.8 |
| 0.1 mm | 10 ± 0.9 | 8 ± 0.7 | 12 ± 1.3 | 14 ± 1.1 |
| pH | 6.3 ± 0.01 | 6.2 ± 0.02 | 6.8 ± 0.02 | 6.4 ± 0.02 |
| | Values in μg/g | | | |
| Total soil N | 680 ± 20 | 740 ± 50 | 860 ± 40 | 2910 ± 20 |
| $NH_4^+ + NO_3^- $-N | 5.8 ± 0.2 | 7.6 ± 0.3 | 15.6 ± 0.2 | 16.4 ± 0.1 |
| $NaHCO_3$-P | 5.0 ± 0.5 | 7.3 ± 0.6 | 8.9 ± 0.9 | 15.0 ± 1.1 |
| Exchangeable Na | 114 ± 6 | 97 ± 10 | 64 ± 4 | 35 ± 4 |
| Exchangeable K | 35 ± 5 | 49 ± 4 | 74 ± 5 | 264 ± 10 |
| Microbial biomass C | 209 ± 9 | 276 ± 10 | 496 ± 14 | 477 ± 20 |
| Microbial biomass N | 20 ± 2 | 23 ± 2 | 36 ± 2 | 75 ± 3 |
| Microbial biomass P | 7 ± 1 | 10 ± 0 | 16 ± 1 | 29 ± 2 |

*Note:* Analysis of variance indicated significant differences at $p < 0.05$ in all soil parameters among the spoil ages and between spoil and native forest.

[a] Mean ± 1 SE.

of low organic matter content, unfavorable pH, droughty from coarse texture, or oxygen deficient due to compaction. Salinity, acidity, poor water-holding capacity, inadequate supplies of plant nutrients, accelerated rate of erosion, and spoil texture are major problems in mine spoils that affect the revegetation process.[49] There is a lack of information on mechanisms of natural plant succession on mine spoils. By accelerating this process, a self-sustained ecosystem may be developed in a short period.

The present study was aimed at understanding the mechanism of natural revegetation on an age series (5, 10, 12, 16, and 20 years) of mine spoils near Jhingurda coal mines (Figure 1) and to investigate the possibility of accelerating the natural revegetation process through directly seeded plants. However, in the present chapter only data for three age groups (5, 10, and 20 years) are considered. In Jhingurda there are two coal seams[8] called "Jhingurda Top" and "Jhingurda Bottom," about 132 m and 10 to 25 m thick, respectively, separated by a massive sandstone layer having a thickness of 31 to 53 m.

## Physicochemical and Biological Properties

The proportion of coarse fragments (>2.0 mm) was higher in mine spoils than in the native forest soil (Table 12). The 5-year-old mine spoil had higher amount of coarse fragments than the 20-year-old mine spoil. The proportion of coarse fragments decreased with the age of mine spoils (r = –0.943, $p<0.005$).[8] Fragmentation, redistribution, and aggregation of particles evidently occurred with time and development of vegetation and soil processes. Generally, coarse fragments reduced the volume of soil for root penetration and caused lower moisture retention. Particle-size distribution influenced erosion rates on reclaimed mine spoils in the Northern Great Plains.[50]

Total soil nitrogen (TSN) in the 5-year-old spoil was only 23% of that in the native forest (Table 12). Total soil nitrogen increased slightly with the age of mine spoil. The recovery of total soil N during 20 years of succession was 0.0012% per year. Total soil nitrogen and available P concentration in 10 to 70-year-old mine spoils were twice as high as in native forest soils in Oklahoma and these were positively and significantly correlated with age of mine spoils.[51] Similarly, in 1 to 50-year-old mine spoils in North Dakota soils, N and available P increased with age.[52] In Alberta levels of available N and P in 26-year-old mine spoils were lower than native soils.[53]

Mineral N ($NH_4^+$-N+ $NO_3^-$-N) was lower in the present mine spoils than in the native forest soil (Table 12). In the present study, total soil N, mineral N, $NaHCO_3$-extractable P, and exchangeable K increased with the age of mine spoils (r = 0.930 to 0.996, $p<0.05$ to 0.01). These parameters in mine spoils were lower than in native forest soil even after 20 years of succession. Exchangeable Na decreased with the age of mine spoils (r= –0.989, p<0.01) but was higher than native forest soil even after 20 years of succession. High exchangeable Na contents result in deterioration of soil structure, reduced infiltration and greater crusting in spoils. Increase in available P with the age of mine spoils was also reported in North Dakota,[52] and in Alberta.[53]

Microbial biomass C, N, and P increased with the age of mine spoils (Table 12). Microbial biomass C was positively and significantly related with spoil age and root biomass. Root biomass provides direct input of C and nutrients to the soil microbial population through organic secretion and following mortality. Mineral N was positively related with microbial N and $NaHCO_3$-extractable P was positively related with microbial P. The soil biomass in the present spoils can be characterized by a mean C:N:P ratio of 29:3:1. The soil biomass carbon (MB-C, µg/g) was related to $NaHCO_3$-extractable soil phosphorus (P, µg/g) and total soil nitrogen (N, %) according to: MB–C = –167.10+69.83 P–104.35 N, ($r^2$= 0.976, $p<0.005$). Soil biomass C, N, and P were lower in the mine spoils than in the native forest soil, except microbial biomass C was about equal in 20 years. The reduced microbial nutrients in the mine spoils are mainly due to the lack of (1) top soil layer with its associated plant components, (2) more favorable nutrient levels, and (3) more active microbial system.

On the basis of results obtained in the present study, microbial biomass can be considered a critical factor in ecological succession on mine spoils in the early stages of soil redevelopment as it aids in the reestablishment of nutrient cycling.[54]

**Table 13**
**Plant Communities in Different Microsites on Coal Mine Spoils of Different Ages**

| Microsite | Age 5 | Age 10 | Age 20 |
|---|---|---|---|
| Slope | Dactyloctenium aegyptium | Bothriochloa pertusa | Aristida adscensionis |
|  | Digitaria setigera | Cassia pumila | Bothriochloa pertusa |
| Coalpatch | Dactyloctenium aegyptium | Aristida adscensionis | E. tenella |
|  | Aristida adscensionis | Dactyloctenium aegyptium | A. adscensionis |
| Undulating surface | D. aegyptium | A. adscensionis | A. adscensionis |
|  | A. adscensionis | B. pertusa | B. pertusa |
| Flat surface | D. aegyptium | A. adscensionis | A. adscensionis |
|  | A. adscensionis | D. aegyptium | B. pertusa |

## General Floristics

Members of Poaceae, Fabaceae, and Convolvulaceae were predominantly represented in the mine-spoil vegetation.[55] On older spoils, *Butea monosperma* was the most frequent species among the woody components. Scattered individuals of *Acacia nilotica, Cassia fistula, Carissa carandas, Dalbergia sissoo, Holoptelia integrifolia, Diospyros melanoxylon, Azadirachta indica,* and one unidentified tree species occurred on the 20-year-old site in addition to *Butea monosperma.* Sparsely distributed isolated individuals of *Calotropis procera, Woodfordia fruticosa* and *Zyzyphus glaberrima* occurred on all sites.

The grass to legume to other forbs ratios were 0.9:0.8:1; 0.9:0.9:1, 0.8:0.8:1, and 1.1:0.8:1 for slope, coalpatch, undulating surface, and flat surface microsites, respectively.[55] The proportion of legumes on the present mine spoils was higher than in the adjacent grassland. The grasses are beneficial in checking erosion and the legumes have ameliorative benefits on both the physical and chemical properties of spoils because most are potential $N_2$-fixers.

## Plant Community Development

The sum of relative frequency, relative density and relative dominance (IVI) was used to identify a total of six plant communities (Table 13). Species with > 100 IVI on slope were *Dactyloctenium aegyptium, Bothriochloa pertusa,* and *Aristida adscensionis;* on coalpatch *Aristida adscensionis* and *Eragrostis tenella,* on undulating surface *Aristida adscensionis,* and on flat surface *D. aegyptium* and *A. adscensionis.*[8] *Xanthium strumarium* occurred abundantly, but in patches on all mine spoils. The early colonizing species were *D. aegyptium, Digitaria setigera,* and *A. adscensionis.*

The early colonizing plant communities were *D. aegyptium-D. setigera,* occurring on the 5-year-old slope microsite and *D. aegyptium-A. adscensionis*

**Table 14**
**Analysis of Vegetation Developed on Mine Spoils**

| Plant Species | Age 5 Relative Shoot Biomass (%) | IVI Index | Age 10 Relative Shoot Biomass (%) | IIVI Index | Age 20 Relative Shoot Biomass (%) | IVI Index |
|---|---|---|---|---|---|---|
| Alysicarpus monolifer (L.) DC. | 3.9 | 14.1 | 0.6 | 6.4 | 0.1 | 5.0 |
| Aristida adscensionis L. | 30.6 | 74.8 | 62.1 | 130.7 | 18.1 | 82.1 |
| Bothriochloa pertusa L. | 1.5 | 6.9 | 24.3 | 64.7 | 68.7 | 102.2 |
| Cassia tora L. | 3.1 | 13.4 | 0.2 | 2.0 | 0.2 | 4.7 |
| Dactyloctenium aegyptium (L.) Willd. | 39.6 | 99.2 | 4.5 | 21.9 | 0.9 | 9.2 |
| Digitaria setigera R. and S. | 9.5 | 28.3 | 0.9 | 7.6 | 0.1 | 1.7 |
| Eragrostis tenella (L.) | 0.0 | 0.0 | 0.1 | 1.6 | 8.7 | 35.2 |
| Heteropogon contortus (L.) P. Beauv. ex R & S. | 0.0 | 0.0 | 0.3 | 1.6 | 0.1 | 1.3 |
| Tephrosia purpurea (L.) Pers. | 2.4 | 8.7 | 1.4 | 8.8 | 0.7 | 8.4 |
| Tridax procumbens L. | 3.8 | 19.6 | 1.8 | 11.0 | 0.2 | 2.7 |
| Xanthium strumarium L. | 1.9 | 8.6 | 1.0 | 4.2 | 0.5 | 7.1 |
| Other species | 3.7 | 26.4 | 2.8 | 39.5 | 1.7 | 40.4 |
| Species richness | 2.26 | | 2.37 | | 4.32 | |
| Pielou's evenness index | 0.708 | | 0.632 | | 0.597 | |
| Shannon-Weiner diversity index | 2.087 | | 2.005 | | 2.188 | |
| Simpson's index of concentration of dominance | 0.192 | | 0.247 | | 0.208 | |

*Note:* Separate values for relative shoot biomass and importance value index are given for those species which contributed >1% shoot biomass on any site. Data for other species are lumped together. Total number of species sampled was 14 on 5 yr old, 25 on 10 yr old and 39 on 20 yr old spoil.

occurring on 5-year-old coalpatch, undulating surface and flat surface microsites. In the present sites, on slope, undulating surface, and flat surface *A. adscensionis-Bothriochloa pertusa* was the endpoint community in 20-year succession, while on coalpatch the endpoint community was *E. tenella-A. adscensionis* (Table 13). Notwithstanding the changes in species composition, only a few species participated in community formation as dominants or codominants (Table 14). This is illustrative of harsh environmental conditions, both climatic and edaphic. The study region is characterized in the annual cycle by high temperatures, high variability in rainfall, and a very long dry period. Therefore, once the propagules have arrived, only those species which are adapted to the special conditions of the sites, or those species which can acclimatize to these extreme conditions, are selected in very distinctive flora of such a drastically disturbed land. Interestingly, *B. pertusa,* a species of high successional order growing under better soil conditions, participated in the community formation on mine spoils with *A. adscensionis,* which is a major species of degraded grasslands.[56] This indicated the possibility of revegetating the mine spoil with desirable species of higher successional order.

Table 15
Area-Weighted Shoot and Root Biomass (g/m$^2$) in Mine Spoils

| | Shoot Biomass | | Root Biomass | | |
| | Monospecific Patches of *Xanthium strumarium* | Other Vegetation | Monospecific Patches of *Xanthium strumarium* | Other Vegetation | Total |
| Spoil | | | | | |
|---|---|---|---|---|---|
| 5 | 185 | 95 | 139 | 280 | 699 |
| 10 | 201 | 184 | 122 | 401 | 908 |
| 20 | 141 | 190 | 89 | 554 | 797 |

Note: Area-weighted values have been calculated on the basis of area occupied by different microsites.

## Community Characteristics

Community characteristics, i.e., species richness, concentration of dominance, species diversity, and evenness, were evaluated by shoot biomass (Table 14). Species richness was calculated as (S-1) log (N) following Margalef,[57] species diversity (H′) as $\sum_{i=1}^{s}$ (ni/N) log (ni/N) following Shannon and Wiener,[58] evenness as H′/log (S) following Pielou,[59] and concentration of dominance as $\sum_{i=1}^{s}$ (ni/N)$^2$ following Simpson.[60] Where, ni = shoot biomass of the ith species, N = sum of ni values and S = number of species. Species richness and species diversity were maximum in the 20-year-old spoil. Evenness of species was maximum in 5-year-old community and concentration of dominance was greatest in 10-year-old community. Evidently, habitat improvement over time facilitates species recruitment. Higher species richness on fine-textured spoil than on coarse-textured spoil in Alberta, Canada was also reported.[53]

## Shoot Biomass

In the present study, using the area of microsites and shoot biomass values, area-weighted biomass values for monospecific patches of *Xanthium strumarium* and of other vegetation were calculated (Table 15). The area-weighted shoot biomass of *Xanthium strumarium* patches decreased with age while that of other vegetation increased. Decline in shoot biomass of *X. strumarium* patches was due to decline in size and number of *X. strumarium* patches with the age of mine spoils.[8] The contribution of *Xanthium* patches to the total area-weighted shoot biomass remained important, however, and this plant may be an important natural agent for habitat improvement of mine spoils in this region.

In southeastern British Columbia, the 15-year-old naturally revegetated coalmine spoil showed low shoot biomass in comparison with the fertilized 2-, 3-and 5-year-old spoils and native grassland.[61] Total biomass increased with increasing water-holding capacity in mine spoils at Colstrip, Montana.[62] Older mine spoils which contained high Ca and N had higher biomass, but on sites with less favorable conditions biomass remained low for several years.[51] Many plant communities were highly productive for 2 to 5 years, but productivity and vigor declined due to large buildup of litter.[63] Direct relationships between increasing spoil age and increased productiviy have been reported for coal mine spoils in the Northern Great Plains.[64] Better productivity and equal to or greater animal performance were reported in revegetated mined land than in nonmined land in North Dakota.[65]

## Root Biomass

In the present mine spoils, roots were concentrated (65 to 85%) in the 0 to 15 cm layer.[66] Similarly concentrated root biomass in upper soil layer has also been reported for mine spoils in Colstrip, Montana,[67] in the Northern Great Plains,[68] and in southeastern British Columbia.[61] Thus, the plants would derive maximum benefit from soil water furnished by light showers.

Root biomass increased with the age of spoils in all microsites.[66] Area-weighted root biomass of *X. strumarium* patches declined with age while that of other vegetation increased (Table 15).

## Effect of Age and Microsites

In order to study the influence of age, the data on spoil features, microbial biomass C, N and P; shoot and root biomass were subjected to Discriminant Analysis.[69] A territorial map was constructed using first two discriminating functions with group centroid means (Figure 2). Function 1 accounted for 82.14% of the variance and exhibited high positive correlations with the proportions of 2.0 to 0.2 mm particle size, microbial biomass P, and exchangeable K. Function 2 accounted for 15.20% of the variance and showed high positive correlations with microbial biomass N and shoot biomass. The scatter plot indicated that the mine spoils of different ages are separated from each other. The 10-year-old and 12-year-old spoils were closer to each other. The 5-year-old spoil and 20-year-old spoil were quite far apart.[8] Thus, spoil of each age was characterized by its own complex physicochemical and biological characteristics with a progressive trend of improvement.

Differences in plant communities and plant biomass indicated the importance of microsite. The effect of microsite was further explored through Discriminant Analysis. Spoil features, microbial biomass C, N and P, and root and shoot biomass values were used in this analysis. By using first two Discriminant Functions a territorial map was constructed with group centroid means (Figure 3). Function 1 accounted for 83.27% of the variance and showed high positive correlation with exchangeable K and a high negative correlation with microbial biomass P. Function 2 accounted for 15.74% of the variance and showed high positive correlations with the proportion of 0.2 to 0.1 mm particle size and microbial biomass P. The scatter plot indicated that the microsites are separated from each other. Undulating surface and flat surface microsites were closer together, while coalpatch and slope microsites were very distinct from each other. This indicated that microsites remain important throughout the 20 years of succession and they have to be considered in any plan

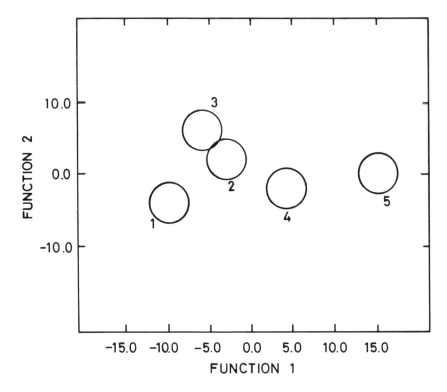

**FIGURE 2.**    Group Centroid means plotted against first and second Discriminant, 1= 5-year-old spoil, 2= 10-year-old spoil, 3= 12-year-old spoil, 4= 16-year-old spoil and 5= 20-year-old spoil. Data were used for five age groups of mine spoils (i.e., 5, 10, 12, 16, and 20). Data are available in Jha.[74]

of revegetation. Generally undulating and flat surface microsites were superior in terms of plant growth compared to slope and coalpatch microsites.

In the present study, although data on species composition satisfy the requirements of "tolerance model"[70] or "initial floristic composition"[71] hypothesis; the resource-ratio hypothesis[72] seems to be operational inasmuch as change in relative proportions of species are concerned.

## Evaluation of Mine Spoil as a Medium for Plant Growth

During mining activities, overburden strata displacement may contain materials that may be unfavorable or potentially unfavorable to plant growth. Soil fertility is a major factor which regulates plant growth; usually mine spoils are deficient in most nutrients.

Mine spoil alone of different ages; mine spoil of different ages+NPK (equivalent to 120 kg/ha N as urea, 60 kg/ha P as $P_2O_5$, and 60 kg/ha K as $K_2O$); and mine spoil of different ages+forest soil (5:1, mine spoil to forest soil) were seeded in triplicate with 13 plant species (tree legumes, leguminous forbs, and grass) in polyethylene pots.

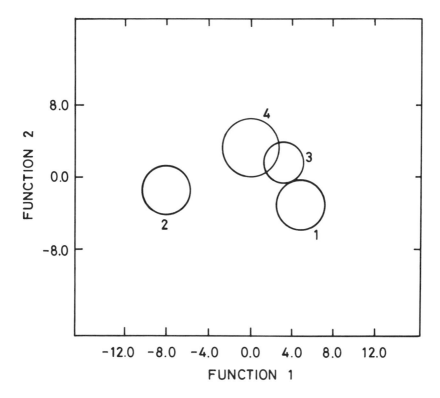

**FIGURE 3.** Group Centroid means plotted against first and second Discriminant, 1= slope, 2= coalpatch, 3= undulating surface and 4= flat surface. Data were used for five age groups of mine spoils (i.e., 5, 10, 12, 16 and 20). Data are available in Jha.[74]

Applications of NPK showed significantly greater shoot growth as compared to mine spoil only and mine spoil+forest soil treatments (Table 16).[73] Mine spoil+forest soil treatment showed significantly greater shoot growth only in three plant species. Forest soil is impoverished, so that shoot growth response in a majority of plant species was not greater after addition of forest soil to mine spoil compared to NPK treatment.[74]

In a majority of plant species, shoot growth was higher in younger ages of spoils than in older ages of mine spoils.[74] This apparently was due to enhanced nutrient content with age of mine spoil.[73] Effect of application of NPK was generally greater in grass compared to leguminous trees and forbs (Table 16). A sevenfold increase in grass yields and a sixfold increase in herbaceous ground cover following nitrogen fertilization of gold-tailing from telluride ores has been reported.[75] A linear effect of NPK fertilizer on yield was obtained when NPK were applied at higher rates in southern Ohio coalmine spoil.[76] In spoil from an opencast bauxite mine at Weipa in Queensland, Australia, P was the most limiting element followed by Cu for the legume *(Stylosanthes gracilis)* and N for the grass *(Brachiaria decumbens).*[77]

**Table 16**
**Effect of Application of NPK and Forest Soil on Maximum Shoot Growth Length (cm) in a Mine Spoil[a]**

| Species | Mine Spoil | Mine Spoil + NPK | Mine Spoil + NPK |
|---|---|---|---|
| Acacia catechu | 26.6 ± 0.04[b] | 38.7 ± 0.02 | 32.4 ± 0.02 |
| Acacia nilotica | 62.8 ± 0.02 | 90.0 ± 0.03 | 77.6 ± 0.07 |
| Albizia procera | 47.6 ± 0.04 | 109.2 ± 0.08 | 63.4 ± 0.06 |
| Dichrostachys cinerea | 47.5 ± 0.05 | 80.6 ± 0.06 | 70.0 ± 0.04 |
| Leucaena leucocephala | 68.4 ± 0.04 | 82.4 ± 0.06 | 74.4 ± 0.03 |
| Pongamia pinnata | 54.2 ± 0.03 | 72.4 ± 0.07 | 58.2 ± 0.02 |
| Prosopis juliflora | 78.0 ± 0.06 | 105.1 ± 0.06 | 92.2 ± 0.01 |
| Clitoria ternatea | 105.8 ± 0.09 | 124.0 ± 0.05 | 112.6 ± 0.05 |
| Desmanthus virgatus | 76.0 ± 0.07 | 107.0 ± 0.04 | 93.6 ± 0.04 |
| Macroptelium atropurpureum | 95.4 ± 0.06 | 118.6 ± 0.03 | 105.8 ± 0.06 |
| Stylosanthus hamata | 48.0 ± 0.04 | 61.3 ± 0.02 | 56.0 ± 0.08 |
| Stylosanthus humilis | 42.8 ± 0.02 | 56.2 ± 0.04 | 47.9 ± 0.07 |
| Cenchrus ciliaris | 27.8 ± 0.01 | 59.4 ± 0.04 | 48.8 ± 0.03 |

Note: Analysis of variance showed significant differences among the treatments at $p<0.001$ in S. hamata and S. humilis (mine spoil, mine spoil + NPK and mine spoil + forest soil) and at $p<0.0001$ in the remaining species.

[a] Values are averaged across spoil ages.
[b] Mean ± 1 SE.

## Growth Performance of Certain Directly Seeded Plants

Herbaceous plant species were used to enhance natural succession and development of wildlife habitat in Appalachian.[78] Seeding or planting of plant species accelerated succession that fulfilled the revegetation goal. It controlled soil erosion and increased plant diversity. Mine spoils can be stabilized through the establishment of beneficial plant cover. Climax and subclimax native and introduced plant species accelerated soil development on mine spoils.

The growth performance of 30 directly seeded leguminous trees, nonleguminous trees, leguminous forbs, grasses, and crops were evaluated on 12-year-old flat surface microsite (Table 17). Among three species of *Acacia,* the native species, *A. nilotica* showed better growth performance ($p<0.002$) than the exotic species, *A. tortolis* and *A. auriculiformis.* Among three species of *Albizia, A. lebbeck* showed better performance ($p<0.01$) than *A. amara* and *A. procera. Dichrostachys cinerea* was superior in growth performance ($p<0.03$) to *Dichrostachys corten. Clitoria ternatea* performed best ($p<0.001$) among the leguminous forbs, *Bothriochloa intermedia* (closely followed by *Cenchrus setigerus) ($p<0.0001$)* among the grasses, and *Cajanus cajan* and *Pennisetum typhoides* ($p<0.001$) among the crops. Among the tree species tried, *Sesbania aegyptiaca* attained maximum (130 cm) shoot height after 1 year of seeding.[74] Fast-growing species and trees that maintain the water balance (e.g., Phraetophyte group) have been recommended for revegetation of spoils.

Leguminous trees and forbs tried on the present mine spoils showed satisfactory growth performance. Legumes accumulate 100 kg of N/ha/year in

Table 17
Performance of Plant Species Seeded on 12-Year-Old Flat Surface Mine Spoil

| Species | Mean Length of Shoot (cm) | Mean Shoot Dry Wt (g/plant) | Mean Root Dry Wt (g/plant) | Total Wt (g/plant) |
|---|---|---|---|---|
| Acacia auriculiformis | 33.5 ± 0.9[a] | 2.0 ± 0.03 | 1.0 ± 0.02 | 3.0 |
| Acacia nilotica | 86.5 ± 1.0 | 5.0 ± 0.02 | 2.5 ± 0.01 | 7.5 |
| Acacia tortolis | 47.5 ± 0.8 | 3.0 ± 0.03 | 2.5 ± 0.02 | 5.5 |
| Albizia amara | 40.0 ± 0.7 | 2.0 ± 0.01 | 1.5 ± 0.02 | 3.5 |
| Albizia lebbeck | 45.0 ± 0.7 | 1.5 ± 0.01 | 3.0 ± 0.01 | 4.5 |
| Albizia procera | 20.0 ± 0.4 | 2.0 ± 0.01 | 1.5 ± 0.01 | 3.5 |
| Dichrostachys cinerea | 66.2 ± 0.8 | 7.0 ± 0.03 | 1.5 ± 0.01 | 8.5 |
| Dichrostachys corten | 28.0 ± 0.3 | 3.5 ± 0.02 | 2.5 ± 0.02 | 6.0 |
| Pongamia pinnata | 46.0 ± 0.6 | 6.1 ± 0.03 | 4.0 ± 0.02 | 10.1 |
| Sesbania aegyptiaca | 130.0 ± 1.0 | 10.0 ± 0.04 | 2.5 ± 0.01 | 12.5 |
| Clitoria ternatea | 90.0 ± 1.2 | 20.0 ± 0.04 | 2.5 ± 0.01 | 22.5 |
| Desmanthus virgatus | 65.0 ± 1.0 | 5.0 ± 0.02 | 1.2 ± 0.01 | 6.2 |
| Desmodium tortosum | 100.0 ± 0.8 | 6.0 ± 0.03 | 3.2 ± 0.02 | 9.2 |
| Macroptelium atropurpureum | 75.0 ± 0.6 | 8.0 ± 0.04 | 2.5 ± 0.01 | 10.5 |
| Stylosanthus hamata | 35.0 ± 0.5 | 8.0 ± 0.05 | 1.2 ± 0.01 | 9.2 |
| Bothriochloa intermedia | 105.0 ± 0.9 | 21.0 ± 0.03 | 4.0 ± 0.02 | 25.0 |
| Bothriochloa pertusa | 110.0 ± 1.0 | 15.0 ± 0.04 | 2.0 ± 0.01 | 17.0 |
| Cenchrus setigerous | 95.0 ± 0.8 | 18.0 ± 0.05 | 6.6 ± 0.02 | 24.6 |
| Chrysopogon fulvus | 105.0 ± 0.7 | 14.0 ± 0.05 | 6.5 ± 0.02 | 20.5 |
| Dendrocalamus strictus | 45.0 ± 0.5 | 5.0 ± 0.04 | 2.5 ± 0.01 | 7.5 |
| Cajanus cajan | 103.3 ± 1.0 | 57.6 ± 0.03 | 10.3 ± 0.02 | 67.9 |
| Pennisetum typhoides | 90.0 ± 1.0 | 12.3 ± 0.04 | 1.9 ± 0.01 | 14.2 |
| Phaseolus mungo | 21.0 ± 0.5 | 2.6 ± 0.02 | 0.8 ± 0.01 | 3.4 |
| Phaseolus aureus | 22.0 ± 0.5 | 2.7 ± 0.01 | 0.7 ± 0.01 | 3.4 |

Note: Data for C. cajan were collected after 150 days of sowing, for P. typhoides P. mungo, and P. aureus after 90 days of sowing, and for all other species after 1 year of sowing.

[a] Mean ± 1 SE.

certain types of waste,[79] in an easily mineralizable form. Use of legumes was more advantageous than fertilizers when energy cost was considered,[80] and an actively $N_2$-fixing legume should supply nitrogen to the root zone of a sward continuously in a way that no fertilizer can do. Legumes have ameliorative benefits on both the physical and chemical properties of spoils.[81–83] The maintenance of even a minor legume component in a plant community coupled with the development of adequate mineralizable organic P would ensure long-term N and P availability.[84] For bauxite mine spoils in Australia, use of legumes inoculated with effective strains of *Rhizobium* to meet the N requirements of pastures, and application of 50 kg/ha of Mo superphosphate to overcome Mo deficiency in N-fixation have been recommended.[85]

Grass species such as *Bothriochloa pertusa, B. intermedia, Chrysopogon fulvus,* and *Cenchrus setigerous,* and the bamboo *Dendrocalamus strictus* performed well on the present mine spoils.[74] Sod-forming grasses protect surface from erosion. Once grass species established, organic matter began to accumulate in the surface layer, and the spoil was equally stable from erosion so long as the land was not heavily grazed.[86]

On Australian mine spoils, potential pastures such as grass-legume pastures, legume shrublands, agro-forestry, and native woodlands with stylos have been developed which have potential for grazing and future improvement.[85] Coal mine spoils in the Bowen Basin of Queensland, Australia, are revegetated with mixed grass-legume pastures. Aerial seeding and fertilizing have been done with mixed success.[87] In opencast bauxite mine spoil at Weipa in Queensland, several introduced legumes and grasses have been found to be productive and persistent.[88] *Panicum maximum, Brachiaria mutica, B. humidicola, B. decumbens, Cenchrus ciliaris, Paspalum plicatulum* cv. Roddy's Bay, *Digitaria decumbens, Macroptelium atropurpureum* cv. Siratro, *Centrosema pubescens, Stylosanthes hamata* cv. Verano, and *S. scabra* have been recommended for poorly prepared ground areas where dry seasons extend more than 6 months without effective rainfall and in soils with limited available nutrients.[85]

Crop plants tried on the present mine spoils also performed well. The growth performance of *Cajanus cajan* was better in mine spoil than in adjacent agricultural farm plots.[74] Cultivation of *Cajanus cajan* increased total P availability, and it can grow and yield well in soils of low available P level without P fertilizer applications because of its ability to tap Fe-P, which cannot be easily utilized by other crops.[89] Plant species, however, may not be equally suited for all environmental conditions in a coal mining region. Only the species or species cultivars best suited for specific postmine land uses and adapted to the post-mine environment should be selected.[74]

## CONCLUSIONS AND RECOMMENDATIONS

The TPPs located in the Singrauli belt have ecologically disturbed the area, and their emissions eventually could prove ruinous for the terrestrial ecosystems of forested areas, agricultural fields, and grasslands. The growth, productivity, and regeneration of remaining forest vegetation under the present level of emissions probably will induce irreversible changes in the community structure.

Fly ash use may be limited in low rainfall areas as it creates trace element buildup in the area, but being alkaline it neutralizes the acidity caused by $SO_2$ emissions from TPPs. This may be why acidic rains are not frequent in the area in spite of significant $SO_2$ levels emitted from the stack. The pH of the first rainfall during 1988-1989 ranged from 4.8 to 5.2 at different sites within an 8 km² area around power plants.

Electrostatic precipitators and wet scrubbers should be used to minimize the stack emission of particulate and gaseous pollutants in the area. Control equipment already installed must be made to perform at maximum efficiency to reduce the emissions at the source. The overburden and the fly ash dumps should not be left dry, but should be covered with plants to check particulate pollution due to surface erosion.

A massive tree plantation program should be undertaken to alleviate the effects of particulate and gaseous pollutants in the area, especially in the vicinity of emission sources. For developing a green belt near the sources, sensitive species like *Mangifera indica* and *Delonix regia* should be avoided. Plant species such as *Acacia nilotica, Eucalyptus hybrid, Phyllanthus distichus, Prosopis juliflora,* and *Tamarindus indica* should be planted.

The mine spoils studied were generic in nature, nontoxic, and suitable (although sub-optimal) media for plant growth. The natural process of revegetation can be accelerated by seeding mixtures of grasses, leguminous forbs, and leguminous tree species which showed satisfactory growth performance in field trial experiments. Direct seeding is easier and more cost-effective than spot seeding or drilling. The three most promising species in each category suggested for direct seeding on mine spoils are

1.  **Leguminous trees:** *Sesbania aegyptiaca, Acacia nilotica,* and *Dichrostachys cinerea*
2.  **Leguminous forbs:** *Desmodium tortosum, Clitoria ternatea,* and *Macroptelium atropurpureum*
3.  **Grasses:** *Bothriochloa intermedia, Bothriochloa pertusa,* and *Chrysopogon fulvus*
4.  **Crops:** *Cajanus cajan, Phaseolus aureus,* and *Pennisetum typhoides*

During 20 years of plant succession, microsite remained important. Undulating and flat surface microsites were better than slope and coal patch for plant growth. Undulating and flat surface microsites recovered more rapidly than other microsites. Therefore, in any revegetation plan the proportion of slope and coalpatch microsites should be reduced in comparison to undulating and flat surface microsites. On slope, plant species such as *Stylosanthes hamata* should be seeded to check soil erosion.

## ACKNOWLEDGMENT

Funding support from the Ministry of Environment and Forests, Government of India is gratefully acknowledged.

## REFERENCES

1. Guo, B. S., Zhang, G. F., and Cai, L., Study on countermeasures for issues of energy and acid precipitation in China, in *Proc. Symp. Effects of Acid Precipitation on Comprehensive Agriculture and Its Countermeasures,* China Association for Science and Technology, 1988, 36.
2. Rao, D. N., Air pollution and plants, *Biol. Memories,* 11(1), 60, 1985.
3. Kumar, S. and Upadhyay, S. N., Atmospheric emissions from thermal power plants in India, in *Proc. Symp. Air Poll. Control,* Air Pollution Control Association, India, 1983, 119.
4. Hougfa, C., Air pollution and its effects on plants in China, *J. Appl. Ecol.,* 26, 763, 1989.
5. Rajgopal, S., Power and environment: on a collision course, in *The Hindu Survey of the Environment,* 1991, 93.

6. Huancheng, G., Dengru, W., and Hongxing, Z., Land restoration in China, *J. Appl. Ecol.*, 26, 787, 1989.

7. Bose, A. K. and Singh, B., Environmental problems in coal mining areas, impact assessment and management strategies — case study in India, in *Man and his Ecosystem*, Vol. 4, Brasser, L. J. and Mulder, W. C., Eds., Elsevier, Amsterdam, 1989, 243.

8. Singh, J. S., Singh, K. P., and Agrawal, M., Eds., Environmental Degradation of Obra-Renukoot-Singrauli Area and Its Impact on Natural and Derived Ecosystems, Ministry of Environment and Forests, Government of India, New Delhi, 1990, 498.

9. Rao, D. N., Agrawal, M., and Singh, J., Eds., Study of Pollution Sink Efficiency, Growth Response and Productivity Pattern of Plants with Respect to Fly Ash and $SO_2$, Ministry of Environment and Forests, Government of India, New Delhi, 1990, 136.

10. Sinha, A. K. and Pradhan, G. K., Bulk handling explosive system — its application in Indian surface mines, *Indian Min. Eng. J.*, 25, 27, 1986.

11. Sharma, H. C., Flue gas conditioning in superthermal power station in India, in *Proc. Semin. Industrial Impact on Eastern U.P's Environment and Ecology*, HINDALCO, Renukoot, India, 1990, 7.

12. Banerjee, S. P. and Hussain, H. F., Level of air pollution in the fire area of Jharia coalfield, *Indian J. Environ. Protection*, 9(7), 499, 1989.

13. Rao, M. V. and Dubey, P. S., Plant response against sulphur dioxide in field conditions, *Asian Environ.*, 3(10), 3, 1988.

14. **Anon.**, The Kwinana Air Modelling Study, Report 10, Department of Conservation and Environment, Perth, Western Australia, 1982.

15. Subrahmanyam, G. V., Rao, D. N., Varshney, C. K., and Biswas, D. K., Air Pollution and Plants: A State-of-Art Report, Ministry of Environment and Forests, Government of India, New Delhi, 1985, 193.

16. Ricks, G. R. and Williams, R. J. H., Effects of atmospheric pollution on deciduous woodland. II. Effects of particulate matter upon stomatal diffusion resistance in leaves of *Quercus pertreae* (Mattuschka) Liebl, *Environ. Pollut.*, 6, 87, 1974.

17. Fluckiger, W., Fluckiger-Keller, H., and Oertli, J. J., Der Enfluss von Strassenstaub auf den stomataren Diffusions-widerstand und die blatt-temperaturein antagonistischer Effekt, *Staub-Reinhalt. Luft*, 38, 502, 1978.

18. Beasley, E. W., Effects of some chemically inert dusts upon the transpiration rate of yellow coleus plants, *Physiol. Lancaster*, 17, 101, 1942.

19. Darley, E. F., Studies on the effect of cement-kiln dust on vegetation, *J. Air Poll. Control Assoc.*, 16, 145, 1966.

20. Davis, T., Grasses more sensitive to $SO_2$ pollution in condition of low irradiative and short days, *Nature (London)*, 284, 483, 1980.

21. Heck, W. W., Blum, U., Reinert, R. A., and Heagle, A. S., Effects of air pollution on crop production, in *Strategies of Plant Production*, Mendt, W. J., Ed., BARC Symp. No. 6, 1981, 333.

22. Steubing, L. and Fangmeier, A., $SO_2$ sensitivity of plant communities in a beech forest, *Environ. Pollut.*, 44, 297, 1987.

23. Malhotra, S. S., Effects of aqueous sulphur dioxide on chlorophyll destruction in *Pinus contorta*, *New Phytol.*, 78, 101, 1977.

24. Agrawal, M. and Agrawal, S. B., Phytomonitoring of air pollution around a thermal power plant, *Atmos. Environ.*, 23, 763, 1989.

25. Lauenroth, W. K. and Dodd, J. L., Chlorophyll reduction in western wheat grass (*Agropyron smithii* Rybd.) exposed to sulphur dioxide, *Water Air Soil Poll.*, 15, 309, 1981.

26. Pandey, S. N., Impact of thermal power emissions on vegetation and soil, *Water Air Soil Poll.,* 19, 87, 1983.

27. Vij, B. M., Trivedi, T., Sharada, A., and Dubey, P. S., Chlorophyll damage in tree species due to air pollution, *Biol. Bull. India,* 3, 193, 1981.

28. Asada, K. and Kiso, K., Initiation of aerobic oxidation of sulfite by illuminated spinach chloroplasts, *Eur. J. Biochem.,* 33, 253, 1973.

29. Evans, C., *The Quantitative Analysis of Plant Growth,* Blackwell Scientific, Oxford, 1972.

30. Knabe, W., The role of tree stands for reducing air pollution, in *Proc. 4th Int. Clean Air Congr.,* Tokyo, 1977, 952.

31. Shin, F. B. and Park, W. C., Dose response relationship between rice plants and atmospheric pollution, in *Man and His Ecosystem,* Vol. 2, Brasser, L. J. and Mulder, W. C., Eds., Elsevier, Amsterdam, 1989, 35.

32. Sisson, W. B., Both, J. A., and Thronenberry, G. O., Absorption of $SO_2$ by pecan *(Caryo illinoensis* [Wang] K. Koch) and alfalfa *(Medicago sativa* L.) and its effect on net photosynthesis, *J. Exp. Bot.,* 32, 523, 1980.

33. Cowling, D. W. and Koziol, M. J., Growth of rye-grass *(Lolium perenne)* exposed to $SO_2$, *J. Exp. Bot.,* 29, 1029, 1978.

34. Liberti, A., Brocco, D., and Possanzini, M., Adsorption and oxidation of sulphur dioxide on particles, *Atmos. Environ.,* 12, 255, 1978.

35. Little, P. and Martin, M. H., A survey of zinc, lead and cadmium in soil and natural vegetation around a smelting complex, *Environ. Pollut.,* 3, 241, 1972.

36. Butler, T. J., Composition of particles dry deposited to an inert surface at Ithaca, New York, *Atmos. Environ.,* 22, 895, 1988.

37. Williams, A. L. and Stensland, G. J., The scavenging of sulfur dioxide by dry deposited road dust, in *Proc. 1989 EPA/ADWMA Int. Symp. Meas. Toxic and Related Air Pollutants,* ADWMA Publication No. VIP. 13, EPA Report No. 600/9-89-060, 1989, 230.

38. Dobrovolsky, V., *Geography of microelements,* Myst. Publ., Moscow, 1983.

39. Norrish, K., The geochemistry and minerology of trace elements, in *Trace Elements in Soil-Plant-Animal Systems,* Nicholas, D. J. D. and Egan, A. R., Eds., Academic Press, New York, 1975, 55.

40. Burton, K. W. and John, E., A study of heavy-metal contamination in the Rhondda Fawr, South Wales, *Water Air Soil Poll.,* 7, 45, 1977.

41. Carlson, R. W., Bazzaz, F. A., and Rolfe, G. L., The effect of heavy metals on plants. II. Net photosynthesis and transpiration of whole corn and sunflower plants treated with Pb, Cd, Ni, Ti, *Environ. Res.,* 10, 113, 1975.

42. Briggs, D., Genecological studies on lead to tolerance in groundsel, *New Phytol.,* 77, 173, 1986.

43. Lagerwerff, J. V., Heavy metal contamination of soils, in *Agriculture and the Quality of Our Environment,* Brady, N. C., Ed., 1967, 343.

44. Ormrod, D. P., *Pollution in Horticulture,* Elsevier, Amsterdam, 1978.

45. Bazzaz, F. A., Carlson, R. W., and Rolfe, G. L., The effect of heavy metals on plants. I. Inhibition of gas exchange in sunflower by Pb, Cd, Ni and Ti, *Environ. Poll.,* 7, 241, 1974.

46. Mukherji, S. and Das Gupta, B., Characterization of copper toxicity in lettuce seedlings, *Physiol. Plant.,* 27, 126, 1972.

47. Dixon, N. E., Gazzola, C., Blakeley, R. L., and Zarer, B., Jack bean urease (E.C. 3.5.1.5.): a metalloenzyme. A simple biological role for nickel, *J. Am. Chem. Soc.,* 97, 4131, 1975.

48. Bertland, D., Microbiologie du sol. le nickel, oligo-element dynamique pour les micro-organisms fixateurs de l'azote de l'azote de l'air, *C. R. Acad. Sci. Ser. D*, 278, 1974, 2231.

49. Jha, A. K. and Singh, J. S., Revegetation of mine spoils: review and a case study, in *Environmental Management of Mining Operations*, Dhar, B. B., Ed., Ashish Publishing House, New Delhi, India, 1990, 300.

50. Toy, T. J., and Shay, D., Comparison of some soil properties on natural and reclaimed hillslopes, *Soil Sci.*, 143, 264, 1987.

51. Johnson, F. L., Gibson, D. J., and Risser, P. G., Revegetation of unreclaimed coal strip mines in Oklahoma. I. Vegetation structure and soil profiles, *J. Appl. Ecol.*, 19, 453, 1982.

52. Wali, M. K., The structure, dynamics, and rehabilitation of drastically disturbed ecosystems, in *Perspectives in Environmental Management*, Khoshoo, T. N., Ed., Oxford Publications, New Delhi, India, 1987, 163.

53. Russell, W. B. and La Roi, G. B., Natural vegetation and ecology of abandoned coalmined land, Rocky Mountain foothills, Alberta, Canada, *Can. J. Bot.*, 64, 1286, 1986.

54. Srivastava, S. C., Jha, A. K., and Singh, J. S., Changes with time in soil biomass C, N and P of mine spoils in a dry tropical environment, *Can. J. Soil Sci.*, 69, 849, 1989.

55. Jha, A. K. and Singh, J. S., Vascular flora of naturally revegetated coalmine spoils in a dry tropical environment, *J. Trop. For.*, 6, 131, 1990.

56. Pandey, C. B. and Singh, J. S., Influence of grazing and soil conditions on derived savanna vegetation in India, *J. Vegetation Sci.*, 2, 95, 1991.

57. Margalef, R., Information theory in ecology, *Gen. Syst.*, 3, 36, 1958.

58. Shannon, C. E. and Weaver, W., The mathematical theory of communication, University of Illinois Press, Urbana, 1963, 117.

59. Pielou, E. C., The measurement of diversity in different types of biological collections. *J. Theor. Biol.*, 13, 131, 1966.

60. Simpson, E. H., Measurement of diversity, *Nature*, 163, 688, 1949.

61. Fyles, J. W., Fyles, I. H., and Bell, M. A. M., Vegetation and soil development on coalmine spoil at high elevation in the Canadian Rockies, *J. Appl. Ecol.*, 22, 239, 1985.

62. Schafer, W. M. and Nielsen, G. A., Soil development and plant succession on 1- to 50-yr old strip mine spoils in southern Montana, in *Ecology and Coal Resource Development, Vol. 2*, Wali, M. K., Ed., Pergamon Press, New York, 1979, 541.

63. DePuit, E. J., Coenenberg, J. G., and Willmuth, W. H., Research on revegetation of surface-mined lands at Colstrip, Montana, *Agricultural Experiment Station Research Report*, Montana State University, Bozeman, 165, 1978.

64. Schafer, W. M., Mine-soil restoration and maturity: a guide for managing mine-soil development, in *Proc. Symp. Surface Coal Mining and Reclamation in the Great Plains*, 172, 1984.

65. Hofmann, L., Ries, R. E., and Gilley, J. E., Relationship of runoff and soil loss to ground cover of native and reclaimed grazing lands, *Agron. J.*, 75, 599, 1983.

66. Jha, A. K., A note on root development in dry tropical naturally revegetated coalmine spoils, *Vegetatio*, 85, 67, 1989.

67. Holechek, J. L., Root biomass on native range and mine spoils in southeastern Montana, *J. Range Manage.*, 35, 185, 1982.

68. Wyatt, J. W., Dollhopf, D. J., and Schafer, W. M., Root distribution in 1 to 48-year-old stripmine spoils in southeastern Montana, *J. Range Manage.*, 22, 101, 1980.

69. SPSS, *SPSS/PC+,* SPSS Inc., Chicago, 1985.

70. Connell, J. H. and Slatyer, R. O., Mechanisms of succession in natural communities and their role in community stability and organization, *Am. Nat.,* 111, 1119, 1977.

71. Egler, F. E., Vegetation science concepts. I. Initial floristic composition — a factor in old-field vegetation development, *Vegetatio,* 4, 412, 1954.

72. Tilman, D., The resource-ratio hypothesis of plant succession, *Am. Nat.,* 125(6), 827, 1985.

73. Jha, A. K., Evaluation of coal mine spoil as a medium for plant growth in a dry tropical environment, *Indian Forester,* 1991.

74. Jha, A. K., A Study on Revegetation of Coal-Mine Spoil, Ph.D. thesis, Department of Botany, Banaras Hindu University, Varanasi, India, 1990, 242.

75. Berg, W. A., Vegetative stabilization of mine wastes, in *1972 Mining Yearbook,* Colorado Mining Association, Denver, 1972, 24.

76. Sutton, P. and Hall, G. F., Fertilization for forage production on calcareous mined land in Ohio, *J. Soil Water Cons.,* 42, 361, 1987.

77. Grundy, M. J., Mineral Nutrient Requirements for Pasture Establishment on Mined Land at Weipa, North Queensland, Master of Agricultural Science thesis, University of Queensland, Austrialia, 1979.

78. Rafaill, B. L. and Vogel, W. G., A guide for vegetating surface-mined lands for wildlife in eastern Kentucky and West Virginia, Fish and Wildlife Service, U.S. Department of Interior, Washington, D.C., 1978.

79. Dancer, W. S., Handley, J. F., and Bradshaw, A. D., Nitrogen accumulation in kaolin mining wastes in Cornwall. II. Forage legumes, *Plant Soil,* 48, 303, 1977.

80. Laidlaw, A. S. and Wright, C. E., The advantages in energy terms of legumes in grassland systems, *Grass Forage Sci.,* 35, 70, 1980.

81. Vimmerstedt, J. P., House, M. C., Larson, M. M., Kasile, J. D., and Bishop, R. L., Nitrogen and carbon accretion on Ohio coal mine spoils: influence of soil forming factors, *Landscape and Urban Planning,* 17, 99, 1989.

82. Alexander, M. J., The long-term effect of *Eucalyptus* plantations on tin-mine spoil and its implication for reclamation, *Landscape and Urban Planning,* 17, 47, 1989a.

83. Alexander, M. J., The effect of *Acacia albida* on tin-mine spoil and their possible use in reclamation, *Landscape and Urban Planning,* 17, 61, 1989b.

84. Roberts, J. A., Daniels, W. A., Bell, J. C., and Martens, D. C., Tall fescue production and nutrient status on southwest Virginia mine soils, *J. Environ. Qual.,* 17, 55, 1988.

85. Lawrie, J. and Gunness, A., Improved pastures on mined land at Weipa, in *Ecology and Management of World's Savannas,* Tothill, J. C. and Mott, J. J., Eds., Australian Academy of Science, Canberra, in conjunction with C.A.B. Farnham Royal, Bucks, 1985, 332.

86. Ries, R. E. and Hofmann, L., Re-establishment and use of grassland in reclaimed soils, in *Can Mined Land Be Made Better Than Before Mining?,* Bismarck, ND, North Dakota Energy Development Impact Office, 1983, 85.

87. Roe, P. A., Rehabilitation of opencut coal mines in the Bowen Basin of Queensland, in *Ecology and Management of World's Savannas,* Tothill, J. C. and Mott, J J., Eds., Australian Academy of Science, Canberra, in conjunction with C.A.B., Farnham Royal, Bucks, 1985, 336.

88. Coaldrake, J. E., Restoration of coastal mining areas, Annual Report (1967-68), Division of Tropical Pastures, C.S.I.R.O., 1968.

89. Ae, N., Arihara, J., Okada, K., Yoshihara, T., and Johansen, C., Phosphorus uptake by Pigeonpea and its role in cropping systems of the Indian subcontinent, *Science,* 248, 477, 1990.

# 4

# Trace Element Concentrations in the Soft Tissue of Transplanted Freshwater Mussels Near a Coal-Fired Power Plant

**C. S. Klusek, M. Heit and S. Hodgkiss**

Environmental Measurements Laboratory, U.S. DOE, New York, NY

## ABSTRACT

Freshwater Unionidae mussels *(Lampsilus radiata)* were collected from the relatively pristine northern basin of Lake George in New York State and transplanted to 13 sites in Cayuga Lake (also in New York) at various distances from the Milliken Station coal-fired power plant. The mussels were retrieved after periods of approximately one and two years. The soft tissues of the mussels were analyzed for 16 elements (Ag, As, Ba, Br, Ca, Cl, Cd, Cr, Cu, Fe, Mn, Pb, Se, Sr, V, and Zn) by atomic absorption and proton-induced X-ray emission. Concentrations significantly greater $(p < 0.05)$ than the baseline values measured in the tissue of the Lake George mussels (in situ) prior to the transplant were seen for As and Cu. The levels of Cu in the mussel tissue appear to be related to the discharge of cooling water from Milliken Station. The mussels accumulated several elements to levels in excess of the concentrations present in the surface sediments in Cayuga Lake.

## INTRODUCTION

The ability of many aquatic organisms to accumulate trace metals has led to their use as indicator organisms of trace metal pollution.[1,2] Bivalve molluscs, such as the common marine mussel *Mytilus edulis,* have been widely used as indicators of environmental degradation in estuarine and coastal aquatic systems.[3–5] Freshwater bivalves, both indigenous and transplanted, have been used to monitor the levels of trace metals.[6–8] Unionidae mussels, which are widely distributed in rivers and lakes of North America, have been found to be good indicators of pollution in freshwater environments.[9–12] Both marine and freshwater mussels can be readily transplanted from nonpolluted to polluted zones for field studies.[5,13,14] However, freshwater mussels are more sensitive than marine species to environmental disturbances, such as rapid temperature, pH, and substrate changes.[15]

The present study is part of an integrated ecosystem study[16–19] to determine the environmental fate of pollutants released from the Milliken Station coal-fired electric generating station located on the shore of Cayuga Lake, in the Finger Lakes region of New York. Cayuga Lake was chosen for this study in part because its limnology, hydrology, meteorology, nutrient loading, and biota have been described.[20] Mass balance calculations for trace metal concentrations of the various coal ash compartments at the plant have been reported.[21] Thus, several environmental factors required to assess the impact of the electrical generating station on the ecosystem were available prior to beginning our investigations.

The present investigation was undertaken to determine whether transplanted Unionidae mussels would be useful indicators of bioaccumulation of trace metals emitted from a coal-fired power plant into an aquatic ecosystem. Three species of Unionidae mussels *(Elliptio complanatus, Lampsilus radiata,* and *Anodonta grandis)* were collected from Lake George and transplanted to sites in Cayuga Lake at various distances from the Milliken Station. These filter feeders are long-lived sedentary organisms that come into contact with both the sediment and water when feeding and respiring. The selection of a suitable organism depends on the ecosystem, the nature of the pollution source, and the pollution load. Factors such as season, size and age of the organism, concentration of free ions in the water column, and the chemical form of the elements in the sediments influence the extent to which aquatic organisms will accumulate trace elements.[1] The molluscs can accumulate trace metals both through the ingestion of particulate- associated pollutants and by the direct absorption of dissolved metals from the water.[12] The analysis of the soft tissue of the mussels revealed detectable levels of 16 elements (Ag, As, Ba, Br, Ca, Cl, Cd, Cr, Cu, Fe, Mn, Pb, Se, Sr, V, and Zn). The elements include both abundant and trace elements, as well as micronutrients and potentially toxic elements.[22,23]

## EXPERIMENTAL

### Study Site

Cayuga Lake is situated in the rural Finger Lakes region of New York (Figure 1). The history and limnology of Cayuga Lake have been described.[18,20,21,24] Selected hydrographic and chemical data have been tabulated (Table 1).

The Milliken coal-fired generating station is located on the east bank of the lake, approximately 21 km northwest of Ithaca, NY. The plant began operation in 1955 and consists of two tangentially fired 135 MW coal combustion units using bituminous coal for fuel. Electrostatic precipitators were installed in 1976 with a reported efficiency of 99.7%. Prior to the installation of the precipitators, there

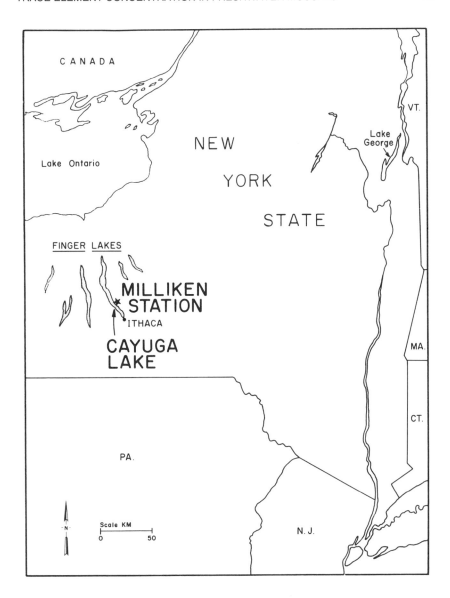

**FIGURE 1.** Map showing locations of Cayuga Lake and Milliken Power Station.

were reports of ash fallout in the vicinity of the plant.[25] A complete description of the plant is given elsewhere.[18,21] The prevailing winds in the Cayuga Lake basin are from the southwest in the summer and the northwest in the winter.[20,26] However, the shape of the basin channels the winds along the long axis of the lake;[20] thus, any emissions from the plant could be dispersed over the entire length of the lake.

**Table 1**
**Characteristics of Cayuga Lake and Lake George**

|  | Cayuga Lake[a] | Lake George[b] |
|---|---|---|
| Hydrographic: |  |  |
| Total catchment area, km$^2$ | 2033 | 606 |
| Surface area, km$^2$ | 172 | 114 |
| Volume, km$^3$ | 9.4 | 2.1 |
| Maximum depth, m | 133 | 57 |
| Mean depth, m | 54 | 18 |
| Thermal class | Warm monomictic | Dimictic |
| Mean residence time, yr | 12.8 | 8.6 |
| Stream runoff, km$^3$/yr | 0.342 | 0.181 |
| Sedimentation rate, mm/yr | 12 | 0.28 |
| Sediment | Clay and silt | Silty clay |
| Bedrock | Shale-limestone | Granitic |
| Soil | Coarse textured and calcareous | Shallow glacial till |
| Water Chemistry: |  |  |
| pH | 7.7–8.4 | 7.0–7.8 |
| Total alkalinity, mg/L | 96–98 | 20–30 |
| Dissolved oxygen, mg/L[c] | 7.2–9.7 | 8.6–10.8 |
| Conductivity, μmhos | 596 | 90–100 |
| Chloride, mg/L | 80–84 | 7–8 |
| Total phosphorus μg/L | 15–19 | 5 |
| Nitrate, μg/L | 300–350 | <10 |

[a] Data from References 20 and 24; water chemistry data for 1972-1973.
[b] Data from References 27, 28, and 29.
[c] Minimum value near bottom of lake.

## Sample Transplanting and Collection

In September 1978, three species of Unionidae mussels *(Elliptio complanatus, Lampsilus radiata,* and *Anodonta grandis)* were collected from the northern basin of Lake George, located in the eastern Adirondack Mountains, approximately 267 km from Cayuga Lake (Figure 1). The northern basin is essentially oligotrophic, receiving little input from local sources. Lake George has been studied extensively [27-29] and some hydrographic and water chemistry data have been published (Table 1).

The mussels were transported directly to Cayuga Lake for transplant within 24 hours of collection. The mussels were placed at nine locations in the littoral zone around Cayuga Lake. Most of the sites (Table 2; Figure 2) were on the east bank of the lake, both north and south of Milliken Station. One site (2.9 W) was located on the west bank of the lake, directly opposite the plant. The site locations are identified by the distance (km) and direction (N, S, W) from Milliken Station. For comparative purposes, mussels were placed in or near two marinas where combustion emissions from power boats or discharges of petroleum-based fuels were likely to occur. Approximately 100 Lake George mussels in situ were used to establish the baseline concentration of the elements in the mussels prior to transplanting.

A complete description of the mussel cage system and transplanting procedures are described in Heit and Klusek.[16] The cage system (Figure 3) utilized commercially available nylon mesh bags that allowed the mussels both lateral and

**Table 2**
**Sampling Locations in Cayuga Lake**

| Distance (km) from Milliken Station | Site Description | Distance (m) from Shore | Placement Depth (m) | No. of Samples Analyzed | |
|---|---|---|---|---|---|
| | | | | 1979 | 1980 |
| Lake Shore Sites: | | | | | |
| 16.2 N | Littoral zone, east side of the lake | 33 | 2.9 | — | 5 |
| 8.1 N | Littoral zone, east side of the lake | 23 | 3.0 | 7 | 4 |
| 4.8 N | Littoral zone, east side of the lake | 23 | 3.0 | 5 | 7 |
| 1.6 N | Littoral zone, east side of the lake | 23 | 3.0 | 5 | 5 |
| 0.1 N | Littoral zone, east side of the lake, in front of northern coal unloader | 35 | 3.0 | 7 | 3 |
| 0.05 S | Littoral zone, east side of the lake, in the discharge plume, in front of southern coal unloader | 10 | 3.0 | 4 | — |
| 0.1 S | Littoral zone, east side of the lake, in the discharge plume | 10 | 3.0 | 6 | 6 |
| 1.6 S | Littoral zone, east side of the lake | 23 | 3.0 | 4 | 6 |
| 4.8 S | Littoral zone, east side of the lake | 23 | 3.0 | 5 | 4 |
| 8.1 S | Littoral zone, east side of the lake | 30 | 3.0 | 4 | 6 |
| 2.9 W | Littoral zone, west side of the lake, directly opposite the plant | 7 | 3.0 | 5 | 5 |
| Marina Sites: | | | | | |
| 28.3 N | Castelli Marina, Union Springs, NY | — | 2.0 | 3 | 4 |
| 20.0 S | Inlet Park Marina, Ithaca, NY | — | 2.0 | 2 | 3 |

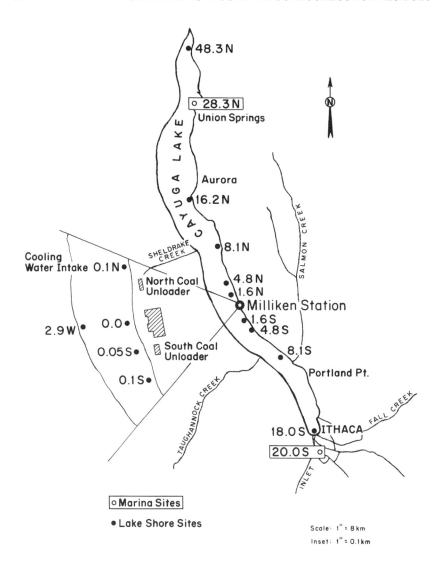

**FIGURE 2.**  Sampling sites with locations identified by distance (km) and direction (N, S, W) from Milliken Station.

vertical movement within the cage. Each cage contained two bags of approximately 35 mussels, with the exception of the cages placed at the marinas, where only a single bag was used. Two cages were placed at each site.

In June 1979, nine months after transplanting, one cage placed at each location was retrieved, except at site 16.2 N, where the cages could not be located. The remaining mussel cage at each site was collected in May 1980, 20 months after transplanting. At this time, the cage at site 0.05 S could not be located, but both cages were retrieved from 16.2 N. All of the mussels were frozen at 0°C prior to shipment to our laboratory for analysis.

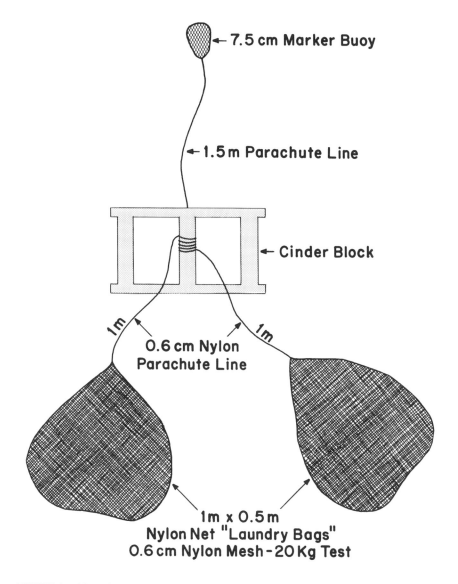

← 7.5 cm Marker Buoy

← 1.5 m Parachute Line

← Cinder Block

0.6 cm Nylon
Parachute Line

1m x 0.5 m
Nylon Net "Laundry Bags"
0.6 cm Nylon Mesh - 20 Kg Test

**FIGURE 3.** Mussel cage system.

The mortality rate at the time of the first retrieval was <7% for the littoral zone sites and 12% for the marina sites. After the second retrieval, <7% mortality was observed at all sites except at 1.6 N, 0.1 N, 0.1 S and 4.8 S, where the mortality was 10 to 25%. These mortality rates are consistent with levels of natural mortality for transplanted freshwater mussels reported by Sheehan.[14]

Because concentrations of metals in water and sediment influence bioaccumulation and/or bioconcentration, samples of both matrices were collected at each site at the time of mussel placement, except for sediment from the marina at 20.0 S and water at sites 0.05 S, 0.1 S and the marina at 28.3 N.

Additional water and sediment samples were collected directly in front of the power plant (0.0) in the discharge water plume and at site 48.3 km north and 18.0 km south of Milliken Station. Metal concentrations in water and sediment from Lake George have been reported.[11] Two liters of surface water were collected using acid-cleaned (1:1 $HNO_3$) polyethylene containers. The samples were immediately preserved with 50 mL of Ultrex grade $HNO_3$. Sediment grab samples were taken with a "petite ponar" grab (Wildco Co., Saginaw, MI), which penetrated to a depth of about 8 cm and had a sampling area of about 225 $cm^2$. The sediment samples were placed in acid-cleaned (1:1 $HNO_3$) 400-mL glass bottles.

## Sample Preparation and Analysis

### Mussels

The frozen mussels were allowed to thaw slightly, prior to excising the entire soft tissue from their shells. The soft tissues of three mussels (of the same species) were usually composited to provide the approximately 30 g of wet tissue needed for analysis. The Lake George mussels in situ were prepared in the same way for analysis. For this report, only samples of *Lampsilus radiata,* the most abundant (54%) of the three species collected, were analyzed. Three to eight samples (composites) were analyzed from each site. All dissections were performed with titanium or stainless steel scalpels. The wet tissue samples were weighed and then placed in polyethylene bottles, which were precleaned with 1:1 $HNO_3$, doubly-deionized $H_2O$. Each sample was then freeze-dried at 10°C.

The entire freeze-dried tissue sample was digested with doubly distilled $HNO_3$ in Teflon®* beakers covered with Teflon® watch glasses. Cadmium, Cr, Pb, and V were analyzed by atomic absorption spectroscopy (AAS). The method of standard additions was used for As, Pb, and V. For the Pb analysis, ammonium phosphate was used as a matrix modifier in conjunction with a L'vov platform.[30] Nickel was used as a matrix modifier for the As and Se analyses by hydride generation. Five elements (As, Cu, Se, Sr, and Zn) were analyzed by proton-induced X-ray emission (PIXE). The PIXE analysis provided information at detectable levels for the following elements: Ag, Ba, Br, Ca, Cl, Fe, and Mn.

### Water and Sediment

The unfiltered water samples were analyzed by flameless AAS for Ag, Al, As, Be, Cd, Co, Cr, Pb, Se, and V. Detectable concentrations of Ba, Br, Ca, Cu, Fe, Mn, Sr, and Zn were found in the water samples analyzed by PIXE.

The sediment samples were homogenized by stirring with a Teflon®-covered spatula. Aliquots of each sample (10 to 40 g) were placed into prewashed polyethylene bottles prepared in the same manner as those used for the mussel tissues. Samples were air dried at 10°C. The dried samples were digested with 4 $HNO_3$ : 1 $HClO_4$. The digestate was then diluted to a known volume with deionized water and mixed. Flame AAS was used to analyze Al, As (hydride generation), Be, Co, Cr, Cu, Fe, Mn, Pb, Se, V, and Zn. High levels of Fe in some sediments were found to suppress the Cr signal. The low recovery

---

*Registered trademark of E.I. duPont de Nemours and Company, Inc., Wilmington, DE.

was corrected by the addition of 2% $NH_4Cl$. Silver concentrations were determined by flameless AAS. Cadmium concentrations were determined by inductively coupled argon plasma (ICAP) spectroscopy. Strontium concentrations were determined by flame emission X-ray spectroscopy (FES).

## Analytical Quality Control

Quality control measures were taken to assess contamination and the reliability of the data and are described in detail in Silvestri et al.[19] Method blanks were run to determine the presence of contamination sources. The concentrations reported for the blanks were either below the limits of detection or at concentrations at least an order of magnitude below the levels found in the mussel tissues.

Results for precision and accuracy were based upon the analysis of National Institute of Standards and Technology (NIST) reference materials: NIST 1566 Oyster Tissue, NIST SRM 1643a Trace Elements in Water, and NIST SRM 1645 River Sediment. Although there was variation in the precision obtained for the different reference materials, the precision of analysis (% coefficient of variation of replicate analysis, CV) was determined to be good (>10% to ≤20%) to excellent (≤10%) for all elements. The accuracy of analysis (% recovery of standard materials) was excellent (100 ± 10%) to good (100 ± 20%) for most elements. Poor accuracy (outside the limits of 100 ± 40%) was reported for Br, Cl, and Cr in oyster tissue, and Se in water. The data for these elements, in the corresponding samples (mussel tissue, sediment, lake water), were not reported.

## Data Analysis

The metal concentrations (per dry weight of tissue) in the transplanted mussels were compared to the concentrations in the Lake George controls using Student's t-test. A two-factor analysis of variance was used to determine the significance of site and year on the metal concentrations at the littoral zone sites. When site was shown to be a significant factor for a given element, the significance of the site means was tested by the method of (Fisher's) least significant difference (LSD).[31] Analysis of variance was used to determine the significance of length of exposure (i.e., year of retrieval) at the marina sites. The correlation coefficient was used to determine the relation between element concentrations in the mussels, water, and sediments. Significant difference was assumed at a probability level of 0.05 unless otherwise indicated.

## RESULTS AND DISCUSSION

### Element Concentrations in Mussels

The average trace element concentration (± standard deviation, SD) for the transplanted mussels were recorded by site and year of retrieval (Table 3). The average trace element concentration (± SD) for the Lake George in situ mussels

**Table 3**
Trace Element Concentrations (μg/g dry wt) in Transplanted Mussels[a]

| Site | Ag 1979 | Ag 1980 | As 1979 | As 1980 | Ba 1979 | Ba 1980 |
|---|---|---|---|---|---|---|
| Lake Shore: | | | | | | |
| 16.2 N | NS[b] | 2.88 ± 0.77 | NS | 7.34 ± 0.48 | NS | 91.20 ± 6.98 |
| 8.1 N | 2.66 ± 0.47(5)[c] | 2.60 ± 1.08 | 6.80 ± 0.74 | 7.68 ± 0.56 | 119.57 ± 46.86 | 71.50 ± 12.66 |
| 4.8 N | 2.80 ± 1.39(4) | 2.73 ± 0.84 | 7.04 ± 0.59 | 7.34 ± 0.65 | 178.40 ± 18.39 | 83.00 ± 31.35 |
| 1.6 N | 3.06 ± 0.40 | 1.90 ± 0.67(4) | 6.56 ± 0.40 | 6.96 ± 0.73 | 121.40 ± 46.53 | 140.40 ± 10.71 |
| 0.1 N | 3.00 ± 0.69 | 2.47 ± 0.40 | 6.61 ± 1.17 | 7.27 ± 0.38 | 85.60 ± 45.20 | 55.67 ± 13.05 |
| 0.05S | 2.63 ± 0.85(3) | NS | 6.83 ± 0.28 | NS | 95.00 ± 30.21 | NS |
| 0.1 S | 3.95 ± 1.26 | 4.16 ± 2.10(5) | 7.13 ± 0.77 | 8.26 ± 1.00 | 112.33 ± 53.14 | 87.70 ± 22.70 |
| 1.6 S | 1.73 ± 0.83 | 3.28 ± 1.00(4) | 6.40 ± 0.22 | 7.63 ± 0.60 | 131.50 ± 55.45 | 90.17 ± 25.09 |
| 4.8 S | 3.10 ± 1.39 | 4.60 ± 2.18 | 6.68 ± 0.61 | 8.93 ± 1.67 | 170.60 ± 58.55 | 79.50 ± 8.35 |
| 8.1 S | 3.63 ± 2.03(3) | 3.80 ± 1.57 | 7.00 ± 1.84 | 8.18 ± 1.09 | 157.00 ± 52.98 | 88.30 ± 36.00 |
| 2.9 W | 3.08 ± 0.48(4) | 2.86 ± 0.76 | 6.23 ± 1.10 | 7.32 ± 1.39 | 132.25 ± 35.81 | 114.00 ± 20.96 |
| Marina: | | | | | | |
| 28.3 N | 1.97 ± 0.29 | 1.63 ± 0.32(3) | 5.40 ± 0.17 | 5.78 ± 0.39 | 85.33 ± 9.07 | 105.75 ± 31.67 |
| 20.0 S | — | 1.13 ± 0.57 | 4.45 ± 0.64 | 4.63 ± 0.23 | 153.00 ± 18.39 | 91.33 ± 52.37 |

| Site | Ca[d] 1979 | Ca[d] 1980 | Cd 1979 | Cd 1980 | Cu 1979 | Cu 1980 |
|---|---|---|---|---|---|---|
| Lake Shore: | | | | | | |
| 16.2 N | NS | 36.46 ± 7.92 | NS | 6.78 ± 1.01 | NS | 16.94 ± 2.01 |
| 8.1 N | 34.27 ± 6.16 | 35.28 ± 8.94 | 9.89 ± 1.52 | 8.38 ± 1.97 | 17.73 ± 2.14 | 16.78 ± 2.02 |
| 4.8 N | 32.08 ± 5.07 | 34.27 ± 7.45 | 8.84 ± 2.24 | 9.73 ± 1.66 | 16.72 ± 2.66 | 20.49 ± 1.28 |
| 1.6 N | 35.42 ± 5.14 | 34.60 ± 4.93 | 7.32 ± 0.46 | 7.34 ± 0.83 | 20.36 ± 1.86 | 17.32 ± 2.52 |
| 0.1 N | 32.67 ± 7.55 | 31.23 ± 5.64 | 7.43 ± 2.93 | 7.83 ± 0.55 | 17.46 ± 3.23 | 17.50 ± 4.18 |
| 0.05S | 37.60 ± 13.42 | NS | 7.63 ± 1.46 | NS | 34.18 ± 4.13 | NS |
| 0.1 S | 41.45 ± 5.19 | 40.55 ± 9.80 | 10.20 ± 2.16 | 10.28 ± 3.32 | 29.33 ± 6.12 | 46.78 ± 11.98 |
| 1.6 S | 33.55 ± 13.90 | 43.28 ± 8.42 | 9.25 ± 2.06 | 8.68 ± 2.51 | 17.05 ± 3.10 | 21.48 ± 2.50 |

**Table 3 (Continued)**
**Trace Element Concentrations (μg/g dry wt) in Transplanted Mussels[a]**

| Site | Ca[d] 1979 | Ca[d] 1980 | Cu 1979 | Cu 1980 | Cd 1979 | Cd 1980 |
|---|---|---|---|---|---|---|
| Lake Shore: (Cont.) | | | | | | |
| 4.8 S | 37.80 ± 14.87 | 42.50 ± 8.10 | 17.24 ± 4.10 | 22.15 ± 4.89 | 8.80 ± 1.59 | 6.46 ± 4.52 |
| 8.1 S | 37.53 ± 8.86 | 40.44 ± 6.04 | 18.53 ± 0.64 | 19.85 ± 1.14 | 10.33 ± 2.99 | 10.22 ± 2.69 |
| 2.9 W | 36.08 ± 5.19 | 41.02 ± 9.27 | 19.86 ± 3.13 | 19.92 ± 2.19 | 8.52 ± 1.02 | 2.48 ± 1.34 |
| Marina: | | | | | | |
| 28.3 N | 32.00 ± 5.29 | 28.65 ± 2.79 | 56.33 ± 6.03 | 121.75 ± 13.07 | 5.73 ± 1.03 | 6.35 ± 2.16 |
| 20.0 S | 26.55 ± 2.19 | 20.70 ± 4.99 | 27.85 ± 0.21 | 20.33 ± 2.02 | 7.95 ± 2.05 | 3.43 ± 0.25 |

| Site | Fe 1979 | Fe 1980 | Pb 1979 | Pb 1980 | Mn 1979 | Mn 1980 |
|---|---|---|---|---|---|---|
| Lake Shore: | | | | | | |
| 16.2 N | NS | 846. ± 173 | NS | 4.54 ± 1.39 | NS | 4288. ± 1114 |
| 8.1 N | 896. ± 321 | 1080. ± 330 | 10.36 ± 1.20 | 5.78 ± 2.67 | 3989. ± 691 | 4095. ± 1153 |
| 4.8 N | 996. ± 359 | 960. ± 143 | 10.94 ± 1.09 | 6.02 ± 1.53 | 3698. ± 656 | 4088. ± 934 |
| 1.6 N | 590. ± 92 | 1598. ± 967 | 9.70 ± 1.41 | 5.68 ± 0.96 | 3564. ± 441 | 3736. ± 472 |
| 0.1 N | 1167. ± 384 | 1190. ± 191 | 10.03 ± 1.75 | 4.53 ± 0.47 | 3707. ± 706 | 3310. ± 509 |
| 0.05S | 1055. ± 353 | NS | 12.98 ± 5.91 | NS | 4428. ± 1800 | NS |
| 0.1 S | 1330. ± 546 | 1567. ± 581 | 15.50 ± 3.18 | 8.40 ± 2.55 | 4893. ± 1104 | 5222. ± 1741 |
| 1.6 S | 1175. ± 461 | 862. ± 197 | 5.45 ± 3.76 | 6.67 ± 1.48 | 3875. ± 875 | 4603. ± 963 |
| 4.8 S | 1148. ± 360 | 1593. ± 656 | 12.92 ± 2.88 | 8.28 ± 2.43 | 4604. ± 1640 | 5028. ± 1538 |
| 8.1 S | 1133. ± 458 | 1332. ± 199 | 10.18 ± 3.71 | 7.07 ± 2.37 | 4180. ± 1549 | 4662. ± 782 |
| 2.9 W | 954. ± 108 | 1034. ± 275 | 10.18 ± 0.99 | 6.78 ± 2.53 | 4404. ± 675 | 4682. ± 1271 |
| Marina: | | | | | | |
| 28.3 N | 1193. ± 250 | 1035. ± 371 | 13.00 ± 2.74 | 8.25 ± 1.66 | 3333. ± 635 | 3710. ± 217 |
| 20.0 S | 1335. ± 262 | 1970. ± 298 | 14.10 ± 1.70 | 11.03 ± 2.67 | 3465. ± 50 | 2843. ± 337 |

**Table 3 (Continued)**
Trace Element Concentrations ($\mu$g/g dry wt) in Transplanted Mussels[a]

| Site | Se 1979 | Se 1980 | Sr 1979 | Sr 1980 | V 1979 | V 1980 | Zn 1979 | Zn 1980 |
|---|---|---|---|---|---|---|---|---|
| Lake Shore: | | | | | | | | |
| 16.2 N | NS | 4.22 ± 0.28 | NS | 82.2 ± 5.40 | NS | 0.50 ± 0.09 | NS | 356. ± 72 |
| 8.1 N | 4.53 ± 0.40 | 4.08 ± 0.26 | 74.0 ± 13.87 | 71.5 ± 11.6 | 0.36 ± 0.07 | 0.81 ± 0.33 | 343. ± 46 | 315. ± 67 |
| 4.8 N | 4.54 ± 0.34 | 3.99 ± 0.27 | 89.0 ± 12.88 | 78.7 ± 13.5 | 0.61 ± 0.14 | 0.65 ± 0.24 | 322. ± 52 | 345. ± 53 |
| 1.6 N | 4.36 ± 0.48 | 3.30 ± 0.33 | 77.8 ± 13.54 | 95.0 ± 9.80 | 0.54 ± 0.17 | 1.30 ± 1.29 | 330. ± 14 | 293. ± 27 |
| 0.1 N | 4.60 ± 0.58 | 4.30 ± 0.31 | 64.0 ± 15.40 | 61.0 ± 1.73 | 0.84 ± 0.20 | 1.04 ± 0.28 | 308. ± 64 | 281. ± 24 |
| 0.05S | 4.50 ± 0.39 | NS | 81.8 ± 16.94 | NS | 0.63 ± 0.30 | NS | 388. ± 131 | NS |
| 0.1 S | 5.05 ± 0.42 | 4.60 ± 0.44 | 85.8 ± 14.90 | 85.3 ± 16.6 | 0.62 ± 0.21 | 1.00 ± 0.71 | 382. ± 62 | 378. ± 87 |
| 1.6 S | 4.78 ± 0.44 | 4.00 ± 0.32 | 54.8 ± 29.92 | 86.0 ± 13.1 | 0.82 ± 0.21 | 0.49 ± 0.27 | 313. ± 111 | 379. ± 61 |
| 4.8 S | 4.62 ± 0.20 | 4.85 ± 0.81 | 94.8 ± 36.4 | 77.5 ± 7.59 | 0.67 ± 0.24 | 1.14 ± 0.15 | 355. ± 126 | 374. ± 84 |
| 8.1 S | 4.48 ± 0.96 | 4.55 ± 0.79 | 86.3 ± 9.32 | 80.7 ± 25.6 | 0.67 ± 0.30 | 0.58 ± 0.23 | 340. ± 69 | 369. ± 69 |
| 2.9 W | 5.10 ± 0.50 | 4.10 ± 0.44 | 91.8 ± 15.50 | 85.6 ± 5.86. | 0.46 ± 0.18 | 0.60 ± 0.14 | 368. ± 71 | 379. ± 97 |
| Marina: | | | | | | | | |
| 28.3 N | 4.20 ± 0.17 | 3.23 ± 0.26 | 110.7 ± 7.10 | 152.8 ± 25.80 | 1.24 ± 0.18 | 0.26 ± 0.14 | 363. ± 40 | 367. ± 66 |
| 20.0 S | 3.30 ± 0.10 | 2.29 ± 0.17 | 57.5 ± 3.54 | 42.5 ± 16.90 | 0.403 ± 0.004 | 1.29 ± 0.11 | 310. ± 28 | 246. ± 33 |

a  Mean ± SD.
b  NS indicates that mussels were not retrieved for this location.
c  ( ) for Ag represent number of detectable values reported. Only detectable values were used in calculated mean.
d  Concentration is reported in mg/g dry wt.

**Table 4**
**Average Trace Element Concentrations (µg/g dry wt) in Lake George Control Mussels**

| Element | Mean[a] | SD | % CV | Estimated Natural Variability,%[b] |
|---|---|---|---|---|
| Ag | 4.18 | 1.43 | 34 | 29 |
| As | 4.06 | 1.05 | 26 | 24 |
| Ba | 100.60 | 15.09 | 15 | 14 |
| Ca | 44,280.00 | 9,340.00 | 21 | 21 |
| Cd | 8.76 | 2.33 | 26 | 26 |
| Cu | 20.70 | 4.07 | 20 | 19 |
| Fe | 999.00 | 406.00 | 41 | 41 |
| Mn | 5,087.00 | 2,021.00 | 40 | 40 |
| Pb | 10.49 | 3.90 | 37 | 37 |
| Se | 5.49 | 0.63 | 11 | 9 |
| Sr | 72.00 | 12.70 | 18 | 18 |
| V | 0.64 | 0.22 | 34 | 34 |
| Y | 0.79 | 0.23 | 29 | 10 |
| Zn | 438.20 | 120.80 | 27 | 27 |

[a] Mean of 8 to 10.
[b] Estimated natural variability was calculated as the total uncertainty minus the contribution from analytical uncertainty.

were considered baseline values or controls (Table 4). Some less than detectable values were reported for the Ag analyses. Although this data represents only one to two samples at any site (Table 3), its exclusion from the calculated mean value of the site results in an overestimated mean.

The percent coefficient of variation about the mean of the Lake George control mussels was greater than 30% for several elements, but the analytical error was a small component of the overall variability. Therefore, natural variability (total variation-analytical uncertainty) in trace element concentrations may be an important factor in the data evaluation for some elements (i.e., Fe, Mn, Pb, Sn and V). Age, size, and body weight of the mussels may also contribute to the variability in the levels of elements in the mussel tissue. Although our previous study of the Lake George mussels[11] showed no correlation between soft tissue concentration and dry body weight, other researchers have reported such an association for some metals in marine molluscs; however, this association appears to be species related.[32-34] Different studies utilizing the same species have also yielded contradictory results for several trace metals.[34] However, a metal-body size relationship is generally not evident when there is less than a tenfold difference in the body weights of the bivalves.[32] For the studies presented here, mussels of similar size were used to minimize such effects.

In the following sections the metal concentrations in the transplanted mussels are discussed in relation to: (1) the concentrations in the Lake George control mussels, (2) the spatial and temporal changes, (3) the Cayuga Lake water and surface sediment concentrations, (4) emissions from the Milliken Station coal-fired power plant, and (5) other sources of metals in Cayuga Lake.

## Comparison of Metal Concentrations in Transplanted Mussels and Lake George Mussels

The northern basin of Lake George, a relatively pristine lake,[35] was chosen as a source of mussels for transplanting. The metal concentrations in the control (in situ) mussels represent a baseline concentration from which bioavailability or bioaccumulation of metals in the transplanted mussels were determined.

A comparison of the metal concentrations in the tissues of the Lake George controls and the mussels transplanted to the littoral zone in Cayuga Lake showed significantly greater than the baseline concentrations: As at all Cayuga Lake sites in both retrieval years; and Cu at sites 0.05 S and 0.1 S (the water discharge sites) in both years (Table 5). Significantly elevated concentrations were also seen for several other elements, but no clear spatial patterns were apparent.

At the marina site located 28.3 km north of Milliken Station, the concentration of As, Cu, and Sr in transplanted mussels retrieved in both years and V in mussels retrieved in 1979 were significantly greater than the concentrations in the Lake George controls (Table 5). At the marina site 20.0 S, which is also impacted by stream runoff, the mussel concentrations of Cu in 1979 and Fe and V in 1980 were also significantly greater than the levels measured in the Lake George controls.

Several of the elements had significantly lower tissue concentrations in the transplanted mussels than in the Lake George controls (Table 5). These elements were (1) Se at most sites for mussels in both years of retrieval; (2) Pb at most sites for mussels retrieved in 1980; and (3) Ag, Ca, Cu, Mn, and Zn at sites north of the plant for both retrieval years. Similarly, at the marinas, the following elements were significantly lower in concentration than the controls: (1) Ca and Mn at both marina sites in both retrieval years; (2) Ag and Se in both years at site 28.3 N; and (3) Zn in both years at site 20.0 S. These lower levels suggest that the mussels may reduce their metal burden by a self- purging mechanism, possibly in response to lower exposure or bioavailable levels in Cayuga Lake.

For some elements (i.e., Cd, Cu, Fe, Mn and Zn) metal behavior in aquatic systems (i.e., bioavailability, toxicity) is explainable by the availability of free metal ions.[36] Biota inhabiting environments with oligotrophic water conditions would generally be more susceptible to trace metals stress than biota in eutrophic systems or waters affected by anthropogenic discharges. Organic materials, which may be more abundant in eutrophic environments, form metal complexes and, thereby, may reduce the amount of free ion concentration and metal bioavailability.[36] Northern Lake George is considered an oligotrophic system,[27] while Cayuga Lake is generally considered to be mesotrophic. However, at specific times and for certain parameters (i.e., nutrient loading and water residence time), Cayuga Lake may be termed eutrophic.[20] The mussels transplanted to the littoral sites (non-marina) had tissue levels similar to the concentrations present in the Lake George control mussels for Cd and Fe at nearly all of the sites, as well as Ca, Mn, and Zn at the site south of Milliken Station. Of these elements, Fe and Mn may display wide ranges of natural variability (Table 4) which would decrease the likelihood of detecting small changes in the tissue concentrations.

**Table 5**
**Comparison of Concentration in Control vs. Transplanted Mussels[a]**

| Site | As 1979 | As 1980 | Ba 1979 | Ba 1980 | Cu 1979 | Cu 1980 | Fe 1980 | Pb 1979 | Pb 1980 | Sr 1979 | Sr 1980 | V 1979 | V 1980 |
|---|---|---|---|---|---|---|---|---|---|---|---|---|---|
| Lake Shore: | | | | | | | | | | | | | |
| 16.2 N | * | * | | | | | | | | | ** | | |
| 8.1 N | * | * | | | | | | | | | | | |
| 4.8 N | * | * | * | | | | | | | ** | | | |
| 1.6 N | * | ** | | ** | | | | | | | * | ** | |
| 0.1 N | * | * | | | | | | | | | | | |
| 0.05S | * | | | | * | | | | | | | | |
| 0.1 S | * | * | | | * | * | ** | ** | | ** | | | |
| 1.6 S | * | * | ** | | | | | | | | ** | | |
| 4.8 S | ** | * | | | | | | | | | | | * |
| 8.1 S | * | * | | | | | ** | | | ** | * | | |
| 2.9 W | * | * | | | | | | | | ** | * | | |
| Marina: | | | | | | | | | | | | | |
| 28.3 N | * | * | | | * | * | | | | * | * | * | |
| 20.0 S | | | | | * | | * | | | | | | * |

*Note:* * = $p < 0.01$; ** = $p < 0.05$.

[a] Student t-test results for mean concentration of the transplanted mussels is greater than the mean of the Lake George controls.

## Spatial and Temporal Variations in Mussel Concentrations

A second approach to evaluating element concentrations in the transplanted mussels is to examine the relationship between the soft tissue concentration and (1) placement location and (2) duration of exposure time (i.e., year of retrieval). The elements which were elevated in concentration in the transplanted mussels compared to the Lake George controls (Group 1) will be considered first, followed by the remaining elements that had concentrations significantly lower than the Lake George controls (Group 2).

### *Group 1 (Elevated)*

For the littoral sites, the elements in this group were As, Ba, Cu, Fe, Pb, Si, and V. The year of retrieval and site location were both found to be significant factors ($p < 0.01$) affecting the concentration of Ba, Cu, and Pb in the mussel tissue. The two-way ANOVA test also indicated that there was significant interaction between the two factors (location and duration) for these three elements, suggesting that the effect of one of the factors depends on the values of the other factor. For As, V ($p < 0.01$), and Fe concentrations, the year of retrieval was a significant factor in the tissue concentrations of the transplanted mussels.

*Year of Retrieval.* The tissue concentration of Ba and Pb decreased in mean concentration in the second retrieval year. The remaining elements (As, Cu, Fe, V) increased in concentration. The amount of change in mean concentration varied among the elements from $\leq 1$ µg/g for As to >100 µg/g for Fe.

*Site.* For each element considered, only one or two of the sites had mean values that were significantly different from the majority of the other sites (Table 6). For example, the mean Cu tissue concentration at the site located 0.1 km S of Milliken Station in 1979 and 1980 was significantly different from all of the other locations (mean difference = –5 to 13 µg/g in 1979 and 25 to 30 µg/g in 1980). Additionally, the mean Cu concentration in 1979 at site 0.05 S was significantly different from all of the other sites (mean difference = 5 to 17 µg/g). Interestingly, these are the same sites where the transplanted mussels had elevated concentrations compared to the Lake George mussels. Other site-element pairs which exhibited elevated concentrations compared to the Lake George (Table 5) were: (1) Pb at site 0.1 S in 1979; and (2) Ba at 4.8 N and 4.8 S in 1979 and 1.6 N in 1980. In addition, the Pb concentrations in the transplanted mussels at site 1.6 S in 1979, where the mean concentration was significantly different from most of the other sites, were significantly lower than those measured in the Lake George controls.

### *Group 2*

For the littoral zone sites, the elements that were lower in concentration in the transplanted mussels compared to the Lake George controls were Cd, Mn, and Se. Year of retrieval and site location were both significant factors in the concentrations of the transplanted mussels of only Cd (Table 6). Location and duration significantly

**Table 6**
**Comparison of Site Mean Tissue Concentrations[a]**

| Element | Year | Site | 16.2 N | 8.1 N | 4.8 N | 1.6 N | 0.1 N | 0.05 S | 0.1 S | 1.6 S | 4.8 S | 8.1 S | 2.9 W | Min. | Max. |
|---|---|---|---|---|---|---|---|---|---|---|---|---|---|---|---|
| Ba | 1979 | 4.8 N | ND[b] | * | — | * | * | * | * | | | | | 57 | 93 |
| | 1979 | 4.8 S | ND | * | | * | * | * | * | | — | | | 58 | 85 |
| | 1980 | 1.6 N | * | * | * | — | * | ND | * | * | — | * | | 49 | 85 |
| | 1980 | 2.9 W | | | — | | * | ND | | | * | * | | 31 | 58 |
| Cu | 1979 | 0.1 S | ND | * | * | * | * | * | — | * | * | * | * | −5 | 13 |
| | 1979 | 0.05 S | ND | * | * | * | * | — | * | | * | * | * | 5 | 17 |
| | 1980 | 0.1 S | * | * | * | * | * | ND | — | * | * | * | * | 25 | 30 |
| Pb | 1979 | 0.1 S | ND | * | * | * | * | * | — | * | * | * | * | 4.6 | 10.0 |
| | 1979 | 1.6 S | ND | * | * | — | * | — | — | | * | * | * | −4.2 | −7.5 |
| | 1980 | 0.1 S | * | * | * | * | * | ND | — | * | * | * | * | 2.4 | 3.9 |
| Sr | 1979 | 1.6 S | ND | | * | — | * | — | — | | * | * | * | −27 | −40 |
| | 1979 | 0.1 N | ND | | | * | — | * | * | | | * | * | −21 | −31 |
| | 1980 | 0.1 N | * | | * | * | — | * | * | | | * | | −24 | −34 |
| Cd | 1979 | 1.6 N | ND | * | | — | | ND | | * | * | * | | −2.6 | −3.0 |
| | 1979 | 0.1 N | ND | * | | | — | | | * | * | * | | −2.5 | −2.9 |
| | 1980 | 0.1 S | * | | * | * | | ND | — | * | * | * | * | 2.9 | 7.8 |
| | 1980 | 2.9 W | * | | * | * | * | ND | | | | | — | −4.0 | −7.8 |

*Note:* * = a significant difference between the column and line sites at $p < 0.05$.
[a] Results of Fisher's LSD test for the difference of site means.
[b] ND indicates no data was available.

interacted for the Cd in mussels. In addition, year of retrieval was a significant factor in mussel tissue concentrations of Se. The mean concentration of Cd and Se decreased for the second year of retrieval. Site location was a significant factor in the tissue concentration of Mn.

At the marina sites, year of retrieval was significant in the mussel concentrations for Cu, Pb, and Se, whereas location of the marina was important for As, Ca, Cu, Fe, Se, Sr, and Zn.

Most data on the relation of metal concentration in molluscs and the length of exposure are ambiguous, probably since the levels in the mussels were studied for less than one year. In molluscs, the rates of excretion and metabolic turnover are slow for some metals (Fe and Mn) and if environmental levels of these elements are high, the mussels should accumulate them. As the animal gets older, the amount of stored metal increases more rapidly than body weight.[37] Additionally, some metals may accumulate in excess of metabolic needs and excretory capacity. Correlation of metal and age may be attributed to a permanent storage of excess metals within the mussels. The sequestered metals are thought to be either tightly bound to proteins or stored in granules which are removed from metabolic circulation. Alternatively, metals not correlated with age are found in a readily exchangeable and mobile pool within the molluscs. When the amount of permanently stored metal is a large proportion of the total metal concentration of the mollusc, the effect may be to reduce the apparent responsiveness of mussels to changes in ambient metal concentrations.[37] Thus, if the Lake George in situ controls had already accumulated metals, the transplant to Cayuga Lake may not show any change in the levels of some elements.

In addition, the interactions of two or more pollutants in the same aquatic system, such as the formation of some chemical complexes, are not well understood and may represent a more complex equation than a linear relationship. Significantly positive correlations for the mussels placed in the littoral zone sites were observed (Table 7) for Ag-Ca-Pb, Ca-Mn-Pb-Zn, and K in both retrieval years (other correlations were seen in individual years).

## Relation of Mussel Tissue Concentrations to Concentrations in Water and Surface Sediments

### Sediment and Water Concentrations

Although the sediment and water samples were collected at the same locations used for transplanting, precise knowledge of the variability in concentration over the nearly two years of exposure cannot be assessed because it is expected that there were fluctuations in the water or sediment concentrations for at least some of the elements.

*Sediment.* Metal concentrations in Lake George sediments are available for only a few metals (Table 8). However, the metal concentrations reported here for the sediment from the Cayuga Lake littoral zone sites were slightly higher than the values reported for Cu, Cr, and Zn in Lake George sediments.[11,35] With some exceptions, the elemental concentrations in the sediments from Cayuga Lake were

**Table 7**
**Correlation Coefficients Among Elements in Transplanted Mussels[a]**

|  | Ag | Ca | Cu | Fe | Mn | Pb |
|---|---|---|---|---|---|---|
| **1979 Retrieval:** |  |  |  |  |  |  |
| Ca | 0.660 |  |  |  |  |  |
| Cd |  |  |  |  |  |  |
| Cu |  | 0.650 |  |  |  |  |
| Fe |  |  |  |  |  |  |
| Mn |  | 0.874 |  |  |  |  |
| Pb | 0.725 | 0.714 |  |  | 0.703 |  |
| S |  |  |  | 0.636 |  |  |
| Se |  |  |  |  |  |  |
| Sr | 0.636 |  |  |  |  | 0.680 |
| Zn |  | 0.817 | 0.795 |  | 0.853 | 0.741 |

|  | Ag | As | Ba | Ca | Cu | Fe | Mn | Pb |
|---|---|---|---|---|---|---|---|---|
| **1980 Retrieval:** |  |  |  |  |  |  |  |  |
| As | 0.950 |  |  |  |  |  |  |  |
| Ca | 0.745 | 0.637 |  |  |  |  |  |  |
| Fe |  |  |  |  |  |  |  |  |
| Mn | 0.856 | 0.750 |  | 0.894 | 0.633 |  |  |  |
| Pb | 0.825 | 0.798 |  | 0.786 | 0.671 |  | 0.876 |  |
| Se | 0.873 | 0.848 |  |  |  |  |  |  |
| Sr |  |  | 0.900 |  |  |  |  |  |
| V |  |  |  |  |  | 0.836 |  |  |
| Zn | 0.743 |  |  | 0.893 |  |  | 0.918 | 0.696 |

[a] Correlation coefficients for 5-10 pairs of observations significant at $p < 0.05$. Elements not shown had no significant correlations.

similar to values reported for remote lakes in the Adirondacks[38] and the Canadian Shield.[2] The sediment concentrations of Cd, Cu, Pb, Sr, and Zn at the marina site located 28.3 km north of Milliken Station were elevated compared to the littoral zone sites.

*Water.* The water concentrations in Cayuga Lake (Table 9) are low compared to many other freshwater systems and resembled levels found in High Sierra lakes.[39] The data reported here agree reasonably well with the range of values reported previously for this lake.[20,39] At the marina site, located at 20.0 km south of Milliken, the water concentrations of Br, Fe, Pb, and Zn were at least a factor of two higher than those at littoral sites. Such elevated concentrations would be expected in a marina since both Br and Pb are common additives to petro-fuels, while Fe and Zn may be associated with engine parts, propellers, etc.

## Sediment Enrichment Factors

The source of trace elements in sediments may be obtained by comparing the element concentration found in the sediment with that found in the earth's crust or the major rock types in the area. If the main source of an element in the sediment is the earth's crust, the ratio will be equal to one (EF=1). Any ratio differing from one is assumed to be due to local geological conditions, anthropo-

**Table 8**
Trace Element Concentrations (μg/g dry wt) in Cayuga Lake Littoral Sediments[a]

| Site | Ag | Al[b] | As | Be | Cd | Co | Cr | Cu | Fe[b] |
|---|---|---|---|---|---|---|---|---|---|
| Lake Shore: | | | | | | | | | |
| 48.3 N | 0.17 ± 0.06 | 15.1 ± 3.4 | 1.6 ± 0.4 | 0.32 ± 0.04 | 0.90 ± 0.15 | 2.8 ± 0.7 | 11. ± 1. | 18. ± 2. | 9.9 ± 0.9 |
| 16.2 N | 0.04 ± 0.01 | 19.2 ± 1.2 | 5.7 ± 1.1 | 0.48 ± 0.03 | 0.18 ± 0.04 | 8.6 ± 1.5 | 15. ± 3. | 22. ± 7. | 18.9 ± 3.2 |
| 8.1 N | 0.04 ± 0.01 | 16.4 ± 0.2 | 3.8 ± 2.1 | 0.40 ± 0.01 | 0.22 ± 0.12 | 8.2 ± 0.9 | 16. ± 1. | 15. ± 1. | 14.3 ± 0.03 |
| 4.8 N | 0.12 ± 0.11 | 19.3 ± 1.4 | 7.5 ± 1.8 | 0.55 ± 0.11 | 0.16 ± 0.18 | 10.8 ± 0.1 | 17. ± 1. | 17. ± 1. | 23.2 ± 1.7 |
| 1.6 N | 0.03 ± 0.02 | 14.5 ± 1.2 | 3.0 ± 0.4 | 0.34 ± 0.03 | 0.08 ± 0.03 | 7.4 ± 0.4 | 15. ± 3. | 14. ± 2. | 14.1 ± 1.7 |
| 0.1 N | 0.05 ± 0.02 | 15.8 ± 0.7 | 4.4 ± 0.9 | 0.57 ± 0.03 | 0.18 ± 0.10 | 12.0 ± 3.0 | 16. ± 4. | 24. ± 2. | 20.6 ± 4.0 |
| 0.0[c] | 0.04 ± 0.02 | 17.7 ± 1.9 | 6.6 ± 1.4 | 0.66 ± 0.05 | 0.26 ± 0.10 | 8.4 ± 0.1 | 15. ± 1. | 24. ± 4. | 18.8 ± 1.4 |
| 0.05S | 0.05 | 18.6 | 5.5 | 0.49 | 0.15 | 10.3 | 18. | 18. | 18.4 |
| 0.1 S | 0.03 ± 0.01 | 17.4 ± 0.2 | 5.3 | 0.61 ± 0.12 | 0.19 ± 0.04 | 12.4 ± 1.5 | 16. ± 3. | 27. ± 2. | 20.4 ± 1.8 |
| 1.6 S | 0.05 ± 0.01 | 25.9 ± 3.0 | 4.8 ± 1.2 | 0.73 ± 0.10 | 0.24 ± 0.08 | 10.7 ± 1.5 | 15. ± 3. | 21. ± 2. | 21.8 ± 2.3 |
| 4.8 S | 0.06 ± 0.02 | 15.6 ± 1.2 | 3.3 ± 0.9 | 0.46 ± 0.06 | 0.25 ± 0.04 | 7.3 ± 1.0 | 13. ± 1. | 15. ± 2. | 14.4 ± 1.0 |
| 8.1 S | 0.05 ± 0.01 | 15.5 ± 1.1 | 4.7 ± 0.2 | 0.51 ± 0.12 | 0.28 ± 0.05 | 7.7 ± 1.5 | 13. ± 1. | 16. ± 2. | 15.8 ± 2.0 |
| 18.0 S | 0.25 ± 0.07 | 19.6 ± 2.3 | 3.0 ± 0.7 | 0.41 ± 0.08 | 0.19 ± 0.08 | 8.0 ± 0.9 | 15. ± 2. | 16. ± 3. | 17.8 ± 2.1 |
| 2.9 W | 0.04 ± 0.02 | 10.6 ± 0.7 | 1.5 ± 0.1 | 0.27 ± 0.02 | 0.06 ± 0.01 | 5.4 ± 0. | 12. ± 2. | 21. ± 16. | 11.1 ± 0.5 |
| Marina: | | | | | | | | | |
| 28.3 N | 0.04 | 7.9 | 2.4 | 0.29 | 0.4 | 2.5 | 8.2 | 50. | 6.5 |
| Lake George:[c,d] | — | — | 3.1 ± 1.3 | — | 0.3 | — | 4 ± 2 | 4 ± 1 | — |

| Site | Mn | Pb | Se | Sr | V | Zn |
|---|---|---|---|---|---|---|
| Lake Shore: | | | | | | |
| 48.3 N | 211. ± 12. | 7.0 ± 1. | 1.01 ± 0.19 | 467. ± 38. | 9. ± 2. | 90. ± 6. |
| 16.2 N | 275. ± 18. | 13.2 ± 0.4 | 0.38 ± 0.07 | 91. ± 1. | 18. ± 0.2 | 91. ± 7. |
| 8.1 N | 370. ± 60. | 9.0 ± 1. | 0.14 ± 0.02 | 38. ± 2. | 16. ± 1. | 59. ± 6. |
| 4.8 N | 410. ± 50. | 17.0 ± 0.5 | 0.19 ± 0.04 | 60. ± 2. | 17. ± 0.3 | 88. ± 2. |
| 1.6 N | 230. ± 20. | 11.0 ± 2. | 0.23 ± 0.07 | 37. ± 7. | 16. ± 2. | 63. ± 6. |
| 0.1 N | 245. ± 6. | 11.0 ± 1. | 0.55 ± 0.50 | 35. ± 12. | 19. ± 4. | 173. ± 106. |
| 0.0[c] | 218. ± 22. | 10.0 ± 4. | 0.85 ± 0.49 | 96. ± 6. | 18. ± 4. | 64. ± 8. |
| 0.05S | 200. | 11.0 | 0.94 | 58. | 20. | 84. |

**Table 8 (Continued)**
**Trace Element Concentrations (μg/g dry wt) in Cayuga Lake Littoral Sediments[a]**

| Site | Mn | Pb | Se | Sr | V | Zn |
|---|---|---|---|---|---|---|
| Lake Shore: (Cont.) | | | | | | |
| 0.1 S | 312. ± 75. | 10.0 ± 2. | 0.55 ± 0.13 | 43. ± 3. | 19. ± 1. | 112. ± 6. |
| 1.6 S | 300. ± 20. | 16.0 ± 3. | 0.36 ± 0.09 | 34. ± 10. | 22. ± 1. | 88. ± 11. |
| 4.8 S | 210. ± 20. | 12.0 ± 2. | 0.36 ± 0.04 | 57. ± 5. | 16. ± 1. | 69. ± 8. |
| 8.1 S | 250. ± 10. | 16.0 ± 3. | 0.42 ± 0.02 | 59. ± 5. | 13. ± 1. | 84. ± 13. |
| 18.0 S | 330. ± 60. | 20.0 ± 5. | 0.16 ± 0.05 | 26. ± 4. | 15. ± 2. | 73. ± 10. |
| 2.9 W | 190. ± 20. | 7.8 ± 0.5 | 0.15 ± 0.04 | 31. ± 1. | 12. ± 2. | 50. ± 1. |
| Marina: | | | | | | |
| 28.3 N | 198. | 147. | 0.47 | 211. | 10. | 114. |
| Lake George:[c][d] | — | — | 0.22 ± 0.06 | — | — | 33. ± 14 |

a Mean and SD of two to five samples.
b Concentration is in mg/g dry wt.
c Directly in front of Milliken Station in the outfall; mussels were not transplanted to this site.
d Ref. 11; samples collected in 1977.

**Table 9**
**Trace Element Concentrations (ng/mL) in Cayuga Lake Surface Water Samples[a]**

| Site | Ag | Al | As | Be | Br | Ca | Cd | Co |
|---|---|---|---|---|---|---|---|---|
| Lake Shore: | | | | | | | | |
| 48.3 N | — | 38. ± 1 | 1.60 ± 0.20 | — | 53. ± 5 | 17. ± 1 | 0.024 ± 0.003 | 0.120 ± 0.008 |
| 16.2 N | — | 59. ± 2 | 0.78 ± 0.05 | — | 37. ± 3 | 31. ± 1 | 0.004 ± 0.001 | 0.074 ± 0.009 |
| 8.1 N | — | 48. ± 2 | 0.78 ± 0.05 | — | 31. ± 2 | 29. ± 1 | 0.007 ± 0.001 | 0.087 ± 0.009 |
| 4.8 N | — | 23. ± 1 | 0.74 ± 0.08 | — | 29. ± 1 | 31. ± 1 | — | 0.072 ± 0.009 |
| 1.6 N | — | 54. ± 2 | 1.10 ± 0.10 | — | 32. ± 5 | 30. ± 1 | 0.005 ± 0.001 | 0.080 ± 0.008 |
| 0.1 N | — | 21. ± 1 | 1.08 ± 0.08 | — | 33. ± 2 | 29. ± 1 | — | 0.171 ± 0.008 |
| 0.0 b | — | 72. ± 3 | 0.45 ± 0.06 | — | 29. ± 1 | 31. ± 1 | — | 0.075 ± 0.008 |
| 1.6 S | — | 28. ± 1 | 0.54 ± 0.10 | — | 42. ± 2 | 29. ± 2 | — | 0.086 ± 0.008 |
| 4.8 S | — | 28. ± 1 | 0.81 ± 0.06 | — | 43. ± 1 | 29. ± 1 | — | 0.044 ± 0.008 |
| 8.1 S | — | 38. ± 1 | 0.97 ± 0.06 | — | 36. ± 2 | 31. ± 1 | — | 0.065 ± 0.008 |
| 18.0 S | 0.135 ± 0.004 | 39. ± 2 | 1.35 ± 0.07 | 0.021 ± 0.001 | 44. ± 1 | 38. ± 1 | 0.025 ± 0.001 | 0.320 ± 0.01 |
| 2.9 W | — | 55. ± 2 | 0.61 ± 0.06 | — | 32. ± 3 | 27. ± 1 | — | 0.093 ± 0.008 |
| Marina: | | | | | | | | |
| 20.0 S | 0.023 ± 0.002 | 29. ± 1 | 1.70 ± 0.2 | 0.010 ± 0.001 | 85. ± 2 | 42. ± 1 | 0.018 ± 0.001 | 0.390 ± 0.02 |

| Site | Cr | Cu | Fe | Mn | Pb | Sr | V | Zn |
|---|---|---|---|---|---|---|---|---|
| Lake Shore: | | | | | | | | |
| 48.3 N | 0.14 ± 0.01 | — | 9. ± 4 | — | 0.42 ± 0.02 | 260. ± 20 | 0.41 ± 0.04 | 3.3 ± 0.6 |
| 16.2 N | 0.39 ± 0.01 | 1.4 ± 0.6 | 44. ± 1 | — | 0.38 ± 0.02 | 189. ± 9 | 0.30 ± 0.02 | 9.7 ± 0.4 |
| 8.1 N | 0.25 ± 0.01 | 6.0 ± 1.0 | 27. ± 2 | — | 0.35 ± 0.02 | 164. ± 2 | 0.51 ± 0.03 | 11.0 ± 1.0 |
| 4.8 N | 0.65 ± 0.02 | 2.5 ± 0.9 | 18. ± 3 | — | 0.43 ± 0.04 | 190. ± 4 | 0.17 ± 0.02 | — |
| 1.6 N | 0.28 ± 0.01 | 1.5 ± 0.5 | 34. ± 1 | — | 0.76 ± 0.05 | 178. ± 3 | 0.17 ± 0.02 | 7.0 ± 0.7 |
| 0.1 N | 0.23 ± 0.01 | 1.0 ± 0.4 | 142. ± 4 | — | 0.39 ± 0.03 | 178. ± 5 | 0.27 ± 0.04 | 1.4 ± 0.2 |
| 0.0 b | 0.53 ± 0.02 | 3.3 ± 0.4 | 29. ± 2 | — | 0.34 ± 0.03 | 191. ± 5 | 0.47 ± 0.04 | 2.2 ± 0.3 |
| 1.6 S | 0.26 ± 0.02 | 0.8 ± 0.4 | 17. ± 3 | — | 0.58 ± 0.06 | 170. ± 10 | 0.17 ± 0.04 | 0.7 ± 0.3 |
| 4.8 S | 0.35 ± 0.02 | — | 14. ± 1 | — | 0.39 ± 0.03 | 180. ± 10 | 0.21 ± 0.04 | 2.5 ± 0.4 |
| 8.1 S | 0.70 ± 0.02 | — | 19. ± 2 | — | 0.37 ± 0.03 | 181. ± 5 | 0.12 ± 0.02 | 2.2 ± 0.6 |
| 18.0 S | 0.98 ± 0.03 | 3.4 ± 1.0 | 45. ± 2 | 40.0 ± 2 | 2.50 ± 0.20 | 152. ± 3 | 0.55 ± 0.03 | 9.2 ± 0.7 |
| 2.9 W | 0.23 ± 0.01 | — | 35. ± 4 | — | 0.25 ± 0.02 | 167. ± 6 | 0.19 ± 0.06 | 2.8 ± 0.3 |

**Table 9 (Continued)**
**Trace Element Concentrations (ng/mL) in Cayuga Lake Surface Water Samples[a]**

| Site | Cr | Cu | Fe | Mn | Pb | Sr | V | Zn |
|------|-----|-----|-----|-----|-----|-----|-----|-----|
| Marina: | | | | | | | | |
| 20.0 S | 0.68 ± 0.04 | 4.6 ± 0.5 | 452. ± 9 | 112.0 ± 4 | 1.77 ± 0.07 | 8.4 ± 0.40 | 6100. ± 200 | 0.022 ± 0.011 |

*Note*: − indicates that values less than the detection limits were reported for all samples at this site. In addition, less than detectable values were reported for all samples at all sites for Cl. Barium was detected only at the marina site: 200 ± 20 ng/mL.

[a]Mean and SD for three analyses.
[b]Directly in front of the Milliken Station in the outfall; mussels were not transplanted to this site.

genic inputs or both. If, for example, the EF>1, additional sources of input besides average surface crustal sources are indicated; while an EF<1 may indicate a local geological depletion in a particular element. This type of analysis is subject to misinterpretation if the natural levels of the elements in the area differ greatly from their average concentration in crustal material.

The bedrock composition of Cayuga Lake shows a dominant shale- limestone mix extending along most of the immediate border of the lake. The major portion of the drainage basin is composed of shales and siltstones.[20] Based on this minerology, the ratios (or enrichment factors) of the sediment to shale ($EF_s$) and to limestone ($EF_L$) were calculated for Cayuga Lake after normalizing both the sediment and average crustal concentrations[22] to their respective Fe concentrations (Table 10).

Enrichment factors greater than 10 were seen at site 48.3 N and at the marina site, 28.3 N, for some elements (Ag, Cd, Co, and Pb). At other sites in the littoral zone of Cayuga Lake, elements with $EF_s$>3 were Ag (4.8 N, 4.8 S, 18.0 S), Se (0.0, 0.5 S), and Zn (0.1 N). The occurrence of elevated Se concentrations directly in front of Milliken Station is likely to have come from the fly ash. In 1977, fly ash buried in a site 3 km east of the plant entered the lake as a result of a broken dike.[40] In addition, depleted levels of Mn and Sr were seen ($EF_L$<1.0) at all sites, indicating possible local depletion.

Other results showed that the $EF_L$ ratio was >2 only for Co at all sites, suggesting that the natural background of the element may be higher than the average crustal concentration. In addition, the marina at 28.3 N had an $EF_S$ and an $EF_L \geq 3$ for Cd, Cu, Pb, Se, and Zn. These results would be expected since all of these elements are associated with combustion engines, boat hardware, or fuel additives.

## Relation Between Mussel and Sediment Concentrations

Sediments represent the most concentrated physical pool of metals in aquatic environments. Detritus-feeding organisms are exposed directly to sediment-bound metals. Mussels are believed to feed on any fine decaying tissue (detritus and zooplankton).[15] Their feeding is specialized to remove these suspended microscopic particles from water. Although some inorganic silt may be mixed with the organic food ingested, a large portion of the inedible material is separated beforehand.[41]

When the metal concentrations in the transplanted mussels were compared to the concentrations in the surface sediments collected at the same littoral sites, a significant positive correlation was obtained only for Cd (r = 0.69) for samples retrieved in 1979 and V (r = 0.69) for samples retrieved in 1980. Other researchers[8,42] have reported no correlation between metal concentration in bivalves and adjacent sediments for Cd, as well as for Cu, Fe, Mn, Pb, and Zn, even if the study area was heavily polluted. In fact, correlations between tissue concentrations and sediment concentrations have been predicted only when the difference in concentrations is two to three orders of magnitude.[36]

Wren et al.,[2] however, have reported that many of the elements highly concentrated in clams also have high sediment-concentration ratios (i.e., the ratio of the concentration in tissue to the sediment concentration) but are not accumulated in

**Table 10**
**Comparison of Sediment Concentrations and Average Crustal Concentrations[a]**

| Site | Ag | Al | As | Be | Cd | Co | Cr | Cu | Mn | Pb | Se | Sr | V | Zn |
|---|---|---|---|---|---|---|---|---|---|---|---|---|---|---|
| Enrichment Factor, $EF_S$: | | | | | | | | | | | | | | |
| 48.3 N | 12.3 | 0.9 | 0.6 | 0.6 | 14.4 | 0.7 | 0.6 | 1.9 | 1.2 | 1.7 | 8.0 | 7.4 | 0.3 | 4.5 |
| 16.2 N | 1.5 | 0.6 | 1.1 | 0.4 | 1.5 | 1.1 | 0.4 | 1.2 | 0.8 | 1.6 | 1.6 | 0.8 | 0.3 | 2.4 |
| 8.1 N | 2.0 | 0.7 | 1.0 | 0.4 | 2.4 | 1.4 | 0.6 | 1.1 | 1.4 | 1.5 | 0.8 | 0.4 | 0.4 | 2.0 |
| 4.8 N | 3.7 | 0.5 | 1.2 | 0.4 | 1.1 | 1.2 | 0.4 | 0.7 | 1.0 | 1.7 | 0.6 | 0.4 | 0.3 | 1.9 |
| 1.6 N | 1.5 | 0.6 | 0.8 | 0.4 | 0.9 | 1.3 | 0.6 | 1.0 | 0.9 | 1.8 | 1.3 | 0.4 | 0.4 | 2.2 |
| 0.1 N | 1.7 | 0.4 | 0.8 | 0.4 | 1.4 | 1.4 | 0.4 | 1.2 | 0.7 | 1.6 | 2.1 | 0.3 | 0.3 | 4.2 |
| 0.0 | 1.5 | 0.5 | 1.3 | 0.5 | 2.2 | 1.1 | 0.4 | 1.3 | 0.6 | 1.2 | 3.6 | 0.8 | 0.3 | 1.7 |
| 0.05 S | 1.9 | 0.6 | 1.1 | 0.4 | 1.3 | 1.4 | 0.5 | 1.0 | 0.6 | 1.4 | 4.0 | 0.5 | 0.4 | 2.3 |
| 0.1 S | 1.0 | 0.5 | 0.9 | 0.5 | 1.5 | 1.5 | 0.4 | 1.4 | 0.8 | 1.2 | 2.1 | 0.3 | 0.3 | 2.7 |
| 1.6 S | 1.6 | 0.7 | 0.8 | 0.5 | 1.7 | 1.2 | 0.4 | 1.0 | 0.8 | 1.7 | 1.3 | 0.2 | 0.4 | 2.0 |
| 4.8 S | 3.0 | 0.6 | 0.8 | 0.5 | 2.8 | 1.3 | 0.5 | 1.1 | 0.8 | 2.0 | 2.0 | 0.6 | 0.4 | 2.4 |
| 8.1 S | 2.3 | 0.6 | 1.1 | 0.5 | 2.8 | 1.2 | 0.4 | 1.1 | 0.9 | 2.4 | 2.1 | 0.6 | 0.3 | 2.6 |
| 18.0 S | 10.0 | 0.6 | 0.6 | 0.4 | 1.7 | 1.1 | 0.4 | 0.9 | 1.0 | 2.6 | 0.8 | 0.2 | 0.3 | 2.0 |
| 2.9 W | 2.6 | 0.6 | 0.5 | 0.4 | 0.8 | 1.2 | 0.6 | 2.0 | 1.0 | 1.6 | 1.1 | 0.4 | 0.4 | 2.2 |
| 28.3 N | 4.4 | 0.7 | 1.3 | 0.7 | 9.8 | 1.0 | 0.7 | 8.1 | 1.7 | 53.4 | 5.7 | 5.1 | 0.6 | 8.7 |
| Enrichment Factor, $EF_L$: | | | | | | | | | | | | | | |
| 48.3 N | 1.3 | 1.4 | 0.6 | | 9.9 | 10.8 | 0.4 | 1.7 | 0.07 | 0.3 | 4.9 | 0.29 | 0.2 | 1.7 |
| 16.2 N | 0.2 | 0.9 | 1.1 | | 1.0 | 17.3 | 0.3 | 1.1 | 0.05 | 0.3 | 1.0 | 0.03 | 0.2 | 0.9 |
| 8.1 N | 0.2 | 1.0 | 1.0 | | 1.7 | 21.8 | 0.4 | 1.0 | 0.09 | 0.3 | 0.5 | 0.02 | 0.2 | 0.8 |
| 4.8 N | 0.4 | 0.7 | 1.2 | | 0.7 | 17.6 | 0.2 | 0.7 | 0.06 | 0.3 | 0.4 | 0.02 | 0.1 | 0.7 |
| 1.6 N | 0.2 | 0.9 | 0.8 | | 0.6 | 20.0 | 0.4 | 0.9 | 0.06 | 0.3 | 0.8 | 0.02 | 0.2 | 0.8 |
| 0.1 N | 0.2 | 0.7 | 0.8 | | 1.0 | 22.1 | 0.3 | 1.1 | 0.04 | 0.2 | 1.3 | 0.01 | 0.2 | 1.6 |
| 0.0 | 0.2 | 0.8 | 1.3 | | 1.5 | 17.0 | 0.3 | 1.2 | 0.04 | 0.2 | 2.2 | 0.03 | 0.2 | 0.6 |
| 0.05 S | 0.2 | 0.9 | 1.1 | | 0.9 | 21.3 | 0.3 | 0.9 | 0.04 | 0.2 | 2.4 | 0.02 | 0.2 | 0.9 |
| 0.1 S | 0.1 | 0.8 | 1.0 | | 1.0 | 23.1 | 0.3 | 1.2 | 0.05 | 0.2 | 1.3 | 0.01 | 0.2 | 1.0 |
| 1.6 S | 0.2 | 1.1 | 0.8 | | 1.2 | 18.7 | 0.2 | 0.9 | 0.05 | 0.3 | 0.8 | 0.01 | 0.2 | 0.8 |
| 4.8 S | 0.3 | 1.0 | 0.9 | | 1.9 | 19.3 | 0.3 | 1.0 | 0.05 | 0.4 | 1.2 | 0.02 | 0.2 | 0.9 |
| 8.1 S | 0.2 | 0.9 | 1.1 | | 1.9 | 18.5 | 0.3 | 1.0 | 0.06 | 0.4 | 1.6 | 0.02 | 0.2 | 1.0 |
| 18.0 S | 1.1 | 1.0 | 0.6 | | 1.2 | 17.1 | 0.3 | 0.8 | 0.06 | 0.5 | 0.4 | 0.01 | 0.2 | 0.8 |
| 2.9 W | 0.3 | 0.9 | 0.5 | | 0.6 | 18.5 | 0.4 | 1.8 | 0.06 | 0.3 | 0.6 | 0.02 | 0.2 | 0.8 |
| 28.3 N | 0.5 | 1.1 | 1.4 | | 6.7 | 14.6 | 0.4 | 7.3 | 0.10 | 9.6 | 3.4 | 0.20 | 0.3 | 3.3 |

[a] Enrichment factor = ratio of the element concentration in the sediment to the Fe concentration in the sediment divided by the ratio of the average element concentration in shale ($EF_S$) or limestone ($EF_L$) to the average Fe concentration in shale or limestone.

proportion to their sediment abundances. Tessier et al.[12] reported that metal accumulation in *E. complanata* is related not to the total metal load in the sediment but to one or more extractable elemental fractions. These investigators also reported that metal accumulation is also influenced by other constituents of sediment, such as organic matter, particle size and clay mineralogy.

*Bioconcentration Relative to Sediment.* Silver, Cd, and Fe accumulated in the mussels to levels 20 to 100 times greater than the concentration in the sediment (Table 11). Manganese and Se occurred at 10 to 30 times the sediment levels, while Zn was seen at 2 to 5 times sediment levels. Elements in mussel tissue below the levels found in sediment were Pb and V, which were elevated in tissues of trans- planted mussels compared to Lake George in situ. This indicates some elimination or control process by the mussels. The remaining elements (As, Cu, and Sr) were present in the mussels at or near the same concentrations in the sediment. Of the elements that showed significant correlation between the concentrations in the mussels and in the sediment (Cd and V), only the concentration ratio (CR) for Cd was found to be elevated.

At the marina site (28.3 km north of Milliken Station), the CRs for Cd and Fe were >100. At site 2.9 W, the Cd CR was >100 for 1979. With the exception of Sr, which did not appear to be bioconcentrated, the remaining elements measured at the marina sites had CRs similar to the range observed for the littoral sites located on the main body of Cayuga Lake.

Even though an element may not be present at significantly elevated levels in the transplanted mussels compared to the Lake George controls (e.g., Ag, Cd, Mn, Se), the occurrence of elevated CRs imply bioconcentration by the mussels. This supports the use of mussels as biological indicator organisms.

Some CRs have previously been reported for Lake George mussels and sediment concentrations.[11] *Lampsilus radiata* was reported to accumulate Cd, Se, and Zn above the levels found in sediment from northern Lake George (CR>10), while Pb and Cu occurred at similar concentrations in both mussels and sediment (CR = 1-2). Arsenic was eliminated or not accumulated (CR<1). Comparing these CRs with the present results for the transplanted studies (Table 11) showed similar results for Cd, Cu, and Se (CR>10), while Zn accumulation was reduced and Pb appeared not to be accumulated. Arsenic, which showed no accumulation by the mussels in northern Lake George, exhibited tissue levels similar to sediment concentrations present in Cayuga Lake after transplanting. The differences be- tween Cayuga Lake transplants and northern Lake George studies suggest that bioconcentration of metals by mussels may reflect differences in the sediment concentrations present in the two lakes, as well as the available chemical form of the metals. These results also emphasize the site-specific nature of data on bioconcentration for aquatic organisms.

## Relation Between Mussel and Water Concentrations

When the metal concentrations in the Cayuga Lake water were compared to the levels in the soft tissues of the retrieved mussels, significant positive correlations were found only for Cd in both years of retrieval. Elemental concentrations in

**Table 11**
**Concentration Ratios, Retrieved Mussels (µg/g) to Surface Sediment (µg/g)**

| Site | Ag 1979 | Ag 1980 | As 1979 | As 1980 | Cd 1979 | Cd 1980 | Cu 1979 | Cu 1980 | Fe 1979 | Fe 1980 | Mn 1979 | Mn 1980 |
|---|---|---|---|---|---|---|---|---|---|---|---|---|
| Lake Shore: | | | | | | | | | | | | |
| 16.2 N | — | 72 | — | 1.3 | — | 38 | — | 0.8 | — | 45 | — | 16 |
| 8.1 N | 66 | 65 | 1.8 | 2.0 | 45 | 38 | 1.2 | 1.1 | 63 | 76 | 11 | 11 |
| 4.8 N | 23 | 23 | 0.9 | 1.0 | 55 | 61 | 1.0 | 1.2 | 43 | 41 | 9 | 10 |
| 1.6 N | 102 | 63 | 2.2 | 2.3 | 92 | 92 | 1.4 | 1.2 | 42 | 113 | 16 | 16 |
| 0.1 N | 60 | 49 | 1.5 | 1.6 | 41 | 44 | 0.7 | 0.7 | 57 | 58 | 15 | 13 |
| 0.05 S | 53 | — | 1.2 | — | 51 | — | 1.9 | — | 57 | — | 22 | — |
| 0.1 S | 132 | 139 | 1.4 | 1.6 | 54 | 54 | 1.1 | 1.7 | 65 | 77 | 16 | 17 |
| 1.6 S | 35 | 66 | 1.3 | 1.6 | 38 | 36 | 0.8 | 1.0 | 54 | 40 | 13 | 15 |
| 4.8 S | 52 | 77 | 2.0 | 2.7 | 35 | 26 | 1.2 | 1.5 | 80 | 111 | 22 | 24 |
| 8.1 S | 73 | 76 | 1.5 | 1.7 | 37 | 36 | 1.2 | 1.2 | 72 | 84 | 17 | 19 |
| 2.9 W | 77 | 71 | 4.2 | 4.9 | 142 | 41 | 1.0 | 1.0 | 77 | 93 | 23 | 25 |
| Marina: | | | | | | | | | | | | |
| 28.3 N | 49 | 41 | 2.2 | 2.4 | 143 | 169 | 1.1 | 2.4 | 184 | 159 | 17 | 19 |

| Site | Pb 1979 | Pb 1980 | Se 1979 | Se 1980 | Sr 1979 | Sr 1980 | V 1979 | V 1980 | Zn 1979 | Zn 1980 |
|---|---|---|---|---|---|---|---|---|---|---|
| Lake Shore: | | | | | | | | | | |
| 16.2 N | — | 0.3 | — | 11 | — | 0.9 | — | 0.028 | — | 3.9 |
| 8.1 N | 1.2 | 0.6 | 32 | 29 | 2.0 | 1.9 | 0.023 | 0.051 | 5.8 | 5.3 |
| 4.8 N | 0.6 | 0.4 | 24 | 21 | 1.5 | 1.3 | 0.036 | 0.038 | 3.7 | 3.9 |
| 1.6 N | 0.9 | 0.5 | 19 | 14 | 2.1 | 2.6 | 0.034 | 0.081 | 5.2 | 4.6 |
| 0.1 N | 0.9 | 0.4 | 8 | 8 | 1.8 | 1.7 | 0.044 | 0.055 | 1.8 | 1.6 |
| 0.05 S | 1.3 | — | 5 | — | 1.4 | — | 0.032 | — | 4.6 | — |
| 0.1 S | 1.6 | 0.8 | 9 | 8 | 2.0 | 2.0 | 0.033 | 0.053 | 3.4 | 3.4 |
| 1.6 S | 0.3 | 0.4 | 13 | 11 | 1.6 | 2.5 | 0.037 | 0.022 | 3.6 | 4.3 |

**Table 11**
**Concentration Ratios, Retrieved Mussels (µg/g) to Surface Sediment (µg/g)**

| Site | Pb | | Se | | Sr | | V | | Zn | |
|---|---|---|---|---|---|---|---|---|---|---|
| | 1979 | 1980 | 1979 | 1980 | 1979 | 1980 | 1979 | 1980 | 1979 | 1980 |
| Lake Shore: (Cont.) | | | | | | | | | | |
| 4.8 S | 1.1 | 0.7 | 13 | 13 | 1.7 | 1.4 | 0.042 | 0.071 | 5.1 | 5.4 |
| 8.1 S | 0.6 | 0.4 | 11 | 11 | 1.5 | 1.4 | 0.052 | 0.045 | 4.0 | 4.4 |
| 2.9 W | 1.3 | 0.9 | 34 | 27 | 3.0 | 2.8 | 0.038 | 0.050 | 7.4 | 7.6 |
| Marina: | | | | | | | | | | |
| 28.3 N | 0.1 | 0.1 | 9 | 7 | 0.5 | 0.7 | 0.124 | 0.026 | 3.2 | 3.2 |

mussels may be more closely related to the chemical form of the element in the water and, hence, may not correlate well with the *total* element concentrations present in the water. In addition, the lack of correlations between the concentrations of metals in the mussel tissue and the water may be due to slow metal bioaccumulation and excretion in the mussels relative to short-term fluctuations in metal concentrations in the water. Therefore, the metals measured in the mussel tissue at any given time may not necessarily reflect the concentrations measured in the water sampled at the same time.[43] The metal content of mussels with long life-spans has been hypothesized to be dependent on the ratio of times of uptake to depletion, especially in systems where the varying water concentrations may result in rapid depuration.[7]

*Bioconcentration Relative to Water.* Because of the low metal concentrations in Cayuga Lake water, all of the elemental CRs appear to be accumulated by a factor of 1000 (Table 12). The CRs range from $1500 \times 10^3$ for Cd to $0.2 \times 10^3$ for Sr. At the marina sites, the CRs were generally lower due to the higher levels of water concentrations. Thus, the concentration of free metal ions, rather than total concentration or the presence of complexed metals, appears to be the most important factor controlling metal uptake from solution for at least Cd, Cu, Fe, Mn, and Zn in some aquatic organisms.[36] For these elements, the CR may be a better indicator of environmental conditions than measurements of the water concentrations.

## Relation of Mussel Tissue Concentrations to Coal-Combustion Products from Milliken Station

Trace elements associated with coal combustion are dependent on the type of coal used and the combustion-process characteristics of the power plant. Some fly ashes have been reported to have high concentrations of minor elements (As, Pb, Se, Sr, V, Zn) relative to soil and coal concentrations, indicating that the coal-combustion process enriches the elements in the waste products.[44] The concentration of the major elements Al, Ca, and Fe in fossil fuel combustion wastes is within the range of their soil content, while the concentration of K, Na, Mg, and S may exceed their soil levels.[44] The concentration of elements in coal varies widely, depending on grade and geographic origin. Enrichment is the result of condensation of the elements that are volatilized during combustion onto solid particles as the gas cools. The elements are not distributed homogeneously throughout the wastes, but smaller particles are enriched with the more volatile elements. Arsenic, Cd, Cu, Pb, Se, V, and Zn are generally strongly volatilized during combustions and tend to become enriched on the particle surface. Elements with low condensation temperatures (e.g., As and Se) may escape to the atmosphere as gases during combustion. Most of the Cu, Mo, Se, Sr, and V contained in fly ash readily solubilize.[45]

### Trace Elements Released from Milliken Station

Milliken Station is a source of potential releases to the environment from (1) gases and particulate emissions from the stacks and (2) leachate from the ash piles. Analysis of the coal and ash compartments at the Milliken plant

**Table 12**
**Concentration Ratios, Retrieved Mussels (µg/g dry wt) to Water (ng/mL), $\times 10^3$**

| Site | As 1979 | As 1980 | Ca 1979 | Ca 1980 | Cd 1979 | Cd 1980 | Cu 1979 | Cu 1980 | Fe 1979 | Fe 1980 |
|---|---|---|---|---|---|---|---|---|---|---|
| Lake Shore: | | | | | | | | | | |
| 16.2 N | — | 9 | — | 1.2 | — | 1695 | — | 12 | — | 19 |
| 8.1 N | 9 | 10 | 1.2 | 1.2 | 1413 | 1197 | 3 | 3 | 33 | 40 |
| 4.8 N | 10 | 10 | 1.0 | 1.1 | — | — | 7 | 8 | 55 | 53 |
| 1.6 N | 6 | 6 | 1.2 | 1.2 | 1464 | 1468 | 14 | 12 | 17 | 47 |
| 0.1 N | 6 | 7 | 1.1 | 1.1 | — | — | 17 | 18 | 8 | 8 |
| 0.05 S | — | — | — | — | — | — | — | — | — | — |
| 0.1 S | — | — | — | — | — | — | — | — | — | — |
| 1.6 S | 12 | 14 | 1.2 | 1.5 | — | — | 21 | 26 | 69 | 51 |
| 4.8 S | 8 | 11 | 1.3 | 1.5 | — | — | — | — | 82 | 114 |
| 8.1 S | 7 | 8 | 1.2 | 1.3 | — | — | — | — | 60 | 70 |
| 2.9 W | 10 | 12 | 1.3 | 1.5 | — | — | — | — | 24 | 30 |
| Marina: | | | | | | | | | | |
| 28.3 N | 3 | 3 | 0.6 | 0.5 | 442 | 191 | 6 | 4 | 3 | 4 |

| Site | Pb 1979 | Pb 1980 | Sr 1979 | Sr 1980 | V 1979 | V 1980 | Zn 1979 | Zn 1980 |
|---|---|---|---|---|---|---|---|---|
| Lake Shore: | | | | | | | | |
| 16.2 N | — | 12 | — | 0.43 | — | 1.7 | — | 37 |
| 8.1 N | 30 | 17 | 0.45 | 0.44 | 0.7 | 1.6 | 31 | 29 |
| 4.8 N | 25 | 14 | 0.47 | 0.41 | 3.6 | 3.8 | 322 | 345 |
| 1.6 N | 13 | 7 | 0.44 | 0.53 | 3.2 | 7.6 | 47 | 42 |
| 0.1 N | 26 | 12 | 0.36 | 0.34 | 3.1 | 3.8 | 220 | 201 |
| 0.1 S | — | — | — | — | — | — | — | — |
| 0.1 S | — | — | — | — | — | — | — | — |
| 1.6 S | 9 | 12 | 0.32 | 0.51 | 4.8 | 2.9 | 447 | 541 |

**Table 12 (Continued)**
Concentration Ratios, Retrieved Mussels ($\mu$g/g dry wt) to Water (ng/mL), $\times 10^3$

| Site | Pb 1979 | Pb 1980 | Sr 1979 | Sr 1980 | V 1979 | V 1980 | Zn 1979 | Zn 1980 |
|---|---|---|---|---|---|---|---|---|
| Lake Shore: (Cont.) | | | | | | | | |
| 4.8 S | 33 | 21 | 0.53 | 0.43 | 3.2 | 5.4 | 142 | 150 |
| 8.1 S | 28 | 19 | 0.48 | 0.45 | 5.6 | 4.8 | 154 | 168 |
| 2.9 W | 41 | 27 | 0.55 | 0.51 | 2.4 | 3.2 | 131 | 135 |
| Marina: | | | | | | | | |
| 28.3 N | 2 | 1 | 0.19 | 0.14 | 0.6 | 2.0 | 8 | 6 |

indicated that Ag, As, and Pb were preferentially partitioned into the fly ash.[21] Mass balance calculations indicated that Sb and Se were possibly discharged from the power plant. Assuming neutral atmospheric dispersal conditions at 90 m (the actual stack height), the distance to the maximum downwind deposition concentrations is about 2 to 3 km.[46] The Cayuga Lake basin opens into a rather flat basin at the northern end of the lake but becomes progressively steeper toward the south. This orientation is an important factor when investigating the dispersal of pollutants by air movement, since the shape of the basin causes the prevailing winds to be shifted along the axis of the lake.[26] Prior to incorporation of electrostatic precipitators in 1976 (2 years prior to the start of this study), ash fallout was reported by local residents up to 4.8 km from the plant.[25]

*Silver.* The sediment concentration was enriched at sites 48.3 km north and 18.0 km south of the power plant (Table 10). The $EF_S$ indicated lesser enrichment at sites 4.8 N and 4.8 S. The CR for sediment (Table 11) was elevated at sites 1.6N and 0.1S. However, site location was not significant in mussel tissue concentration of Ag. Also, the mussel concentrations of Ag were not significantly greater than the Lake George mussel concentrations at any of the sites. Thus, Ag was probably not released from the plant.

*Arsenic.* Arsenic was elevated in the mussel tissue (relative to Lake George controls) at all sites. Because of the wind patterns over the lake, this may represent uptake of deposited emissions from the power plant. However, the source of the As could be partially due to the use of pesticides in this highly active agricultural area.

*Copper.* The Cu concentrations in mussel tissues near the plant (sites 0.05S and 0.1S) were significantly elevated compared to the Lake George controls. Both locations received discharge water from the power plant's cooling system. The elevated Cu concentrations could be ascribed to leaching from the Cu containing pipes in the condensation system of the power plant cooling system. Similar results have been reported elsewhere for marine bivalves.[13]

*Lead.* Site location was a significant factor in mussel tissue concentration, particularly at site 0.1S where the levels were 2 to 10 μg/g higher than at other sites. Surprisingly, there is no indication of elevated Pb in surface sediment at this site. Nevertheless, the proximity of the power plant and the elevated elemental tissue concentration suggest some relationship.

*Antimony.* Antimony was below the detection limit for mussel tissue, sediment, or water.

*Selenium.* The mussel concentrations of Se were 10 to 35 times higher than the sediment levels; however, the mussel concentrations were not elevated compared to the Lake George controls. Interestingly, the sediment sampled in the immediate vicinity of the power plant (0.0 and 0.5 S) had an $EF_S$=4. The EFs declined at the sampling locations further north or south of the plant. This could be due to either fly ash fallout from the power plant or from fly ash spilled from the broken dike adjacent to Milliken Station.[40]

## Relation of Metal Concentrations and Distance from the Power Plant

*Concentration in Sediment and Water.* Because of the stack height at Milliken Station and wind patterns along the lake, it is difficult to establish with certainty a pattern of element deposition related to the plant. The concentrations in sediments at sites north of Milliken Station showed a significant positive correlation with distance for Ag, Cd, and Sr (i.e., the concentrations increased as distance *from* the plant increased) and a negative correlation for Co, Cr, and V (i.e., the concentration increased as distance *to* the plant decreased). For sites south of the plant, concentration in sediment was positively correlated with distance from Milliken Station for Ag and Pb, while As was negatively correlated. Water concentrations showed no correlation with distance from Milliken Station for any of the elements.

*Concentrations in Mussels.* The correlation of element concentration with mussel distribution is not necessarily a direct relationship but could result from a common variable, whether a physical process (i.e., silting) or the interaction with another element. In fact, when the concentrations in the littoral mussels (excluding site 2.9 W) were compared to distance from the Milliken Station, only Zn in samples retrieved in 1980 was significantly correlated (r = –0.96) with distance south of the plant.

## Other Possible Sources of Pollution

Although the present study is investigating pollutants related to the coal-fired power station, there are several additional potential sources of trace elements to the Cayuga Lake aquatic environment. Marinas and agricultural runoff will be discussed in the following sections.

### Marinas

Marinas represent a local source of metals from several sources including corrosion preventive undercoatings on boats and trailers (Zn, Cr); outercoating paints (Pb); antifouling boat paints (As, Cu), petroleum based fuels (Cd, Cu, Se, Sr, Pb, V, and Zn); and engines and boating hardware (Al, Fe).[8,47] Enriched levels of most of these elements were observed in either the mussels, sediments, or water from the marina sites. For example, significantly greater concentrations of As, Cu, Fe, Sr, and V were measured in the mussel tissue from Cayuga Lake than from Lake George. Other researchers have previously reported increased levels of Cd, Cu, Pb, and Zn in mussels from marina areas.[8,47] The enrichment factor ($EF_s > 5$) calculated for the marina site also indicated an anthropogenic input of Cd, Cu, Pb, Se, Sr, and Zn. Water levels of Fe, Pb, and Zn were at least a factor of two higher than the water concentrations measured in the main littoral zone of the lake.

### Agricultural Land Use

The dominant land use categories in the Cayuga Lake watershed are active agriculture (48%) and forest (31%). Active agriculture includes both cultivated land

(45%) and cropland permanently or temporarily removed from agricultural use (16%).[20] Pesticides and fertilizers are important sources of metals in agricultural areas.

Pesticides (herbicides, insecticides, fungicides) are widely used in production of fruits, vegetables, and other crops. Compounds of Cu have extensive applications as agricultural pesticides, and runoff of these substances from the land may have been responsible for elevated concentrations in mussels in rural areas.[48] Other elements frequently used in pesticides include As, Pb, and to some degree Mn and Zn.[49] In the U.S., agriculture accounts for 46% of the As used in 1981. About one half of the mobile As inventory in the U.S. comes from pesticides. In addition, soils from orchards have As levels about ten times higher than non-treated soils.[49] In addition to macronutrients (K, N, S) and micronutrients (B, Cu, Fe, Mn, Mo, Zn), commercial fertilizers also contain small amounts of potentially toxic trace elements (Cd, Cr, Pb, and Se).[49] The use of Cu and Co as dietary supplements for milk cows can result in elevated levels of these elements in manure and subsequently add to the levels present in soils.[49]

Of the elements associated with agricultural processes, the concentrations of As in the transplanted mussels at all the sites were elevated above the Lake George controls. Therefore, local agricultural practices probably contributed to the concentration of As in the mussels transplanted to Cayuga Lake.

## CONCLUSIONS

1. The freshwater mussel *L. radiata* accumulated Ag, Cd, Fe, Mn, Se, and Zn in excess of the concentrations in the surrounding sediments.

2. Mussels transplanted to the Cayuga Lake system had higher soft tissue concentration of As and Cu than in situ at the pristine northern Lake George. The elevated Cu levels appears to be related to the water discharge from the Milliken power plant. The elevated As concentrations were attributed to local agricultural practices (pesticides).

3. Except for Cu in a small confined area near the water discharge, the Milliken coal-fired power plant had no discernible effect on the metal levels in Cayuga Lake. The stack emissions from the plant were either masked by other sources or were atmospherically transported and deposited elsewhere.

4. Boating marinas appear to be sources of several metals, including As, Cd, Cu, Fe, Pb, Se, Sr, V, and Zn, in Cayuga Lake. The levels of some of the elements in mussels held at marinas were at least two times higher than concentrations seen in the littoral zone of the lake.

5. The lack of clear correlations between the metal concentrations present in the mussels and the Cayuga Lake sediment and the lake water strongly indicates a complex unexplained interrelationship between the biota and the physical environment of this system, which remains poorly understood.

## ACKNOWLEDGMENTS

We would like to acknowledge the assistance of Drs. N. Clesceri and M. Kobayashi of the Rensselaer Fresh Water Institute, Troy, NY, for securing the Lake George mussels, and Dr. C. Schofield of Cornell University, Ithaca, NY, for use of his refrigerated storage facilities.

## REFERENCES

1. Newman, M. C. and McIntosh, A. W., The influence of lead in components of a freshwater ecosystem on molluscan tissue lead concentrations, *Aquat. Toxicol.*, 2, 1, 1982.
2. Wren, C. D., MacCrimmon, H. R., and Loescher, B. R., Examination of bioaccumulation and biomagnification of metals in a Precambrian shield lake, *Water Air Soil Poll.*, 19, 277, 1983.
3. Krieger, R. I., Gee, S. J., and Lim, L. O., Marine bivalves, particularly mussels, Mytilus sp., for assessment of environmental quality, *Ecotoxicol. Environ. Safety*, 5, 71, 1981.
4. Farrington, J. W., Goldberg, J. D., Risebrough, R. W., Martin, J. H., and Bowen, V. T., The U.S. mussel watch 1976-1978: an overview of the trace-metal, DDT, PCB, hydrocarbon and artificial radionuclide data, *Environ. Sci. Technol.*, 17, 490, 1983.
5. Goldberg, E. D., The mussel watch concept, *Environ. Monitoring Assess.*, 7, 91, 1986.
6. Foster, R. B. and Bates, J. M., Use of freshwater mussels to monitor point source industrial discharges, *Environ. Sci. Technol.*, 12, 958, 1978.
7. V.-Balogh, K., Comparison of mussels and Crustacean plankton to monitor heavy metal pollution, *Water Air Soil Poll.*, 37, 281, 1988.
8. Chu, K. H., Cheung, W. M., and Lau, S. K., Trace metals in bivalves and sediment from Tolo Harbour, Hong Kong, *Environ. Int.*, 16, 31, 1990.
9. Clarke, A. H., The freshwater molluscs of the Canadian interior basin, *Malacologia*, 13, 1, 1973.
10. Curry, C. A., The freshwater clam *(Elliptio complanata)*, a practical tool for monitoring water quality, *Water Poll. Res. (Can.)*, 13, 45, 1977.
11. Heit, M., Klusek, C. S., and Miller, K. M., Trace element, radionuclide, and poly-nuclear aromatic hydrocarbon concentrations in Unionidae mussels from northern Lake George, *Environ. Sci. Technol.*, 14, 465, 1980.
12. Tessier, A., Campbell, P. G. C., Auclair, J. C., and Bisson, M., Relationship between the partitioning of trace metals in sediments and their accumulation in the tissues of the freshwater mollusc *Elliptio complanata* in a mining area, *Can. J. Fish. Aquat. Sci.*, 41, 1463, 1984.
13. Stephenson, M. D., Gordon, R. M., and Martin, J. H., Biological monitoring of trace metals in the marine environment with transplanted oysters and mussels, in *Bioaccumulation of Heavy Metals by Littoral and Pelagic Marine Organisms*, Martin, J. H., Ed., U.S. EPA Report EPA-600/3-79-038, 1979, 12.
14. Sheehan, R. J., Neves, R. J., and Kitchel, H. E., Fate of freshwater mussels transplanted to formerly polluted reaches of the Clinch and North Fork Holston rivers, Virginia, *J. Freshwater Ecol.*, 5, 139, 1989.

15. Fuller, S. L. H., Clams and mussels (Mollusca: Bivalvia), in *Pollution Ecology of Freshwater Invertebrates,* Hart, C. W. and Fuller, S. L. H., Eds., Academic Press, New York, 1974, 215.

16. Heit, M. and Klusek, C. S., The release of trace substances by a coal-fired power station into an aquatic ecosystem. I. Environmental sampling in Cayuga Lake, NY, in Environmental Quarterly, U.S. DOE Report EML-363, New York, 1979, 1.

17. Heit, M., The relationship of a coal-fired power plant to the levels of polycyclic aromatic hydrocarbons (PAH) in the sediment of Cayuga Lake, *Water Air Soil Poll.,* 24, 41, 1985.

18. Heit, M., Tan, Y. L., Miller, K. M., Quanci, J., Marinetti, C., Silvestri, S., Swain, A. M., and Winkler, M. G., The sediment chronology and polycyclic aromatic hydrocarbon concentrations and fluxes in Cayuga Lake, NY, U.S. DOE Report EML-451, New York, 1986.

19. Silvestri, S., Heit, M., and Klusek, C. S., Trace element concentrations in transplanted and naturally occurring Unionidae mussels, water, sediment, and macrophytes in Cayuga Lake, U.S. DOE Report EML-514, New York, 1988.

20. Olgesby, R. T., The limnology of Cayuga Lake, in *Lakes of New York State, Ecology of the Finger Lakes,* Vol. 1, Bloomfield, J. A., Ed., Academic Press, New York, 1978.

21. Klusek, C. S., Miller, K. M., and Heit, M., Trace element and radionuclide mass balance at a coal-fired electric generating station, *Environ. Int.,* 9, 139, 1983.

22. Bowen, H. J. M., *Trace Elements in Biochemistry,* Academic Press, New York, 1966.

23. Heit, M., A Review of Current Information on Some Ecological and Health Related Aspects of the Release of Trace Metals into the Environment Associated with the Combustion of Coal, U.S. ERDA Report HASL-320, New York, 1977.

24. Godfrey, P. J., The eutrophication of Cayuga Lake: a historical analysis of the phytoplankton's response to phosphate detergents, *Freshwater Biol.,* 12, 149, 1982.

25. Travis, A., personal communication, 1979.

26. New York State Electric and Gas Corporation, Air quality and meteorology — Cayuga site, in Cayuga Station Application to the New York State Board on Electric Generation Siting and the Environment, Vol. 2, 1974, 73.2-1.

27. Madsen, J. D., Sutherland, J. W., Bloomfield, J. A., Roy, K. M., Eichler, L. W., and Boylen, C. W., The Lake George Aquatic Plant Survey, Department of Environmental Conservation Report, 1989.

28. Aulenbach, D. B. and Clesceri, N. L., Sources and sinks of nitrogen and phosphorus: water quality management of Lake George (NY), FWI Report 72-32, Rensselaer Fresh Water Institute at Lake George, Troy, NY, 1972.

29. Eichler, L., personal communication, 1990.

30. May, T. W. and Braumbaugh, W. G., Matrix modifier and L'vov platform for elimination of matrix interferences in the analysis of fish tissue for lead by graphite furnace atomic absorption spectrometry, *Anal. Chem.,* 54, 1032, 1982.

31. Snedecor, G. W. and Cochran, W. G., *Statistical Methods,* Iowa State University Press, Ames, IA, 1971.

32. Boyden, C. R., Trace element content and body size in molluscs, *Nature,* 251, 311, 1974.

33. Boalch, R., Chan, S., and Taylor, D., Seasonal variation in the trace metal content of *Mytilus edulis, Mar. Poll. Bull.,* 12, 276, 1981.

34. Hinch, S. G. and Stephenson, L. A., Size- and age-specific patterns of trace metal concentrations in freshwater clams from an acid-sensitive and a circumneutral lake, *Can. J. Zool.,* 65, 2436, 1987.

35. Smith, R. M., Metal Ion Complexing in Lake Sediment, Ph.D. thesis, Rensselaer Polytechnic Institute, Troy, NY, 1977.

36. Luoma, S. N., Bioavailability of trace metals to aquatic organisms — a review, *Sci. Total Environ.*, 28, 1, 1983.

37. Jones, W. G. and Walker, K. F., Accumulation of iron, manganese, zinc and cadmium by the Australian freshwater mussel *Velesunio ambiguus* (Phillipi) and its potential as a biological monitor, *Aust. J. Mar. Freshwater Res.*, 30, 741, 1979.

38. Heit, M., Klusek, C. S., and Baron, J., Evidence of deposition of anthropogenic pollutants in remote Rocky Mountain lakes, *Water Air Soil Poll.*, 22, 403, 1984.

39. Mills, E. L. and Oglesby, R. T., Five trace elements and vitamin $B_{12}$ in Cayuga Lake, New York, in *Proc. 14th Conf. Great Lakes Research,* 1971, 256.

40. Risk and power, *Ithaca Journal,* p. 14, July 27, 1977.

41. Pennak, R., *Fresh-Water Invertebrates of the United States,* Ronald Press, New York, 1953, 699.

42. Seagle, S. M. and Ehlmann, A. J., Manganese, zinc and copper in water, sediments and mussels in north central Texas reservoirs, *Trace Substances Environ. Health*, 8, 101, 1974.

43. Fowler, S. W. and Oregioni, B., Trace metals in mussels from the N. W. Mediterranean, *Mar. Poll. Bull.*, 7, 26, 1976.

44. Mattigold, S. V., Rai, D., Eary, L. E., and Ainsworth, C. C., Geochemical factors controlling the mobilization of inorganic constituents from fossil fuel combustion residues: I. Review of the major elements, *J. Environ. Qual.*, 19, 188, 1990.

45. Eary, L. E., Rai, D., Mattigold, S. V., and Ainsworth, C. C., Geochemical factors controlling the mobilization of inorganic constituents from fossil fuel combustion residues. II. Review of the minor elements, *J. Environ. Qual.*, 19, 202, 1990.

46. Turner, D. B., Workshop on Atmospheric Dispersion Estimates, U.S. HEW Public Service Publ. No. 999-Ap-26, 1969.

47. V.-Balogh, K., Heavy metal pollution from a point source demonstrated by mussel *(Unio pictorum L.)* at Lake Balaton, Hungary, *Bull. Environ. Contam. Toxicol.*, 41, 910, 1988.

48. Manly, R. and George, W. O., The occurrence of some heavy metals in populations of the freshwater mussel *Anodonta anatina (L.)* from the river Thames, *Environ. Pollut.*, 14, 139, 1977.

49. Adriano, D. C., *Trace Elements in the Terrestrial Environment,* Springer-Verlag, New York, 1986, 46.

*Tests for and
Monitoring of Fossil
Fuel Dispersion and
Ash Disposal*

# 5

# Strontium and Lead Isotopes as Monitors of Fossil Fuel Dispersion

R. W. Hurst,[1] T. E. Davis,[2] A. A. Elseewi,[3] and A. L. Page[4]

[1]Chempet Research Corporation, Moorpark, CA
[2]California State University, Los Angeles, CA
[3]Southern California Edison Co., Rosemead, CA
[4]University of California, Riverside, CA

## ABSTRACT

The need to monitor the dispersion of fossil fuel residues such as fly ash and coal tar in the environment requires new methodologies to be developed which utilize naturally occurring isotopes of the elements strontium (Sr) and lead (Pb) as tracers. Both of these isotopic systems do not exhibit isotopic fractionation, a characteristic of the light stable isotopes (e.g., hydrogen, oxygen). This removes one potential complication in fate/transport studies, the alteration of a contaminant's or leachate's isotopic signature via mass dependent isotopic fractionation. Our research began with detailed sampling of soils, groundwater, flora, and coal residues (when available) in the vicinity of both operational and abandoned fossil fuel sites. In order to evaluate these isotopic systems' effectiveness as tracers of fossil fuel leachate, we developed chemical extraction techniques designed to preferentially extract: (1) fly ash-derived Sr from biogeologic samples in the vicinity of an operating coal-fired power plant and (2) coal tar Pb from soils at an abandoned coal tar site. The Sr and/ or Pb isotopic compositions of the fossil fuels at both sites were significantly different (>10 times the standard error of the mean of our analyses) from those of the natural background. Our research indicates that Sr and Pb isotopes are effective tracers of fossil fuel leachates in ecosystems and may be used to place constraints on models developed to assess the impact of fossil fuels on the environment.

## INTRODUCTION

In view of the nation's increasing need to reduce its dependence on foreign sources of fuels to produce electricity, the use of coal as an alternative energy

0-87371-890-9/93/$0.00+$.50
© 1993 by Lewis Publishers

source is expected to increase substantially. There are concerns, however, pertaining to the potential impact of coal-burning power plan residues on the environment. Although more than 99% of stack emissions of fly ash are retained by various control devices, the remainder becomes airborne and is dispersed to the environment. In addition, other fossil fuel residues, such as coal tars, may be stored on-site, thereby creating a situation in which groundwater quality may be degraded.

Monitoring the dispersion of airborne particulate emissions and stored residue (e.g., coal tar) has been difficult due to problems in determining the origin of selected trace elements.[1-3] The isotopes of Sr and Pb have been effective tracers of airborne emissions such as fly ash.[4-6] For this reason, these isotopic systems were chosen to evaluate their application as tracers of different fossil fuel residues in the environment. This work focuses on both Sr and Pb isotopes.

## The Geochemistry of Sr

Strontium is an alkaline earth element which readily substitutes for calcium. Thus any mineral or fluid with calcium enrichments will also have high Sr concentrations. For example, the mineral calcite ($CaCO_3$), contains 500 to 1500 ppm Sr as compared to minerals with little or no calcium, where Sr concentrations rarely exceed 100 ppm.[7]

There are four naturally occurring isotopes of Sr: $^{88}Sr$, $^{87}Sr$, $^{86}Sr$, and $^{84}Sr$. The relative abundances of these four isotopies are 83%, 7%, 9.5%, and 0.5%, respectively. The Sr isotopic abundances, however, vary in geologic materials because $^{87}Sr$ is produced by the radioactive decay of $^{87}Rb$, whose half-life is 48.8 Ga. Thus, the amount of $^{87}Sr$ in geologic materials is controlled by the age and amount of $^{87}Rb$ in the sample. A stable isotope of Sr, $^{86}Sr$, has been designated as the reference isotope. Hence, it is the $^{87}Sr/^{86}Sr$ ratio (Sr isotopic composition, hereafter referred to as the IC), which is used as an isotopic tracer. The IC of a sample can be measured both accurately and precisely $\pm$ 0.005%, which makes this method an excellent fingerprint of biogeochemical processes.

## Strontium in Mixtures

The Sr geochemistry of geologic materials can be controlled by mixing. For example, if two groundwaters with different Sr concentrations and ICs are mixed, the IC of the resultant mixture will be determined by the relative proportions of the two groundwaters (endmembers) contributing to the mixture. If the endmembers can be identified, the relative contribution of each can be assessed. The technique is depicted in Figure 1. Detailed discussions of binary mixing and related isotope systematics may be found in Faure.[7] The example that follows was modelled from the equations found in this reference.

In this example, river water (CR) or soil extract (SE), each with an IC of 0.7102 and Sr concentration of 1 ppm, is mixed with the leachate from fossil fuel residue (FFR) which has an IC between 0.7091 and 0.7094 and Sr concentration of 260 ppm.

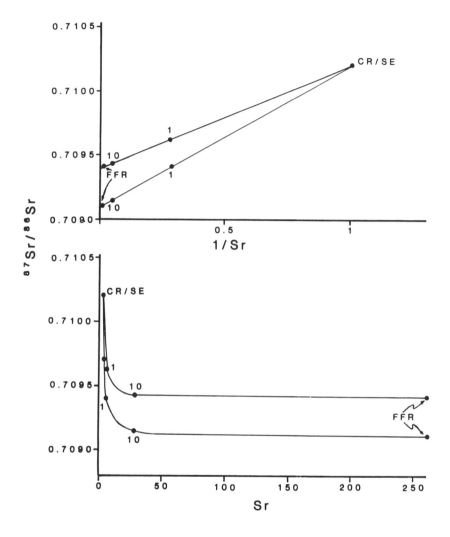

**FIGURE 1.**   Mixing relationships in a two-component system. Endmember components include fossil fuel residue (designated as FFR) and either a soil extract or Colorado River water (designated as SE and CR, respectively).

Hyperbolic mixing curves result when the IC is plotted against Sr concentration; mixing lines result when the IC is plotted against 1/Sr (Figure 1). Since one endmember (FFR) exhibits an IC range, mixtures of the two endmembers (CR or SE and FFR) would plot between the hyperbolic mixing curves (IC vs. Sr concentration) or mixing lines (IC vs. 1/Sr). For reference, mixtures containing 1% and 10% FFR are plotted.

A small contribution from a Sr-enriched endmember, in this case FFR, can significantly control the IC of the mixture. Such plots allow contributions from each endmember to be quantified when endmember compositions are known. An additional benefit arises when mixing curves or lines are produced from Sr isotopic and concentration analyses; in this situation, endmember compositions may be mathematically generated and compared to possible sources of Sr. This approach was used[8] to identify three to four endmembers contributing Pb to an aquifer in the vicinity of a uranium ore body.

## Isotopic Fractionation

Unlike the behavior of light stable isotopes (C, H, O, N. S), Sr and Pb isotopes do not fractionate.[7] Thus, neither Sr nor Pb isotopes are preferentially separated from their isotopic counterparts by biogeologic processes. The lack of fractionation has been demonstrated by controlled greenhouse experiments,[4,5] and in geologic environments.[9] This is an important result since it allows us to tag each endmember in a system and track its migration or interaction without the complications introduced by temperature/biologic dependent fractionation.

## Strontium Isotopes as Tracers of Fly Ash

Elseewi et al.[2,10,11] concluded that plants grown on fly ash-amended soils contained more Sr than those grown on the same soils without fly ash. Strontium concentrations were also shown to correlate with the amount of fly ash present in the soil. However, the source of the Sr in the plant could not be positively identified as having been derived from the fly ash. The concentration data were equivocal, whereas Sr isotopes were not.[4,5]

## Lead Isotopic Evolution in the Crust

Lead isotopic evolution in the earth is depicted in Figures 2A and 2B. The evolution of lead in the earth is based upon a two-stage model with the following parameters:[12]

- Age of the Earth = 4.550 Ga (or aeons, AE)
- Second Stage Pb Evolution Begins = 3.70 Ga (or aeons, AE)
- $^{238}U/^{204}Pb$ (Mu value) = 9.735

The Pb evolution curve becomes more linear as the present (0 AE) is approached. This is due to the decay of $^{235}U$, the parent of $^{207}Pb$, which has experienced 6.5 half-lives since the earth formed. Figure 2B expands upon the region of the curve between 1.0 AE ago and 1.0 AE in the future (1.0 AE).

The particular Pb evolution curve plotted in Figures 2A and 2B represents an average. In reality, there are an infinite number of these curves each with a different Mu-value. The Mu-value is the ratio of $^{238}U$ to $^{204}Pb$; higher Mu-values yield more radiogenic Pb over a given amount of geologic time. Lead evolution curves with Mu-values greater than 9.735 plot above this curve while those with

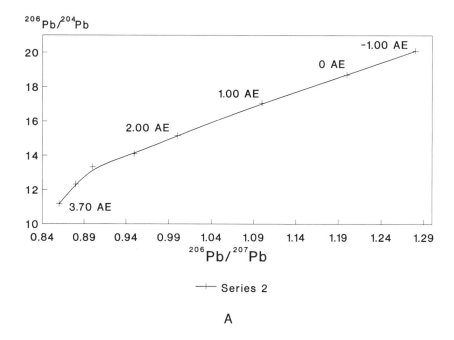

A

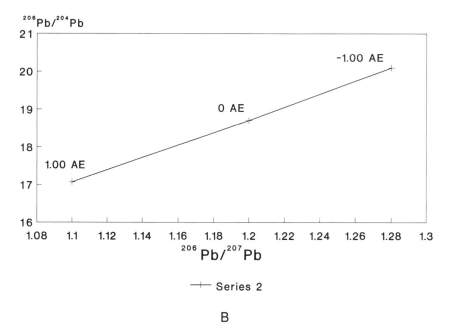

B

**FIGURE 2.**  Lead isotopic evolution in the Earth's crust (A) from 3.7 AE through 0 AE (1 AE = 1 Ga), and (B) from 1.0 AE in the past through −1.0 AE, i.e., 1 Ga into the future. Crustal materials with elevated U/Pb ratios, such as coal fly ash, will have more radiogenic Pb isotope ratios and negative (future) model Pb ages.

lower Mu-values plot below this curve. The significance of this will be discussed later. At this time, it suffices to say that all fossil fuel Pb isotopic compositions plot in the region of these curves represented by future (negative) ages.

## MATERIALS AND METHODS

### Coal-Fired Power Plant

Samples of FFR, soils, monitoring well water, evaporation pond water, river water, brittlebush, and rocks were analyzed for Sr IC and concentration. Collections of samples included both on-site and off-site locations (up to 20 km from the power plant).

Samples were collected using a concentric grid system centered on the Mohave Generating Station. Soil samples consisted of both near surface (0 to 10 cm depth) and subsurface samples collected to depths of 15.5 m; brittlebush leaves and river water were also collected. Samples were placed in acid-cleaned CPE polyethylene bottles or zip-lock bags.

Sample chemistry and cleaning were performed in an ultraclean laboratory environment in order to minimize contamination. Soil samples were ultrasonically cleaned in quartz-distilled water to remove organic debris. Brittlebush leaves were subjected to the same cleaning procedures as soils in order to remove airborne particulates including fly ash. Soils and brittlebush were then air dried to constant weight and weighed into acid-fluxed TFE Teflon© beakers.

An ultrapure (>99.9%) $^{84}Sr$ spike solution was added to determine Sr concentration by isotope dilution. The addition of the tracer was performed prior to the digestion of the brittlebush leaves in 5:1 $HNO_3$-$HClO_4$ mixtures. Soil samples were leached for 10 min with 0.1 $N$ HCl, after which the leachate (carbonate fraction) was spiked with $^{84}Sr$. Leachates were filtered through acid-cleaned Whatman filter paper to remove particulate matter. Soil residues were then dried, weighed, spiked with $^{84}Sr$, and digested in 10:1 $HF$-$HNO_3$ mixtures.

Once digested, the sample solution was evaporated to dryness and converted to a chloride. Dried samples were picked up in 1 mL of 2.5 $N$ HCl and placed on an ion exchange column to separate Sr. The eluant was collected in quartz beakers, evaporated to dryness, and transferred to clean 3 mL quartz beakers for mass spectrometric analysis.

The mass spectrometer used in this work is a thermal ionization, 90 degree sector, 30 cm radius instrument with digital data acquisition. Samples are loaded onto outgassed rhenium filaments. Mass fractionation is corrected by normalizing the $^{86}Sr/^{88}Sr$ ratio to its accepted value of 0.11940. Repeated analyses of NBS SRM 987, a Sr isotopic standard, yielded values between 0.71022 and 0.71024 (accepted IC of 0.71023). Typical precisions and accuracies during this work ranged from ± 0.00002 to 0.00008 at the 95% confidence level (2 sigma standard error of the mean).

## Coal Tar Site

Because the soils at the coal tar site were expected to contain Pb from many sources, we elected to employ a stepwise selective extraction method whereby each successive extraction would remove Pb from a different pool. The method we chose is a modified version of a technique used by soil scientists[13] to selectively extract metals in sediments. The extractants and the species or compounds extracted are as follows:

- 0.01 $M$ DTPA   Soluble, adsorbed, and organically complexed species
- 0.1 $M$ HCl   Acid-soluble monosulfides, carbonates, paracrystalline iron oxides
- Acidified $H_2O_2$   Nonacid-soluble mono/disulfides, organic complexes
- $HF/HNO_3$   Residual silicate minerals
- $HNO_3/HClO_4$   Residual organic complexes

The duration of each extraction was 1 week in order to allow enough time for the sample and extractant to achieve chemical equilibrium. The extracted solution was removed using an ultraclean pipette, filtered, and aliquotted for Sr/Pb spiking. The filtered particulates were washed back into the original sample beaker so that the residue could be weighed in order to determine the mass of sample extracted. Over the course of the five extractions, less than 1% of the original sample was lost because of transferring the pipetting. Both Sr and Pb were separated using standard ion exchange chromatography; mass spectrometric analyses were performed on a 30 cm radius, 90° sector thermal ionization mass spectrometer.

## RESULTS AND DISCUSSION

### Coal-Fired Power Plant

The ICs of the FFR are plotted (Figure 3). Within our analytical limits, the coal (MC) and fly ash (MFA) had identical ICs. The bottom ash (MBA) ICs were not only distinct from each other but also differed from the coal and fly ash. The reason for this variation is not clear, but may be related to differences in the on-site handling and/or temporal variations in the coal IC. Fly ash ICs are known to vary over time. In 1982, the IC of a fly ash sample was 0.70964 as compared to 0.70928 for a 1978 sample. This suggests that temporal variations in the IC of FFR may occur, and that closer monitoring of these coal combustion by-products will be required.

### Criteria for Tracing FFR Dispersion

In order to assess the potential of Sr isotopes as tracers of FFR in the environment, isotopic criteria must be established that allow us to identify impacted soils, waters, or plants. Since Sr concentration is not an unequivocal indicator of fly ash,

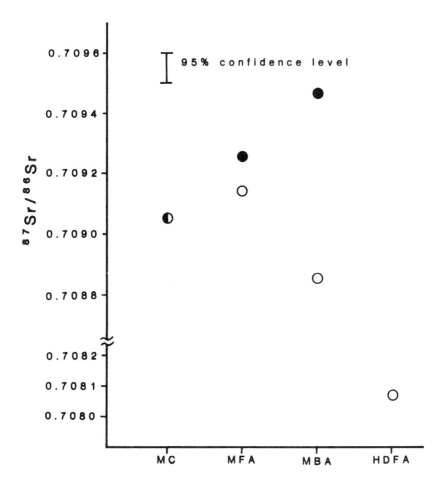

**FIGURE 3.** Strontium isotopic compositions of fossil fuel residues analyzed in this investigation. Solid circles = 24-hr leaches; open circles = 30-min leaches. MC = Mohave coal; MFA = Mohave fly ash; MBA = Mohave bottom ash; HDFA = Mohave fly ash from 1981 investigation.

IC data must be used.[4,5] The observed range of the FFR (0.70883 to 0.70972) was chosen to identify biogeologic samples that may have received Sr from FFR. This range is depicted in each figure as a diagonally ruled field or as two bounding lines. Background ICs of soils, groundwater, and river water differed significantly from those of the FFR.

Local river water ICs (n=21) ranged from 0.70915 to 0.71035. The less radiogenic ICs generally occur in samples located in the vicinity of the power plant. This suggests that river water may receive Sr from FFR.

Local soil and rock samples (n=43) collected off-site were analyzed to determine background IC ranges of the naturally occurring, biogeologically available Sr and to determine if the FFR ICs were distinct from those of the background. Silicate residue ICs and the ICs of local rocks ranged from 0.71260 to 0.79514. Both have ICs that are far more radiogenic than those of the FFR. Based upon the analyses of local waters, silicate residues, and rocks, we conclude that the ICs of the FFR and background are very distinct (>10 times our analytical error).

## On-Site Soil Leachates

Soil samples were analyzed from four on-site locations: by a waste evaporation pond (soil profile 3); 1200 m northwest of the evaporation pond (soil profile 4); near the power plant cooling towers (soil profile 5); and below a waste evaporation pond (soil profiles E, H, and K) (Figures 4 and 5).

Five of the nine analyses of soil Sr leachates from soil profiles 3, 4, and 5 lie within the FFR range (Figure 4). All of the soil Sr leachate ICs from soil profiles E, H, and K below the waste evaporation pond lie within the FFR field (Figure 4). These samples were collected after draining this pond prior to the replacement of the cracked liner. Given that this pond receives power plant wastewater and that the ICs of the subsurface soil Sr leachate lie within the FFR field defined earlier in this work, our selection of the IC interval from 0.70883 to 0.70972 appears valid. Strontium concentrations alone cannot be used to assess the impact of FFR on the environment. This last statement may be generalized to include the use of any chemical concentration data as a tracer of FFR.

Soil leachate ICs that lie above the FFR field in Figure 4 suggest mixing with background Sr sources. One anomalous sample from soil profile 5 has a lower IC (0.70830). This could be the result of an unknown background source or a contribution from fly ash with a less radiogenic IC. The latter is a possibility because fly ash analyzed from this power plant during the 1970s had an IC or 0.70807 ± 0.00025,[5] which is identical, within analytical error, to the IC of this sample.

The variation in on-site soil leachate ICs as a function of depth (Figure 5) is intriguing because all curves, exclusive of soil profile 4, exhibit decreases in IC between 3 and 8 m. With the exception of soil profile 5, all curves appear to converge at 15.5 m to an IC range consistent with the FFR range (0.70883 to 0.70972).

## Off-Site Leachates and Brittlebush

The effectiveness of Sr isotopes as tracers of FFR, predominantly airborne fly ash, distal to the power plant was tested by collecting soils and brittlebush from 13 sites at various distances and directions from the power plant (Figure 5).

Surface soil samples (0 to 1 cm) have ICs that lie well within the FFR range with the exception of two samples. The two anomalous samples lie to the northwest of the power plant and may indicate less deposition of fly ash in this direction over time, more mixing with background Sr and/or a contribution from a less radiogenic Sr source.

Samples collected between 1 and 5 cm have slightly more radiogenic ICs than the surface samples. This trend continues in the 5 to 10 cm depth samples. The

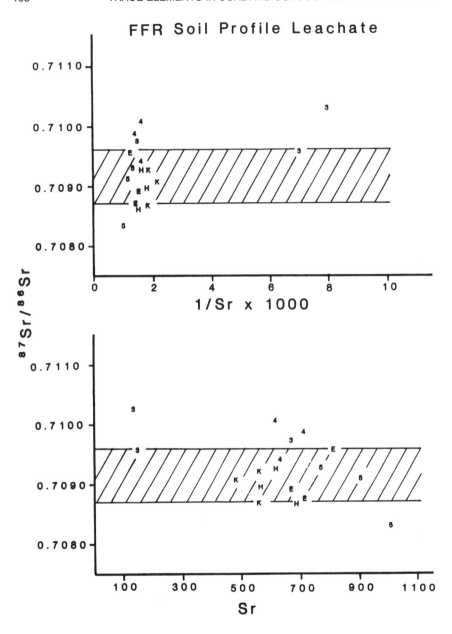

**FIGURE 4.**  Evaluation of mixing relationships of on-site soils. Note that the majority of the data points lie within the fossil fuel residue field (diagonally ruled) indicating the presence of Sr derived from the fossil fuel residue.

average IC increases with depth as follows (errors are 2 sigma standard error of the mean): 0 to 1 cm, $0.70952 \pm 0.00024$; 1 to 5 cm, $0.70992 \pm 0.00020$; 5 to 10 cm, $0.71012 \pm 0.00020$. The percentage of samples that have ICs within the FFR range decreases as a function of depth (0 to 1 cm, 82%; 1 to 5 cm, 40%; 5 to 10

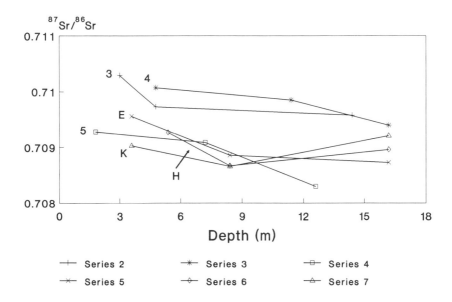

**FIGURE 5.**    Variation of the Sr isotopic composition in soil leachates at the Mohave Generating Station as a function of depth. The numbers and letters designate different
soil borings at the site. The E, H, and K soil-boring samples were taken below
a leaky evaporation pond where fossil fuel residue was stored; note the clustering
of the Sr isotopic compositions of these soil leachates around the measured Sr
isotopic compositions of the Mohave fossil fuel residues (approximately 0.709).

cm, 14%). These data demonstrate that mixing (mechanical, diffusive) is occurring between fly ash deposited by atmospheric fallout and soils at distances up to
12.5 km from the power plant. The mechanical mixing or diffusion of Sr is not
complete; the soils are isotopically stratified.

Brittlebush collected in the vicinity of the soil samples (Figure 6) exhibit ICs
that lie within the IC range of soil-fly ash mixtures or within the FFR range. Hurst
and Davis[5] noted that mixing should be observed when brittlebush are growing on
soil-fly ash mixtures where the fly ash content is <0.25%. This is significant
because the brittlebush with ICs >0.70972 may indicate a soil fly ash content of
<0.25%. If this approach is at least semiquantitative, the brittlebush could serve
not only as a tracer of fly ash, but also as an indicator of soil fly ash content. This
would provide another constraint of airborne fly ash dispersion for atmospheric
models.

### Monitoring Well and Evaporation Pond Waters

Strontium concentrations of monitoring well waters do not exceed 10 ppm
and have ICs that range up to 0.71170 (Figure 7). Most of the monitoring well
water ICs (72%) lie outside of the FFR range, with the more radiogenic values
approaching those of the nonimpacted local soil residues (0.712 to 0.716).
Monitoring well waters whose ICs lie within the range of the FFR are located
next to waste evaporation ponds. The data indicate that local groundwater is

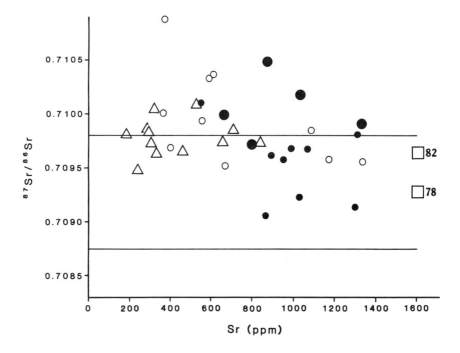

**FIGURE 6.** Evaluation of mixing relationships in off-site soils and brittlebush. The two horizontal boundary lines enclose the fossil fuel residue field; fly ash analyses from 1982 and 1978 (numbered squares) are shown for comparison.

not always impaired by FFR, perhaps due to the volume of fresh river water moving through the system. River water samples are included in Figure 7 for comparison.

Evaporation pond waters have higher Sr concentrations (10 to 40 ppm) and less radiogenic ICs than either river water or monitoring well water (Figure 7). This is expected because power plant wastewater is stored in the waste evaporation ponds. Waste evaporation pond water Sr geochemistry is best explained by mixing river water (1 ppm Sr; IC> 0.710) with FFR (100 to 1000 ppm Sr; IC = 0.70883 to 0.70972). The specific mixing systematics between these two endmembers were discussed earlier (Figure 1). The waste evaporation pond water Sr geochemistry fits the mixing model and supports the selected FFR IC range: 11 of 14 samples lie within or near the FFR field.

## Coal Tar Site

### Lead Concentrations

The data from each extraction are summarized below. All Pb concentrations are based upon the weight of sample extracted.

*DTPA Extraction:* The amount of Pb in the DTPA extract was very high, 62 to 2282 ppm; these concentration ranges accounted for 7 to 43% of the total Pb

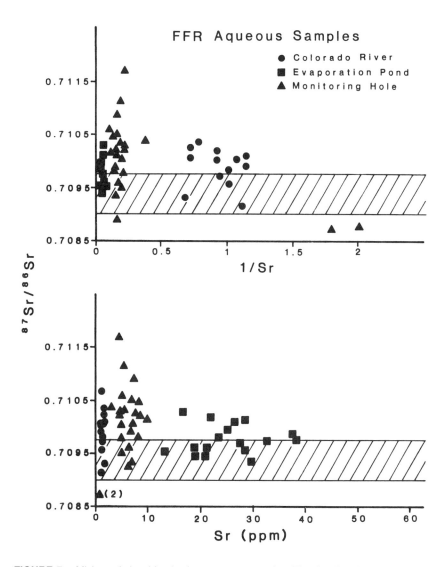

**FIGURE 7.** Mixing relationships in the aqueous samples. The fossil fuel residue field is diagonally striped; the data indicate mixing between Colorado River water and the fossil fuel residue.

in the samples. There appears to be no correlation between the type of soil (e.g., sandy, clayey) and the amount of Pb extracted by the DTPA.

*HCl Extraction:* Lead concentrations in this extract were generally lower than those in the DTPA extract, ranging from 26 to 103 ppm. Much of the Pb removed by the HCl appears to have been adsorbed by clays because clay-rich soils yielded more HCl extractable Pb than the sandier soils.

*$H_2O_2$ Extraction:* Organic Pb complexes were adsorbed by clays in the soils at the site because the Pb concentrations in the acidified $H_2O_2$ extract

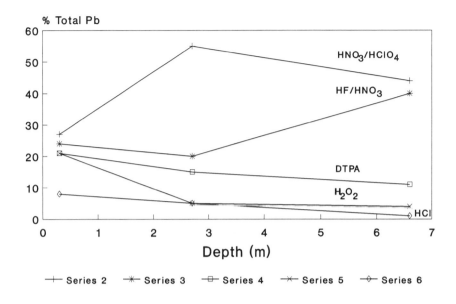

**FIGURE 8.** Percent total Pb in each soil extract vs. depth for the BPW 1 Series core. The redistribution of Pb with depth indicates variations in Pb speciation as a function of depth at the coal tar site.

of clay-rich soils exceeded those of sandy soils (overall range, 33 to 439 ppm).

*HF/HNO₃ Extraction:* Even though the silicate fraction accounted for 56 to 83% of the total sample, it contained the least Pb. Lead concentrations ranged from <1 to 11 ppm in this fraction. This result is extremely significant because it indicates that there is not enough Pb in the mineral matter at the site to account for the total Pb observed in the soils; hence, other Pb sources such as coal tar and organic Pb are required.

*HNO₃/HClO₄ Extraction:* The HNO₃/HClO₄ extraction had Pb concentrations ranging from 24.3 to 244 ppm. This fraction also accounted for the majority of the Pb in most of the soil samples.

*Contribution to the Total Pb as a Function of Depth:* We received one BPW series core with three samples at different depths: BPW 1 at 0.3, 2.6, and 6.5 m. The contribution from each extraction to the total Pb budget of the soil as a function of depth is plotted (Figure 8). The deepest sample at 6.5 m is relatively sandy, whereas the other two tend to be much more clay-rich.

The DTPA, acidified $H_2O_2$, and HCl extractable Pb contributions decrease as a function of depth; more significant drops are observed in the DTPA and HCl extractable Pb than in the acidified $H_2O_2$ extractable Pb. These decreases are concomitant with an increase in the $HNO_2/HClO_4$ extractable Pb contribution. This suggests that Pb is residing in different reservoirs in the soils at different depths. As a consequence, Pb mobility may vary throughout the site

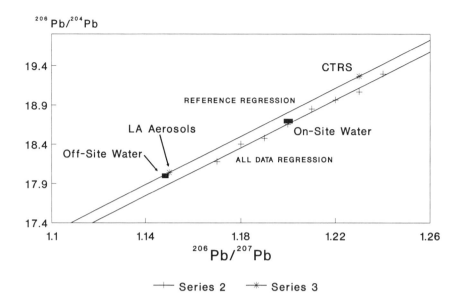

**FIGURE 9.** Reference Regression vs. the All Data Regression at the coal tar site. The All Data Regression (r = 0.999) is defined by the coal tar site soil extract Pb isotopic compositions. The least radiogenic sample (Off-Site Water) and coal tar reference standard (CTRS) lie off the All Data Regression, indicating their Pb is derived from sources other than those located at the coal tar site. Off-site Pb migration has probably not occurred, based upon the absence of coal tar-derived Pb in the off-site water.

as a function of its reservoir. The relative amount of Pb in the silicate residue ($HF/HNO_3$ extract) increases nearly twofold between 2.6 and 6.5 m even though the total Pb concentration only changes by 5% (19.4 to 18.6 ppm, respectively). Redistribution of the Pb as a function of depth is apparent when we compare the sample at 0.3 m to those at 2.6 and 6.5 m. With the exception of the acidified $H_2O_2$ extractable Pb, the Pb in sample BPW 1/0.3 m extracts (DTPA, HCl, $HF/HNO_3$, and $HNO_3/HClO_4$) ranges from 21 to 27% of the total Pb budget in the sample. These percentages change markedly with depth and indicate a redistribution of the Pb during migration and subsequent uptake by deeper soils.

## Lead Isotopic Composition

The Pb isotopic composition results are plotted (Figure 9). One observation to be made prior to the discussion of the isotopic data is the lack of any meaningful correlation between Pb concentrations and the radiogenicity of each extract; i.e., higher Pb concentrations are not necessarily associated with higher $^{206}Pb/^{204}Pb$ or $^{206}Pb/^{207}Pb$ ratios. The background Pb isotopic composition determined from analyses of rocks in the vicinity of the coal tar site does not exceed a $^{206}Pb/^{204}Pb$

value of 18.5. The Pb isotopic composition of any extract that exceeds this value is interpreted to have received Pb from coal tar.

## Soil Samples

The Pb isotopic compositions of all the extracts (DTPA, HCl, $H_2O_2$, HF/$HNO_3$, $HNO_3$/$HClO_4$) exhibit the following ranges (Figure 9):

- $^{206}Pb/^{204}Pb$ = 18.1 to 19.3
- $^{206}Pb/^{207}Pb$ = 1.17 to 1.24

The comparable background ranges from this area are $^{206}Pb/^{204}Pb$ = 18.1 to 18.5 and $^{206}Pb/^{207}Pb$ = 1.17 to 1.19. These are at the less radiogenic end of the values observed at the site. Gasoline and other anthropogenic sources of Pb also have Pb isotopic compositions that do not exceed those of the background. Hence, we conclude that a significant portion of the Pb in the soils at the site must be derived from a more radiogenic source such as coal tar, a fossil fuel.

A linear regression through all of the soil extract Pb isotopic compositions (All Data Regression, Figure 9) is very well correlated (r=0.999) and is interpreted to indicate mixing between the less radiogenic background Pb (soil + anthropogenic source) and fossil fuel-derived Pb (e.g., coal tar).

The soil extract Pb isotopic data indicate that a radiogenic, fossil fuel-derived Pb is present at the coal tar site. The coal tar was removed prior to our study and we did not have a sample of this endmember to analyze. However, by knowing the background Pb isotopic range and the radiogenic nature of fossil fuels, the presence of coal tar in soils is possible.

## Water Samples

Two groundwater samples were analyzed during this study; one sample, MW1 2B, was acquired off-site and the other, TWA P3B, was sampled on-site. The Pb isotopic data are also plotted (Figure 9).

The off-site sample exhibits a lower $^{206}Pb/^{204}Pb$ ratio (17.961) than the on-site sample (18.731). The less radiogenic Pb isotopic signature of MW1 2B is well within the range of gasoline Pb in the Los Angeles Basin ($^{206}Pb/^{204}Pb$ = 17.9 to 18.04.[2,14] Whether this is coincidental or is due to gasoline Pb impacting this groundwater is unknown. The Pb isotopic composition of MW1 2B is similar to Los Angeles aerosol Pb, which is presumed to be gasoline-Pb related.

The on-site sample, TWA P3B, lies above the local background Pb isotopic compositions and within the Pb isotopic compositional range of the soil extracts (Figure 9). A Reference Regression was calculated using a coal tar reference standard and the two groundwater samples (Figure 9). The groundwaters are potentially more representative of a particular site's Pb due to the higher mobility of groundwater relative to solid-state diffusion. If the soil samples are mixtures of components represented by coal tar and a local Pb background similar to MW1 2B, the soil extract data should plot on this regression.

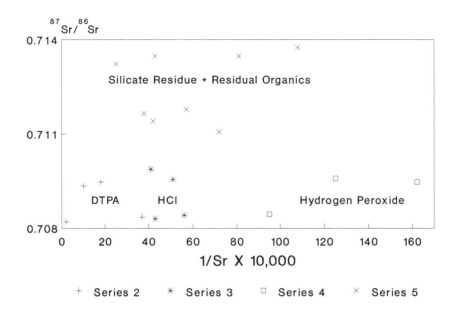

**FIGURE 10.** Strontium isotopic results from soil extracts at the coal tar site. Note the absence of trends but presence of groups on this diagram. The Sr system is not a diagnostic indicator of coal tar. The reciprocal Sr concentration (1/Sr) has been multiplied by 10,000, hence the designation × 10,000.

The Reference Regression was compared to the All Data Regression which is the best-fit through all of our soil extract data (Figure 9). The stars are the Pb isotopic data for MW1 2B, TWA P3B, and the coal tar reference standard. The soil Pb extract data do not fit on the Reference Regression. The on-site groundwater is closely associated with the Pb extraction data and is therefore considered to be genetically related to the mobile Pb pool, i.e., the coal tar-derived Pb. The specific coal tar reference standard is also not directly related to the soil extract Pb; this is to be expected because it originated in Canada. However, note the proximity of the two most radiogenic data points to the coal tar reference standard; these two data points derived from analyses of residual organics, lie well into the fossil fuel field, and may approximate the Pb isotopic composition of the fossil fuel component at the site.

Based on the All Data Regression analysis, we conclude that the on-site water sample contains Pb from the coal tar. Also, the on-site groundwater sample Pb and Los Angeles aerosol Pb are either not part of this Pb system or the relationship is not apparent; at this time, the former interpretation is more clearly substantiated by the data.

### Strontium Concentration and Isotopic Results

The Sr isotopic and concentration results are plotted (Figure 10) ($^{87}$Sr/$^{86}$Sr vs. reciprocal Sr concentration).

The Sr isotopic data demonstrate that there are two populations of Sr: A less radiogenic population which consists of the DTPA, HCl, and acidified $H_2O_2$ extractable Sr, and a more radiogenic population represented by the silicate and residual organic phases. The data indicate distinct isotopic differences between the two groups, but the $^{87}Sr/^{86}Sr$ ratios are not significantly different from the local background, which would allow Sr to be used as a specific tracer of coal tar.

From an isotopic standpoint, the DTPA HCl and acidified $H_2O_2$ extracts have quite similar Sr isotopic compositional ranges. This does suggest an attempt to homogenize the Sr isotopes by a process such as mixing. Unfortunately, the Sr isotopic composition of the coal tar could not be analyzed because this sample was not available from the site.

## CONCLUSIONS

### Coal-Fired Power Plant

Strontium isotopic analyses of FFR, soils, water, rocks, and brittlebush were performed in order to assess the potential use of Sr isotopes as tracers of FFR and airborne fly ash in the environment.

Analyses of FFR such as coal, fly ash, and bottom ash were used to establish Sr isotopic criteria for monitoring the dispersion of FFR in the environment. The Sr isotopic range of the FFR (0.70883-0.70972) was significantly different from that of the natural background materials, which generally exceeded 0.710 and in many cases exceeded 0.712.

On-site soil Sr analyses indicate that FFR leachate is present in soils to depths of 15.5 m. Soil leachate ICs of core samples below a waste evaporation pond lie within the rage of the FFR, indicating that some wastewater has reached the subsurface. Strontium variations as a function of depth may have applications as tracers of hydrogeologic processes if coupled with other available data.

Off-site soil leachate results indicate the presence of fly ash to distances of 20 km from the power plant. The relative contribution of fly ash to the Sr budget is greatest at the surface, but may be detected to depths of at least 2.5 m in some areas.

Brittlebush collected at distances to 10 km from the power plant demonstrates the effectiveness of this plant as a bioindicator of fly ash, and possibly as a semiquantitative monitor of fly ash content in the soils. More detailed mixing models may enable us to utilize brittlebush data to quantify the relative amount of fly ash in soil.

Strontium data from monitoring well water indicate that local groundwater is not significantly impaired by FFR. Waste evaporation pond water does come in contact with FFR; the Sr isotopic and concentration data are consistent with this fact, since the ICs of these waters tend to lie in the FFR field.

Strontium isotopes can be effective tracers of coal-combustion residue in the environment. It is imperative, however, that great care be taken during the sample selection process and that the application is site specific. The ICs of the coal combustion residue to be monitored must be significantly different from that of the local background in order for the method to be effective.

## Coal Tar Site

The total Pb concentration in most of the soils analyzed during this study is within the range of normal background levels (10 to 30 ppm); two soils have high Pb concentrations which exceed normal background levels (95 and 175 ppm). The Pb in the soils originates from two sources: a less radiogenic, presumably background, component and a more radiogenic fossil fuel component which we interpret to be coal tar.

Concentration data suggest that Pb has been mobile at the site. Silicate phases in the soils are not the major Pb reservoir in the soils; in fact, the silicate Pb alone cannot account for more than 38% of the Pb in any sample. The major Pb reservoirs are a residual organic phase and the DTPA extractable Pb. The amount of Pb in any given extraction varies with depth, which indicates different Pb specification at these different depths.

The mobility of Pb at the site is controlled by soil conditions (Eh/pH, microbial interactions, water/soil ratios) and the location of the Pb in a particular soil. Because there are different Pb reservoirs in the soils, it is not possible to conclude that one extractant removes coal tar-derived Pb. In some cases DTPA, for example, extracts nonradiogenic background Pb; in other soils, it extracts radiogenic Pb derived from the coal tar.

Lead isotopic ratios of the soil extracts are very well correlated; the correlation coefficient is 0.999. All of the Pb isotopic data indicate that mixing has taken place in the soils at the site. The Pb isotopic compositional ranges of each separate extraction are similar and support mixing between nonradiogenic and radiogenic Pb components at the site (background and coal tar Pb, respectively). The isotopic composition of the Pb background based upon our data and modelling at the site is well within the observed Pb isotopic compositional range of rocks and soils analyzed in the region. The more radiogenic source must, therefore, be foreign to the site; coal tar is the most obvious choice for this component because of its known presence at the site.

Two groundwater samples were analyzed: one originated off-site and the other on-site. The off-site sample has a relatively nonradiogenic Pb isotopic signature and is not related to the coal tar site samples. The on-site sample is more radiogenic than the local background Pb and lies within the coal tar site sample field. We conclude that this sample contains coal tar Pb; however, we do not know the geographic extent of the coal tar Pb at this time. The Pb isotopic data do indicate that Pb derived from coal tar has been mobile at this particular site.

## REFERENCES

1. Bradford, G. R., Page, A. L., Straughan, I. R., and Phung, H. T., U.S. DOE Symposium Series 45 CONF-760, 429, 383, 1978.
2. Elseewi, A. A., Straughan, A. L., and Page, A. L., Sequential cropping of fly ash-amended soils: effects on soil chemical properties and yields and elemental composition of plants, *Sci. Total Env.,* 15, 247, 1980.
3. El-Amamy, M. M., Fox, C. A., Page, A. L., Bradford, G. R., and Elseewi, A. A., Deposition of Trace Elements on Soils and Vegetation Surrounding the Mohave Generating Station, Southern California Edison Company, Report 83-RD-107, 1983.
4. Straughan, I. R., Elseewi, A. A., Page, A. L., Kaplan, I. R., Hurst, R. W., and Davis, T. E., Strontium as an index to monitor fallout from coal-fired power plants, *Science,* 1267, 1981.
5. Hurst, R. W. and Davis, T. E., Strontium isotopes as tracers of airborne fly ash from coal-fired power plants, *Environ. Geol.,* 3, 363, 1981.
6. Hurst, R. W. and Davis, T. E., Strontium and lead isotopes as monitors of coal combustion residue in the environment, in *Proc. 28th Int. Geol. Cong.,* 2-86, 1989.
7. Faure, G., *The Principles of Isotope Geology,* 2nd ed., John Wiley & Sons, New York, 1986.
8. Hurst, R. W. and Davis, T. E., Lead and Strontium Isotopes in Cores and Groundwater: San Juan Basin Research Site, New Mexico, U.S. DOE Open File Report, 1981.
9. Ericson, J. E., Hurst, R. W., and Davis, T. E., The use of Sr isotopes to characterize prehistoric catchments, *Geol. Soc. Am., Cordilleran Sect.,* March 1986.
10. Elseewi, A. A., Bingham, F. T., and Page, A. L., *J. Environ. Qual.,* 7, 69, 1978.
11. Elseewi, A. A., Bingham, F. T., and Page, A. L., U.S. DOE Symposium Series 45 CONF-760, 429, 568, 1978.
12. Stacey, J. S. and Kramers, J. D., Approximation of terrestrial lead isotopic evolution by a two-stage model, *Earth Plant Sci. Lett.,* 26, 207, 1975.
13. Griffin, T. M., Rabenhorst, M. C., and Fanning, D. S., Iron and trace metals in some tidal marsh soils of the Chesapeake Bay, *Soil Sci. Soc. Am. J.,* 53, 1010, 1989.
14. Chow, T. J. and Johnstone, M. S., Lead isotopes in gasoline and aerosols of the Los Angeles Basin, California, *Science,* 147, 502, 1965.

# 6

# Baker Soil Test™ Applications for Land Reclamation, Animal Health and Food Chain Protection

D. E. Baker, F. G. Pannebaker, J. P. Senft, and J. P. Coetzee

Pennsylvania State University, University Park, PA

## ABSTRACT

This chapter includes the theory and evolution of the Baker Soil Test (BST), with a summary of the experience gained by the authors on the applicability of the method to land management problems. The method has been used with respect to land application of sewage sludges, reclamation of strip mine spoils, fly ash disposal areas, soils contaminated by smelters, and and especially for the production of synthetic soils from coal refuse at disposal sites. The BST was developed from the theory that ion uptake or availability is related to chemical potentials or relative partial molal free energies of ions in the soil solution. The conclusion from the theory and experimental data presented is that plants respond indirectly to the solid phase of soils. Therefore, nonsoil materials or wastes can be used to produce synthetic soils for the production of vegetative cover. The BST approach involves the concept of "small exchange" of ions to reflect the relative partial molal Gibbs free energy, or pi, for any ion, i, in a soil, spoil, or synthetic soil system. A brief description of the method is included. Most of our experience with the method involves its application to soils treated with municipal sewage sludges. While it is possible to calibrate results for any soil test using one soil, our results show that it is not possible to predict plant uptake of Cd and other elements from amounts extracted from different soils. The theory and applications have been applied to many soil, plant, animal and human health related problems involving different wastes.

™Baker Soil Test is a trademark for data interpretation software of Land Management Decisions, Inc., 3048 Research Drive, State College, PA for use by franchised laboratories.

## INTRODUCTION

Many coal refuse piles have remained bare of vegetation for many years in spite of considerable reclamation efforts to revegetate mine spoils and other drastically disturbed sites in Pennsylvania. The Baker Soil Test (BST) and the program initiated by Baker and Pannebaker in 1976 have been used with much success for various mine land and spoil pile reclamation sites.[1-7] The objective of this manuscript is to briefly describe the theory and evolution of the methods, the procedures, and the experiences which have resulted in the development of a database and the technological innovations that are successful for land reclamation without the use of soil cover on mine spoils, coal ash piles, and coal refuse piles.

Soil tests developed recently and calibrated for cropland soils are not suitable for making land reclamation decisions using nonsoil materials — including fly ash, sewage sludge, sludge compost, mine spoils, and other wastes — to produce substrates that substitute for soil. Plant growth rates relate to the nutrient, water, and air relationships that reflect the aqueous solution chemistry of soil water (the soil solution), which forms a continuous film around plant roots and soil or spoil particles which make up the solid phase of any soil. The BST determines the pH, pK, pCa, pCd, etc. using the concept of small exchange, which relates soil solution intensive properties (chemical potentials) to chemical activities and microbial activity (cell growth and metabolism) of ions in solid phase-water systems. Plants respond indirectly to desirable soil properties, but it is not necessary that these properties be supplied by natural soils.

Land reclamation requires attention to plant-water-air and plant-nutrient relations which are affected adversely by too low or too high water-holding capacities of materials, very low pH and associated pAl and pMn, and toxic levels of B, other elements, and soluble salts in some spoils and fly ash samples. Plant species vary greatly in their tolerances to conditions considered adverse for crops normally produced on soils. The main goal of the research-demonstration program discussed here was to provide assistance for the development of "user initiated" systems for land reclamation, which means that the owner/operator ("user") plays a major role in planning, implementing, and paying the direct expenses of each project.

## THEORY AND EVOLUTION OF THE BST

A century and a half ago in the U.S., it was recognized that crop yields could be improved by use of marl or limestone to increase soil pH. Later it was found that the increase in pH was associated with decreases in the availabilities of Al and Mn. In arid regions, high concentrations of soluble salts limit plant growth. The first principle discovered by Baker[8] was that a small effective volume of soil per plant can adversely affect plant growth when the availability of essential elements is adequate. The volume of soil per plant affects air-water relations as well as the availability of ions. Concentrations of essential elements found at deficient or toxic levels in some soils may not be equally deficient or toxic in another soil.

There is a need for soil tests that measure the soil solution availabilities of ions important for plant growth.

Low et al.[9] and Baker and Low[10] provided conclusive evidence that corn seedling accumulations of Na from a clay gel, or the same material in a sol state, were linear functions of the relative partial molar free energy of NaCl at a constant concentration of Na in the substrates. Since the $Cl^-$ was at a constant concentration of $10^{-4}$ $M$, it was concluded that plants were responding inversely to pNa, which is analogous to pH. From this work it was evident that for all essential ions, the availability or biological activity is a function of the activity (passive uptake) of ion i, or the relative chemical potential, $(u - u^0)_i$ (for active uptake) as defined by the relative partial molar, Gibbs free energy:

$$\left(\overline{G} - \overline{G}^0\right)_i = RT\ln a_i = RT\ln\gamma\, c_i = \left(u - u^0\right)_i = 1364 \text{ pi cal / mol at } 25°C$$

where $\left(\overline{G} - \overline{G}^0\right)_i$ is the relative partial molar free energy of ion i, R is the universal gas constant, T is the temperature in °K, ln is the logarithm to the base e, $a_i$ is the activity of ion i in mol/L, $c_i$ is the ionic concentration in mol/L, $\gamma$ is the dimensionless ion activity coefficient of ion i, and pi is the negative logarithm of the activity of ion i, which is analogous to pH. With this evidence, Baker[1] was convinced that the concepts and data of Marshall[11] regarding the efficacy of using ion activity measurements in soil testing were correct. The concept of Vanselow[12] that exchangeable ion activity in a suspension was best approximated by its mole fraction on the exchanger, the concept of energies of exchange,[13] and the lime potential[14] provided the evidence required for one to conclude that both the intensive and relative intensive properties of soil solutions are important parameters to be added to the labile quantity, or existing soil test amounts, which is referred to as an extensive property of matter in chemistry. Scientists have searched for appropriate methods for measuring extensive properties of soils for centuries in order to predict how biological systems respond directly to intensive properties of aqueous systems.

Baker[1] reported "A New Approach To Soil Testing" which defined soil solution compositions that reflect the desired relationships among quantity, intensity, and relative intensity with respect to $H^+$, $K^+$, $Ca^{2+}$ and $Mg^{2+}$ in soils. With additional assumptions relating to the concept that a "small exchange" of ions can reflect relative partial molar Gibbs free energies or pi for any ion i, a diagnostic approach to soil testing was developed for H, Ca, Mg, K, Mn, Fe, Cu, Zn, Cd, Na, P, and S in soils.[11,15] This method, with the laboratory reports prepared by the computer software developed for the "user," is referred to as the Baker Soil Test.[15]

## DESCRIPTION OF THE BST

Details of the test solution composition and methods which are now known as the Baker Soil Test or the Baker Small Exchange Method have been included in the *Handbook on Reference Methods for Soil Testing*.[16] A more theoretical treatment of the method was published by Baker and Amacher.[17] The method enables the user to maintain soils and nonsoils within ranges for optimum growth of most plant species. The soil test solution which is allowed to equilibrate with the soil for 24 hr contains the following in a 1:10, soil:solution ratio on a weight:volume basis: $50 \times 10^{-4}$ M $CaCl_2$, $10 \times 10^{-4}$ M $MgCl_2$, $2.5 \times 10^{-4}$ M KCl, $4 \times 10^{-4}$ M DTPA (diethylenetriaminepentaacetate), pH adjusted to 7.3 with TEA (triethanolamine), 0.1% Superfloc-127 (prevents dispersion and hastens filtering), and preservative (phenylmercuric acetate and dioxane in water)

The test solution provides the desired theoretical attributes, and the use of the preservative and superfloc prevents microbial changes of the soil during equilibration and provides for very rapid filtration of samples — both of which are limitations of many existing soil quick-test methods.[16]

For the macronutrient cations (K, Ca, and Mg), the equilibrating solution contains ions at activities equivalent to those for many soils studied with saturations of the cation exchange capacity equivalent to 2 and 10% for K and Mg, respectively. The method determines a relative pH, pK, pCa, pCd, etc. using the theory of small exchange and the chelation chemistry of DTPA which relates soil solution chemical potentials to chemical and biological activities of ions. With the method and other protocols developed since 1974, desired plant species have been grown on fly ash disposal sites, coal refuse disposal sites, strip mine spoils, mixtures of nonsoil materials, and soils treated with or contaminated by sewage sludges and smelter emissions. Interpretation of the BST for corrective treatments of soils has also been reported earlier at a symposium on soil testing.[18]

## APPLICATIONS OF THE BST

While soils within an area are generally comparable with respect to total composition and other properties affecting the availability of different ions, this cannot be assumed for soils from substantially different parent materials within any area or for soils of a different state or region. This fact has been well documented for P in studies by Wolf et al.[19] and Baker,[20] and for other ions.[18,21–25] It has been observed that the naturally occurring fungus mycorrhiza in association with plant roots enhances the uptake of P and alters interactions with Zn, Cu, and other elements.[26] The buffering capacity of soil is expected to prevent a change in the relative chemical potential of elements by mycorrhiza, but the association could substantially alter the diffusion gradient at the root surface.

From both theory and experimental data, it may be concluded that plants respond to the chemical potential or activity and ionic activity ratios of ions and changes in activity over time resulting from crop removal, fertilizer additions, soil adsorption of ions, and mineral weathering. Ionic diffusion rates, mass flow, and root interception that help explain plant uptake of ions are all related to chemical and biological kinetic factors which depend upon the chemical potentials of the ions at equilibrium and the gradients produced as a result of ion uptake, water tension, aeration, etc.[27,28] For land reclamation, existing species and cultivars best adapted to soil or nonsoil substrate air-water relations must be selected, and then chemical treatments may be incorporated to meet the nutrient requirements, which can be qualitatively predicted from some existing methods. But for nonsoil materials such as fly ash, mixtures of fly ash with coal refuse and/or sewage sludge, paper mill sludges, and other materials that can provide support, water, and nutrients for plants; the substrate can adequately be evaluated by the BST. The following examples are offered as illustrations.

## Sewage Sludge Treated Soils

With the emphasis on land application of wastes in the 1970s, especially sewage effluent and sludges, Baker became concerned that the "hidden hungers" concern of Albrecht[29] could change to a "hidden toxicity" within the food chain. Early work on this concept was published by Baker and Chesnin[27] and Doty et al.[30] In 1974, Baker started the Soil and Environmental Chemistry Laboratory at Pennsylvania State University to assist users of sewage sludges to maintain acceptable levels of macroelements, and to avoid potentially toxic essential and nonessential trace elements in field soils. Most soils of Pennsylvania treated with sewage sludge have been monitored by the BST method. This application includes Bray-1 P, lime requirement by the Shoemaker, McLean, Pratt (SMP) method, and exchangeable cations and ratios by the 1 M $NH_4OAc$ method, the parameters of the BST and an optional analysis for Total Sorbed Metals. For productive agricultural soils of Pennsylvania, the relative chemical potential, available amount, and total concentration of many elements are related to a remarkable degree. However, for soils with variable pH, parent material, and sludge treatments, the negative logarithm of the activity of the ion (pi) in solution in association with the DTPA is the more reliable indicator of plant composition and animal health effects.[30–33]

Within each soil the availability of Cd to plants is predictable from pCd and pH, but at a constant pH and PCd the amount of extractable Cd over soils is variable and not predictable.[24,34] For 16 soils of the Northeast U.S., the loading capacity of the soil to a pCd of 6.5 in solution 6.5 ranged from 1.4 to 134.9 mg Cd/kg soil. The pCd of 6.5 in solution corresponds to a $pCd_{BST}$ of 12, which relates to a field corn leaf concentration of 1 mg Cd/kg dry matter (Figure 1). While a safe loading capacity of the soil for Cd could be determined by relating pCd vs. Cd loading, the soil composition and mineralogical properties do not provide a reliable explanation of the variable loading capacities. Of 22 soil parameters measured, only the organic carbon percentages, zinc concentrations, and surface area of the clay fraction explained a significant amount of the variation in soil loading capacities:[34]

$$CD \text{ Loading} = -0.99 + 9.3\%C - 0.10 \text{ Zn} + 0.27 \text{ SA}_{clay} \quad R^2 = 0.29$$

Similar soil chemistry and mineralogical relationships important for solid waste management decisions have been reported by Baker and Wolf[35] and Wolf and Baker.[36]

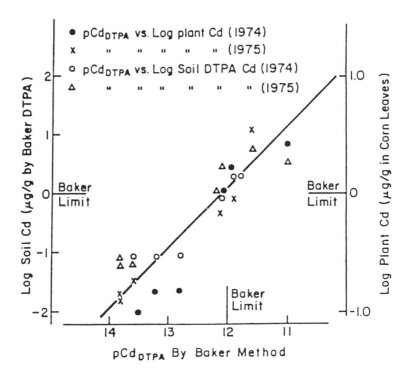

**FIGURE 1.** Interrelationships among pCd-DTPA, the logarithm of soil Cd extracted with DTPA by the Baker method, and logarithm of Cd concentration in corn leaves taken at silking time from plants grown in 1974 and 1975 in field plots receiving different levels of Cd and sewage sludge.

## Strip Mine Spoils

In general, strip mine spoils can be managed as soils, especially where the soil cover material is separated and returned to the site as a part of the reclamation process. However, the pH and lime requirement determined by routine soil test methods for cropland soils are not always sufficiently accurate; and at pH <5.5, the BST results for trace elements are not all reliable. For these samples, a double buffer titration method has been developed for the lime requirement determination.[37] The soil or spoil sample is treated overnight (16 to 24 hrs) with potassium acid phthalate at pH 3.17 to remove reactive Al from interlayers of clays but not to dissolve amorphous $Al(OH)_3$. Then, a Tris-Buffer is added to bring the pH to 8.2, to neutralize complexes and accurately account for all forms of soil acidity. For acid soils (pH <7), the mixture is titrated with standard NaOH to a pH of 8.2, providing a direct measure of the lime requirement.

At a pH of <5.5 the BST results for trace elements are of limited value; therefore, soil or spoil samples are then treated and equilibrated with the required amount of precipitated $CaCO_3$ prior to further testing by the BST. A saturation extract of the material using distilled water is used to measure soluble salt levels by the electrical conductivity method. Using this approach, it is possible to predict the suitability of the site for various plant species and predict plant and animal health effects.[6]

## Fly Ash Disposal Areas

The first application of the BST to nonsoil materials involved experiments for establishing vegetation on fly ash disposal sites in Pennsylvania. For these sites, electric utility companies were required to add 60 cm (2 ft) of soil cover for the establishment of vegetation. With pH reduction from weathering to reduce the free $Ca(OH)_2$ present from hydrolysis of CaO, or by mixing 7.5 to 15 cm (3 to 6 in.) of soil material to the fresh ash, the ash or soil-ash mixtures could be tested by the BST and other cropland soil-testing programs, computerized to provide fertilizer and seeding recommendations. With this protocol, the fly ash can be reclaimed very efficiently.

Because of high erodability of the fly ash, water management to prevent erosion is also very important.[3] Slopes of 15% with and without soil cover have been successfully reclaimed without noticeable erosion when the slope length did not exceed 55 m (180 ft). Rill erosion was observed at 61 m (200 ft), but was stabilized by the vegetation within a few weeks. Fly ash is more erodable than soils with more than about 15% clay; however, the water infiltration rate is much higher than for soils. Erosion associated with new seedings can be protected by adding mulch. The K factor for the universal soil loss equation (USLE) of 0.122 Mg/ha determined by Lehrsch and Baker[3] appears reasonable for sites protected initially by a mulch.

## Soils Contaminated by Smelters

Molybdenosis was observed in cattle within a few kilometers of a Mo smelter in southwest Pennsylvania in 1970.[38] While high soil Mo has no adverse effect on plant growth, high Mo forages lead to a depletion of Cu in the liver and blood of cattle and sheep. Because molybdenosis is a Mo-induced Cu deficiency, injections of copper glycinate and/or additions of copper sulfate to ruminant rations are considered essential for cattle within a few kilometers of the smelter stack where mild symptoms of faded hair color over the withers and/or excessive scouring is observed by cattle on pasture. After several years of monitoring and the development of more sensitive atomic absorption methods for Mo, we have found that the BST provides the best correlation between soil and plant Mo.[39] The BST equilibrium solution is analyzed for Mo using a Perkin-Elmer Model 5100/Zeeman atomic absorption spectrophotometer equipped with a Perkin-Elmer HGA-600 graphite furnace and an AS-60 autosampler. The Mo concentrations in red clover (*Trifolium pratense* L.) were correlated with BST results:

$$\ln[Mo]_{Plant} = 2.99 + 0.63 \ln Mo_{BST} \quad R^2 = 0.82$$

Because the BST equilibrating solution provides a constant ionic strength and a near optimum suite of ions in solution, it is expected to reflect the solubility of soil molybdates. These results support the theory of the method which attempts to predict the effect of the true relative chemical potential of each element.

Home gardens were studied in the vicinity of zinc smelters in both eastern and western Pennsylvania in 1987.[25] Over all soils at 45 locations, the BST results including the negative logarithm of the activity $pi_{BST}$, BST [i] (the amount of ion i, recovered from the soil) and pLi (the negative logarithm of the activity of the free ligand) provided the more reliable regressions with the composition of lettuce plants:

$$[Cd]_{Plt} = -623.8 + 19.89\ pCd_{BST} + 1.39\ BCd + 24.61 pLi_{BST} \quad R_2 = 0.74$$

$$[Zn]_{Plt} = 1604 - 104.9\ pZn_{BST} + 0.568\ BZn - 67.7\ pH_{BST} \quad R^2 = 0.52$$

The plants over locations ranged in dry weight from 0.1 to 37.3 g, in Cd from 0.5 to 155 mg/kg, and in Zn from 38 to 1427 mg/kg. Plant growth was not sufficient for practical gardening when the soil level of total sorbed Zn was greater than 2,000 mg/kg, BST Zn was greater than 275 mg/kg, and $pZn_{BST}$ was less than 11.5. At these Zn levels, the Cd levels were 20 to 30 mg/kg for plants, total sorbed Cd in the soil was 35 to 50 mg/kg; BST Cd was 5 to 20 mg/kg; and pCd was always <11.5. The desired minimum value is 12.0. However, the high Zn concentrations and the inability of the growers to produce a substantial part of their total diet led the toxicology team to the conclusion that garden vegetable production does not pose a health threat for the residents of these areas. These results provided conclusive evidence of the safety of sludges on land where the monitoring program and land management guidelines prepared for the Northeast U.S. are followed.[40]

## Coal Refuse Disposal Sites

Coal from refuse disposal piles is utilized economically by cogeneration plants for steam and electricity production which is then used by the industrial, commercial, and residential sectors of Pennsylvania. Demonstration projects to establish vegetation on these sites were started in 1989 in cooperation with the U.S. Department of Interior, Bureau of Surface Mining, Greenley Energy Holdings, Inc., Greensberg, PA. and Land Management Decisions, Inc., State College, PA.

### Experiment No. 1

The objective of the first experiment was to develop a nonsoil cover and establish vegetation which could prevent further erosion and contamination of surface water leaving the site. The coal refuse material at this site had a pH in water (1:1, w:v) of about 2, a very low water holding capacity because of the coarse texture, a dark color which absorbs the maximum amount of radiant energy, and, of course, has very toxic levels of Al, Mn and in some cases B, which

Table 1
pH, Bray-1 Phosphorus, Exchangeable Cations, Cation Exchange Capacity, and Percent
Saturation of Treated Mine Waste

| Sample No. | pH | P (kg/ha) | cations (meq/100 g) | | | Exchange Capacity | Saturation (%) | | |
|---|---|---|---|---|---|---|---|---|---|
| | | | K | Mg | Ca | | K | Mg | Ca |
| 1 | 6.1 | 662 | 0.57 | 0.9 | 19.8 | 18.5 | 3.1 | 4.9 | 81.1 |
| 2 | 6.4 | 1000 | 0.48 | 1.5 | 19.3 | 19.0 | 2.5 | 7.9 | 78.9 |
| 3 | 6.9 | 758 | 0.53 | 1.6 | 17.8 | 17.1 | 3.1 | 9.4 | 87.7 |
| 5 | 6.8 | 1000 | 0.56 | 2.0 | 19.5 | 17.5 | 3.2 | 11.4 | 85.7 |
| 6 | 6.6 | 1000 | 0.61 | 1.6 | 20.5 | 19.2 | 3.2 | 8.3 | 78.1 |
| 7 | 6.6 | 800 | 0.25 | 1.5 | 16.3 | 18.8 | 1.3 | 8.0 | 79.8 |
| 9 | 6.5 | 1000 | 0.70 | 2.1 | 20.3 | 19.8 | 3.5 | 10.6 | 75.8 |
| 11 | 6.1 | 195 | 0.15 | 4.0 | 24.5 | 21.2 | 0.7 | 18.9 | 70.8 |
| 14 | 6.1 | 449 | 0.29 | 1.7 | 16.8 | 19.0 | 1.5 | 8.9 | 78.9 |
| 16 | 6.0 | 328 | 0.34 | 2.5 | 20.5 | 19.9 | 1.7 | 12.6 | 75.4 |
| 18 | 6.8 | 122 | 0.58 | 1.1 | 23.8 | 16.7 | 3.5 | 6.6 | 89.8 |

may be toxic to some plant species. The nonsoil surface layer or synthetic soil developed for the establishment of vegetation consisted of a mixture of the coal refuse, fly ash from a coal-burning power plant, agricultural grade limestone, and commercial fertilizers alone and in combination with composted sewage sludge. This first test plot was established at the Greenley Energy Mine site 40 on an area 12 m (40 ft) by 91 m (300 ft) ranging in slope from 4 to 30%. A plow depth of 15 to 20 cm was prepared by additions of, on a $m^2$ basis, 45 kg (200 tons/acre) of fly ash, 2.0 kg (9 tons/acre) of agricultural grade limestone, 0.0067 kg N (60 lb/acre), 0.022 kg of $P_2O_5$ (200 lb/acre), and 0.044 kg of $K_2O$ (400 lb/acre). The site was split to provide two $6 \times 90$ m ($20 \times 300$ ft) plots. One plot received 0.013 kg N/$m^2$ (120 lb/acre) as sulfur-coated urea; the other plot received composted sewage sludge to supply 0.056 kg N/$m^2$ (500 lb/acre), 0.022 kg $P_2O_5$/$m^2$ (200 lb/acre) and 0.045 kg of $K_2O$/$m^2$ (400 lb/acre). All surface-applied materials were completely mixed to the 15 to 20 cm depth using a small tractor with a field cultivator, then to the 40 to 50 cm depth with a 40 cm moldboard plow. A seed mixture of spring oats, tall fescue, crown vetch and birdsfoot trefoil was seeded, and the entire plot area was mulched with grass hay at a rate of 0.22 kg/$m^2$ (3 tons/acre) on April 17, 1989. Except for a few small areas where high concentrations of pyritic spoil outcrops were encountered and at the edges where acid runoff entered the plot area, the vegetative cover was complete throughout the 1989 growing season.

*Plant Composition and Substrate "Soil" Analysis Results:* Plant samples and matching soil samples from the synthetic soil substrate described above were taken on June 8, 1989. Table 1 lists the synthetic soil data for 11 sampling points throughout the plots and treatments where the growth of oats was excellent (35 to 45 cm tall). The pH, macronutrient levels, and cation exchange capacity were found to be within the range considered excellent for agricultural soils. Availability data from the BST indicated acceptable ranges for plant growth (Table 2). Table 3 lists the element levels found in plant samples from the areas of the plots where soil

**Table 2**
**Relative Availability of Ions in Soil Samples as Measured by BST**

| Sample No. | pK | pMg | pCa | pAl | pMn | pFe | pNi | pCu | pZn | pCd | pPb |
|---|---|---|---|---|---|---|---|---|---|---|---|
| 1 | 3.3 | 3.2 | 2.1 | 12.1 | 7.3 | 18.8 | 12.8 | 13.5 | 9.8 | 12.2 | 11.1 |
| 2 | 3.4 | 3.1 | 2.1 | 11.7 | 6.8 | 18.3 | 12.3 | 13.0 | 9.2 | 11.8 | 10.6 |
| 3 | 3.3 | 3.1 | 2.2 | 12.7 | 7.5 | 19.0 | 13.0 | 13.6 | 9.9 | 12.6 | 11.1 |
| 5 | 3.3 | 3.0 | 2.2 | 11.9 | 7.0 | 18.5 | 12.5 | 13.1 | 9.3 | 12.1 | 10.6 |
| 6 | 3.3 | 3.1 | 2.1 | 12.1 | 7.3 | 18.8 | 12.7 | 13.4 | 9.6 | 12.4 | 10.8 |
| 7 | 3.6 | 3.1 | 2.2 | 12.4 | 7.3 | 19.2 | 13.0 | 13.7 | 9.8 | 12.3 | 10.9 |
| 9 | 3.3 | 3.0 | 2.3 | 12.3 | 7.4 | 19.1 | 13.3 | 13.5 | 9.7 | 12.4 | 10.9 |
| 11 | 3.7 | 2.9 | 2.0 | 12.0 | 7.2 | 18.9 | 12.7 | 13.5 | 10.1 | 13.0 | 11.0 |
| 14 | 3.6 | 3.1 | 2.2 | 12.0 | 7.6 | 19.2 | 13.1 | 13.7 | 10.2 | 13.1 | 11.4 |
| 16 | 3.4 | 2.9 | 2.1 | 11.7 | 7.5 | 19.1 | 12.8 | 13.7 | 9.9 | 12.4 | 10.6 |
| 18 | 3.2 | 3.1 | 2.1 | 12.0 | 7.9 | 19.3 | 13.2 | 13.7 | 9.9 | 12.4 | 10.6 |
| Normal Range | 3.8– 3.4 | 3.3– 2.8 | 2.7– 2.5 | 12.0– 10.5 | 8.0– 6.0 | 21.0– 18.0 | 14.0– 12.5 | 14.5– 11.5 | 11.5– 10.5 | 13.5– 12.5 | 12.5– 11.5 |

**Table 3**
**Essential Nutrient Elemental Concentration of Plant Material Grown on Soil Samples**

| Sample No. | Concentration Level (g/kg) | | | | | | Concentration Level (mg/kg) | | | | |
|---|---|---|---|---|---|---|---|---|---|---|---|
| | N | P | K | Ca | Mg | S | Mn | Fe | Cu | B | Zn |
| 1 | 46.8 | 5.9 | 38.9 | 4.3 | 1.2 | 4.7 | 93 | 103 | 8 | 24 | 40 |
| 2 | 47.6 | 7.3 | 42.5 | 4.4 | 1.5 | 4.0 | 119 | 96 | 12 | 41 | 57 |
| 3 | 43.8 | 6.8 | 42.2 | 4.2 | 1.6 | 4.3 | 111 | 95 | 10 | 23 | 45 |
| 5 | 47.7 | 6.7 | 44.0 | 4.9 | 1.5 | 3.6 | 107 | 95 | 11 | 24 | 56 |
| 6 | 45.9 | 5.9 | 43.2 | 4.1 | 1.6 | 4.1 | 107 | 87 | 12 | 39 | 42 |
| 7 | 38.5 | 6.4 | 40.6 | 4.0 | 1.7 | 4.5 | 108 | 81 | 11 | 35 | 42 |
| 9 | 48.1 | 6.6 | 42.0 | 4.1 | 1.7 | 3.8 | 108 | 93 | 12 | 39 | 52 |
| 11 | 30.8 | 4.4 | 41.1 | 2.1 | 2.4 | 4.5 | 120 | 85 | 9 | 62 | 21 |
| 14 | 41.9 | 5.3 | 39.0 | 3.4 | 1.5 | 6.2 | 103 | 93 | 10 | 42 | 33 |
| 16 | 38.8 | 4.8 | 42.0 | 2.2 | 2.0 | 3.4 | 115 | 110 | 9 | 56 | 24 |
| 18 | 39.1 | 4.5 | 39.1 | 4.1 | 1.1 | 6.3 | 119 | 93 | 10 | 67 | 21 |

samples were taken. These levels are typical of those found for plants on fertile soils. The results confirm the hypothesis that the BST and protocol developed for establishing vegetation on mine spoils, fly ash, and other materials are applicable to the most adverse sites represented by this very acid, heterogeneous, coal refuse. Symptoms of an Al toxicity were observed on the upper end of the plot area where no agricultural lime or composted sludge had been applied.

While the methods used on the upper 15 to 20 cm of the coal refuse pile were successful for the establishment of vegetation, it was not considered feasible to count on this depth for the long term. Comparison data (Table 4) show the pH, electrical conductivity, and B levels for two sampling dates. During the growth period following heavy rains in May and June, dying off of oats and other species was observed on the upper nonlimed and zero compost area that was also adjacent to the uphill slope which caused the acid runoff to flow into the plot. These

**Table 4**
**Soil pH, Electrical Conductivity, and Boron Levels at Two Time Intervals After Plot Establishment**

| Sample No. | pH | | Conductivity (mmho/cm) | | Boron (mg/kg) | |
|---|---|---|---|---|---|---|
| | (6 wk) | (17 wk) | (6 wk) | (17 wk) | (6 wk) | (17 wk) |
| 1 | 7.5 | 4.1 | 3.8 | 4.9 | 0.35 | 0.9 |
| 2 | 7.5 | 7.2 | 4.1 | 4.8 | 0.73 | 1.8 |
| 3 | 7.6 | 7.5 | 3.2 | 4.5 | 0.62 | 0.6 |
| 5 | 7.4 | 7.1 | 3.7 | 4.2 | 0.96 | 0.7 |
| 6 | 7.4 | 7.7 | 4.7 | 3.6 | 0.84 | 0.5 |
| 7 | 7.6 | 6.8 | 7.0 | 3.3 | 0.48 | 0.5 |
| 9 | 7.4 | 7.5 | 5.0 | 9.0 | 0.94 | 2.6 |
| 11 | 7.0 | 3.6 | 2.9 | 9.5 | 0.41 | 0.7 |
| 14 | 7.1 | 5.2 | 4.4 | 11.5 | 0.58 | 0.9 |
| 16 | 6.7 | 6.6 | 4.3 | 13.0 | 0.54 | 0.7 |
| 18 | 7.7 | 7.4 | 4.4 | 13.0 | 1.73 | 2.6 |

**Table 5**
**Potentially Toxic Elements in Plant Material**

| Sample No. | Concentration Level (mg/kg) | | | | | | | |
|---|---|---|---|---|---|---|---|---|
| | Al | Sr | Pb | Cd | Ni | Mo | Se | As |
| 1 | 48 | 4 | 4 | 0.13 | 2.4 | 2.02 | 1.05 | 0.58 |
| 2 | 47 | 5 | 4 | 0.18 | 2.3 | 1.56 | 0.61 | 0.17 |
| 3 | 39 | 4 | 4 | 0.11 | 2.1 | 1.82 | 1.23 | 0.33 |
| 5 | 40 | 4 | 5 | 0.14 | 1.9 | 1.32 | 0.50 | 0.46 |
| 6 | 42 | 4 | 4 | 0.13 | 1.7 | 1.37 | 0.80 | 0.54 |
| 7 | 48 | 5 | 5 | 0.12 | 2.5 | 1.95 | 0.85 | 0.62 |
| 9 | 45 | 4 | 5 | 0.12 | 1.5 | 1.98 | 0.42 | 0.43 |
| 11 | 81 | 7 | 4 | 0.13 | 9.6 | 1.63 | 2.34 | 0.60 |
| 14 | 53 | 5 | 4 | 0.10 | 2.5 | 0.82 | 1.23 | 0.83 |
| 16 | 60 | 4 | 4 | 0.15 | 8.6 | 1.39 | 1.58 | 0.87 |
| 18 | 60 | 7 | 5 | 0.15 | 4.6 | 2.73 | 1.65 | 0.63 |

sampling sites are represented by numbers 11 to 18 in Tables 1 to 5. The pH drop was associated with an increase in electrical conductivity (formation of $CaSO_4$) and soluble B in saturation extracts of the synthetic soil. The beneficial effect of the compost on soil pH buffering was substantial, but the high levels of N and other elements resulted in excessive growth of the oats, which lodged and greatly reduced the stand of grasses and legumes. The 10% per year value for N release from compost is much too low for these applications.

## Experiment No. 2

Excitement over the excellent results obtained for the first season of Experiment No. 1, combined with the uncertainty regarding the external vs. internal effects on the pH of the plot receiving no composted sewage sludge, led to a desire to develop methods to incorporate the materials added to the surface of the refuse pile to a greater depth. A desirable minimum depth of treated materials is 40 to 60 cm for grass-legume mixtures and up to 1.2 m for trees. A plot area half the size of the original site was established on August 24, 1989. For this site the treatments of the original site were doubled with respect to fly ash and ground limestone. All other treatments on a per acre basis were the same, except that no compost was included. A large tractor and a 210 cm (7 ft) chisel plow were taken to the site to develop a seed bed and incorporate the fly ash and fertilizer to a depth of 40 to 60 cm. It was found that at most locations on the plot area, the chisel plow could not penetrate the coal refuse, and where it did, the tractor could not pull it. A large bulldozer with a ripper was used to loosen the coal refuse to a depth of 60 to 80 cm and then the tractor with the chisel plow was used to break up the large chunks of refuse to enable the final mixing using a small tractor and a 40 cm moldboard plow. The mixing was successful in producing a surface layer 40 to 60 cm deep.

The plot was seeded to winter rye, tall fescue, reed canary grass, and a half seeding of birdsfoot trefoil, and mulched with straw on August 24, 1989. Birdsfoot trefoil and crown vetch were seeded in the spring of 1990. When the plots were visited on September 21, 1989, the vegetative cover was complete and uniform. A visit to the site on April 3, 1990, found complete vegetative cover for all areas of both experimental areas that had a 100% cover in 1989. The grazing of the plots by deer left the new area looking like a freshly mowed lawn.

From the experience to date it has been concluded that even these most severe sites with respect to acidity, slope (length and steepness), fertility, and color (black for maximum heat absorption) can be constructed, treated, and seeded success-fully with no use of our valuable soil resources. If this site remains successful, the 90 kg/m$^2$ (400 tons/acre) of fly ash and 4 kg/m$^2$ (18 tons/acre) of ground limestone incorporated to a depth of 40 to 60 cm, equivalent to 448 to 560 kg/m$^2$ (2,000 to 2,500 tons/acre) of soil, is an economically sound approach. Even if soil were available at a purchase and hauling cost of $22/metric ton, ($20/ton), this system could save the operators $100,000 to $125,000/ha ($40,000 to $50,000/acre). In Pennsylvania, especially in the coal mining regions, there is not sufficient field soil to cover these coal refuse sites. The fly ash provides the needed water-holding capacity, increases the infiltration rate, does not seem to be erosive when mixed with the refuse, and helps to reduce the black color. The BST and operations protocol provide the chemical treatments and seed bed preparation required for successful establishment of the vegetation.

Sites must be prepared to a depth of at least 40 to 60 cm and have the following physical and chemical properties, which are listed in order of their establishment using the developed protocol:

1.   The pH must be adjusted to a range of 6.5 to 7.5 using fly ash or fly ash-limestone mixtures. The use of fast-reacting hydrated lime or calcium oxide is not recommended. The dual buffer titration method developed by Harper et al.[37] is reliable for this determination. However, if sulfide sulfur is present, additional limestone must be added to neutralize this potential acidity over time. More work is needed to determine

the added benefit from compost on the rate of weathering of pyritic materials. Preliminary data from the experimental site and evidence in the literature indicate that the lime equivalency of the compost is greater than its $CaCO_3$ equivalency.

2. The electrical conductivity of saturation extracts of the spoil, ash, lime mixture must be less than 4 mmho/cm. For sites above this level, some more salt-tolerant plant species might be used until the excessive salt levels leach out of the plant root zone. Otherwise, enough soil (5 to 7.5 cm [2 to 3 in.] in Pennsylvania) can be used on the surface to establish vegetation as was done at the fly ash disposal areas.

3. The refuse, spoil, fly ash mixtures must contain enough fine materials to provide moisture levels of 18 to 20% on a dry weight basis or 40 to 50% of the pore space full of water at field capacity.

4. The chemical treatments should be adjusted to bring essential and potentially toxic elements (Table 5) into appropriate ranges using the BST.

## SUMMARY AND CONCLUSIONS

Today we know that soil tests based on extractants calibrated for similar cropland soils are not reliable for making decisions for other soils or lands that are reclamation using spoil materials, other wastes, and combinations to produce synthetic soils. Soil test methods that extract a relatively large amount of a nutrient from soils are not suitable for soils other than those for which the method has been calibrated.[40,41]

After 30 years of research and experience gained in searching for appropriate parameters to explain plant responses to nutrients in different substrates, the BST is considered ready and appropriate for application to any substrate when combined with knowledge of the relative requirements of the plant species with respect to pH, soil aeration, soil-air-water relations, and climatic factors.

The BST provides a diagnostic approach to soil testing that is theoretically based on chemical principles relating intensive (chemical potentials) and relative intensity parameters to extensive parameters (the available and total concentrations) of ions in soils, spoils, and nonsoil materials.

## ACKNOWLEDGMENTS

Partial support of this research by the U.S. Department of Interior (USDI), Bureau of Surface Mining, Pittsburgh, PA; Greenley Energy Holdings, Inc., Greensburg, PA; and Land Management Decisions, Inc., State College, PA is gratefully acknowledged.

## REFERENCES

1. Baker, D. E., A new approach to soil testing, *Soil Sci.,* 112, 381, 1971.
2. Pannebaker, F. A., Chemical Monitoring of Soils and Wastes for the Establishment of Vegetative Cover on a Fly Ash Disposal Area, M.S. thesis, Pennsylvania State University, University Park, 1988.

3. Lehrsch, G. A. and Baker, D. E., Fly Ash Erodability, *J. Soil Water Cons.*, 44, 624, 1989.

4. Buck, J. K., Direct Revegetation of Anthracite Waste Using Coal Fly Ash as a Major Soil Amendment, in *Proc. 8th Int. Ash Util. Symp.*, 1, 1987.

5. Buck, J. K. and Houston, R. J., Direct Revegetation of Four Coal Waste Sites — Four Approaches, in *Innovative Approaches to Mine Land Reclamation*, Carlson, C. L. and Swisher, J. H., Eds., Illinois University Press, Southern Illinois University, Carbondale, 1987, 385.

6. Baker, D. E. and Buck, J. K., Using computerized expert systems, unique soil testing methods, and monitoring data, in Land Management Decisions — Mine Drainage and Surface Mine Reclamation, Vol. 2, U.S.D.I. Bureau of Mines Circular 9184, 246, 1988.

7. Buck, J. K. and Houston, R. J., Direct Revegetation of Anthracite Refuse Using Coal Fly Ash as a Major Soil Amendment, in Proc. Conf. Surface Mining and Reclamation, U.S.D.I. Bureau of Mines, 236, 1988.

8. Baker, D. E. and Woodruff, C. M., Influence of volume of soil per plant upon growth and uptake of phosphorous by corn from soils treated with different amounts of phosphorous, *Soil Sci.*, 94, 409, 1963.

9. Low, P. F., Davey, B. G., Lee, K. W., and Baker, D. E., Clay sols versus clay gels: biological activities compared, *Science,* 161, 1968.

10. Baker, D. E. and Low, P. F., Effect of the sol-gel transformation in clay-water systems on biological activity. II. Sodium uptake by corn seedlings, *Soil Sci. Soc. Am. Proc.,* 34, 49, 1970.

11. Marshall, C. E., *The Physical Chemistry and Mineralogy of Soils,* Vol. 1, John Wiley & Sons, New York, 1964.

12. Vanselow, A. P., Equilibria of the base exchange reactions of bentonites, permutites, soil colloids and zeolites, *Soil Sci.,* 33, 95, 1932.

13. Woodruff, C. M., The energies of replacement of calcium by potassium in soils, *Soil Sci. Soc. Am. Proc.,* 19, 167, 1955.

14. Schofield, R. K. and Taylor, A. W., The measurement of soil pH, *Soil Sci. Soc. Am. Proc.,* 19, 164, 1955.

15. Baker, D. E., A new approach to soil testing. II. Ionic equilibria involving H, K, Ca, Mg, Mn, Fe, Cu, Zn, Na, P, and S, *Soil Sci. Soc. Am. Proc.,* 37, 537, 1973.

16. Baker, D. E., Baker Method, in *Handbook on Reference Methods for Soil Testing*, Council on Soil Testing and Plant Analysis, Athens, GA, 100, 1980.

17. Baker, D. E. and Amacher, M. C., Development and Interpretation of a Diagnostic Soil Testing Program, Bulletin No. 826, Pennsylvania Agricultural Experiment Station, University Park, 1981.

18. Baker, D. E., Ion activities and ratios in relation to corrective treatments of soils, in *Soil Testing: Correlating and Interpreting the Analytical Results,* American Society of Agronomy, Madison, WI, 55, 1977.

19. Wolf, A. M., Baker, D. E., Pionke, H. B., and Kunishi, H. M., Soil tests for estimating labile, soluble, and algae-available phosphorous in agricultural soils, *J. Environ. Qual.,* 14, 341, 1985.

20. Baker, D. E., Phosphorous pollution of water from agricultural practices, in Proc. Phosphorous Symp. Rep. S. Africa, 198, 1988.

21. Baker, D. E., Soil chemical constraints in tailoring plants to fit problem soils. I. Acid soils, in *Plant Adaptation to Mineral Stress in Problem Soils,* Wright, M. J., Ed., Cornell University Press, Ithaca, NY, 127, 1976.

22. Stout, W. L. and Baker, D. E., A new approach to soil testing. III. Differential adsorption of potassium, *Soil Sci. Soc. Am. J.,* 42, 307, 1978.

23. Stout, W. L. and Baker, D. E., Effect of differential adsorption of potassium and magnesium in soils on magnesium uptake by corn, *Soil Sci. Soc. Am. J.,* 45, 996, 1981.

24. Baker, D. E., Rasmussen, D. S., and Kotuby, J., Trace metal interactions affecting soil loading capacities for cadmium, *Hazardous and Industrial Waste Management and Testing: 3rd Symp.,* ASTM STP 851, Jackson, L. P., Rohlik, A. R., and Conway, R. A., Eds., American Society for Testing and Materials, Philadelphia, 1984, 118.

25. Baker, D. E. and Bowers, M. E., Human health effects of cadmium predicted from growth and composition of lettuce in gardens contaminated by zinc smelters, *Trace Substances Environ. Health,* 22, 281, 1988.

26. Lambert, D. H., Baker, D. E., and Cole, H., Jr., The role of mycorrhizae in the interactions of phosphorous with zinc, copper and other elements, *Soil Sci. Soc. Am. J.,* 43, 976, 1979.

27. Baker, D. E. and Chesnin, L., Chemical monitoring of soils for environmental quality, and animal and human health, *Adv. Agron.,* 27, 305, 1975.

28. Barber, S. A., *Soil Nutrient Bioavailability,* John Wiley & Sons, New York, 1984.

29. Walters, C., Jr., Ed., *Albrecht Papers, Vol. 1,* Acres, U.S.A., Kansas City, Mo., 1982.

30. Doty, W. T., Baker, D. E., and Shipp, R. F., Heavy metals in Pennsylvania sewage sludges, *Compost Sci. Land Util.,* 19, 26, 1978.

31. Williams, P. H., Shenk, J. S., and Baker, D. E., Cadmium accumulation by meadow voles *(Microtus pennsylvanicus)* from crops grown on sludge-treated soil, *J. Environ. Qual.,* 7, 450, 1978.

32. Baker, D. E., Amacher, M. C., and Leach, R. M., Sewage sludge as a source of cadmium in soil-plant-animal systems, *Environ. Health Pers.,* 28, 49, 1979.

33. Leach, R. M., Jr., Wei-Li Wang, K., and Baker, D. E., Cadmium and the food chain: the effect of dietary cadmium on tissue composition in chicks and laying hens, *J. Nutr.,* 109, 437, 1979.

34. Rasmussen, D. S., Laboratory Measurement of Cadmium Loading Capacities for Soils of the Northeast, M.S. thesis, Pennsylvania State University, University Park, 1984.

35. Baker, D. E. and Wolf, A. M., Soil chemistry, soil mineralogy and the disposal of solid wastes, in *Solid and Liquid Wastes: Management, Methods, and Socioeconomic Considerations,* Majumdar, S. K. and Miller, E. W., Eds., Pennsylvania Academy of Sciences, Philadelphia, 1984, 238.

36. Wolf, A. M. and Baker, D. E., Cadmium and other trace elements in the environment in *Solid and Liquid Wastes: Management, Methods, and Socioeconomic Considerations,* Majumdar, S. K. and Miller, E. W., Eds., Pennsylvania Academy of Sciences, Philadelphia, 1984, 140.

37. Harper, W. S., Baker, D. E., and McCormick, L. H., A dual buffer titration for lime requirement of acid mine soils and forest soils, *J. Environ. Qual.,* 17, 452, 1988.

38. Hornick, S. B., Baker, D. E., and Guss, S. B., Crop production and animal health problems associated with high soil molybdenum, in *Molybdenum in the Environment,* Vol. 2, Chappell, W. R. and Peterson, K. K., Eds., Marcel Dekker, New York, 1976, 665.

39. Murray, M. R. and Baker, D. E., Monitoring and assessment of soil and forage molybdenum near an atmospheric source, *Environ. Monitoring Assess.,* 15, 25, 1989.

40. Wolf, A. M. and Baker, D. E., Comparisons of soil test phosphorous by Olsen, Bray-P1, Mehlich I, and Mehlich III methods, *Commun. Soil Sci. Plant Anal.,* 16, 467, 1985.

41. Baker, D. E., McMackin, R. L., and Wolf, A. W., Developing an expert system for phosphorous soil test interpretation, in Proc. Phosphorous Symp. Rep. of S. Africa, 244, 1988.

*Transport and Leachability of Metals from Coal and Ash Piles*

# 7

# Multicomponent Transport Through Soil Subjected to Coal Pile Runoff Under Steady Saturated Flow

M. A. Anderson,[1] P. M. Bertsch,[2] and L. W. Zelazny[3]

[1]University of California, Riverside, CA
[2]Savannah River Ecology Laboratory, Aiken, SC
[3]Virginia Polytechnic Institute and State University, Blacksburg, VA

## ABSTRACT

Runoff and leachate from coal storage facilities can be very acidic and may possess high concentrations of transition metals and other components. This study sought to evaluate the transport of dissolved components as coal pile runoff percolated through subsoil. Samples were collected from a site adjacent to a coal storage facility on the U.S. Department of Energy's (DOE) Savannah River Site. The subsoils were packed in 25 $\times 2.5$ cm columns to uniform bulk densities of 1.5 metric ton/$m^3$ and subjected to steady, saturated flows of 0.2 and 1.3 cm/hr. Effluent was collected and solute transport through the subsoils was evaluated. Observed transport was then related to soil chemical and mineralogical properties. Sulfate mobility was found to be governed principally by ion exchange. Initial breakthrough of most cationic components was coincident with sulfate breakthrough and a result of the necessary condition of electroneutrality within solution. Ion exchange was also important to the mobility of the alkaline earth and a number of the divalent first row transition metals, whereas precipitation-dissolution reactions regulated Fe(III) mobility. Coprecipitation with Fe apparently controlled Cr and Cu mobility. Reduction in Darcy velocity from 1.3 to 0.2 cm/hr had little influence on observed transport of components subject to ion exchange, though it did influence components regulated by precipitation or coprecipitation reactions. Solubility relations based upon speciation of column effluent using MINTEQA2 were, in general, of limited utility in describing element fluxes.

## INTRODUCTION

Acid mine drainage has been recognized for some time as a serious threat to surface waters. More recently, concerns have been expressed as to the potential

0-87371-890-9/93/$0.00+$.50

impacts of acid mine drainage, and coal mining operations in general, on groundwater quality,[1] and also the related influences that leachate and runoff from coal storage facilities may have on local ground and surface water quality.[2] To assess the potential impacts of coal mining on groundwater resources, the Bureau of Mines, U.S. Department of the Interior, supported the National Research Council's Committee on Groundwater Resources in Relation to Coal Mining.[1] In its report, the committee's recommendations included further research on: (1) water movement in the unsaturated zone of both undisturbed and reclaimed lands, (2) geochemical interactions between groundwater and spoil material, and (3) the potential of natural hydrogeologic processes in ameliorating adverse effects of coal mining operations.

An issue which the committee did not directly address, but which is intimately related with coal mining operations, is the storage of coal by utilities and industries dependent upon the fossil fuel. Just as in the case of acid mine drainage, coal pile leachate can be highly acidic and heavily laden with toxic metals and organics.[3] As a result, the same concerns regarding acid mine drainage can be expressed for coal stockpiles, with the further considerations that coal stockpiles are relatively abundant and frequently not controlled.

While the general chemistry of acid mine drainage and coal pile runoff is well established, natural geochemical attenuation processes remain incompletely understood. Alkaline neutralization reactions have been shown to effectively reduce the pernicious character of the metalliferous acid leachates[4] and are frequently employed as a treatment of acid mine drainage prior to release to receiving streams. Similarly, Wangen and Jones[5] demonstrated that carbonatic soils were highly effective at reducing acidity and retaining a number of elements when acidic leachate generated from coal-minerals wastes percolated through soil samples. A naturally acidic, non-carbonatic soil provided relatively little attenuation, however.[5] As a significant portion of the U.S. coal reserves and coal stockpiles are situated on moderately to highly weathered geologic materials with limited carbonate alkalinity, direct neutralization of acid mine drainage via carbonate dissolution may be of secondary importance relative to more complex and heterogeneous adsorption, dissolution, and precipitation reactions.

The objectives of this research, then, were to improve current understanding of the environmental impacts of the coal industry by evaluating the hydrogeochemical interactions of acidic runoff from coal stockpiles with soil and groundwater. This chapter summarizes results from an evaluation of the relative mobility of components within coal pile runoff upon infiltration of this highly acidic effluent through a weathered, naturally acidic subsoil-aquifer unit.[6]

## MATERIALS AND METHODS

### Study Site

Soil samples were collected from a site adjacent to the D-Area coal pile runoff containment basin within the DOE Savannah River Site near Aiken, SC (Figure 1).[6,7] The coal stockpile contains approximately 131,000 metric tons of low-sulfur

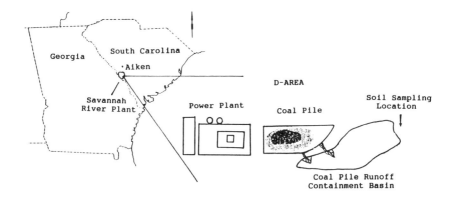

**FIGURE 1.** Study site.[7] (Reprinted with permission from the American Chemical Society.)

(1 to 2% S) coal and covers a 3.6 ha area. The basin has a volume capacity of approximately $3.8 \times 10^{10}$ L and is 5.1 ha in areal extent.[7] Monitoring-well data indicate degradation of a shallow water table aquifer (at approximately 1 to 3 m depth) downgradient (south) of the coal pile and basin.

## Soils

The runoff containment basin was created by excavation and removal of the surficial 1 to 2 m of a Fuquay soil (a thermic, sandy, kaolinitic Plinthic Paleudult). Subsoil samples were collected by augering adjacent to the basin, with samples collected every 25 cm (depth) until the sandy, unconfined aquifer was reached (1.25 m). Bulk densities were determined on samples extruded from cores of known volume[8] which were weighed prior and subsequent to oven drying to 105°C. Bulk densities were calculated from the oven dry weights and core volumes. Porosities were calculated from bulk density data assuming a particle density of 2.65 $Mg/m^3$.[9] Particle size distributions were determined via the modified pipette method of Miller and Miller.[10] Bulk soil samples were air-dried, and passed through a 2 mm sieve. Soil pH was determined on 1:1 soil to water equilibrations.[11] Noncrystalline and free oxide contents of triplicated 1 g (oven dry weight basis) subsamples were estimated via the 4-hr ammonium-oxalate in the dark (Ox) and dithionite-citrate-bicarbonate (DCB) procedures, respectively.[12] Soil cation exchange capacities were determined by summation of native cations exchanged from triplicate 5 g oven-dry-weight-basis subsamples with three 30-min extractions of $M$ $NH_4Cl$. Soil organic matter contents were determined using the Walkley-Black method.[13]

The mineralogical compositions of the subsoils were also determined. Silt- and clay-sized materials were separated from whole soils after Fe removal by DCB[12] via wet sieving and centrifugation.[14] Magnesium and K-saturated oriented clay mounts were prepared using a tile mounting technique.[15] Powder mounts of the silt fractions (Na-saturated) were prepared by packing oven-dry samples into a powder holder. The mounts were X-rayed with a Diano X-ray spectrometer from 2 to 32° 2 θ at a step speed of 0.075° 2 θ/4 s. A Cu source was used (γ = 1.542 × 10$^{-10}$ m). The K-saturated tile mounts were X-rayed at room temperature and after heating to 110°, 300°, and 550°C. The Mg-gly tile mounts were X-rayed at room temperature and after heating to 110°C. Powder mounts were X-rayed at room temperature. Samples of 5 mg from the K-saturated clay mounts were removed for analysis by differential scanning calorimetry. Each sample was placed in a gold pan and heated to 625°C at a rate of 20°C/min with a DuPont 1090 Thermal Analyzer.

## Column Leaching Experiments

Air-dried, < 2 mm soil was packed into 25 × 2.54 cm lucite tubes to a uniform bulk density of 1.50 Mg/m$^3$. A Dionex APM pump module was used to supply a constant flow of deionized water, equivalent to a Darcy velocity (q) of 1.3 cm/hr, until steady saturated (upward) flow was attained. The columns then were allowed to equilibrate overnight (no flow). Coal pile runoff collected from the basin was then supplied and the columns leached under steady saturated flow (q = 1.3 cm/hr). Effluent was collected via an LKB linear fraction collector. Column dispersivities were determined via pulse injections of Br$^-$ using the parameter estimation method of Parker and van Genucthen.[16]

## Analytical

The pH of the soil:water extracts and column effluent fractions was determined using a GK2401C combined glass-calomel electrode and a Radiometer PHM 84 pH meter. Bromide concentrations were determined using a Fisher 13-620-520 Br$^-$ ion-specific electrode and a Radiometer K401 reference electrode. Effluent fractions and extracts were analyzed for total dissolved major and trace element concentrations via inductively coupled plasma emission spectroscopy (ICPES). Sulfate analysis was performed using a Dionex System 2010i ion chromatograph (IC). Ferrous iron concentrations were determined via the o-phenanthroline method.[17]

## Speciation

Direct analytical speciation of elements within the coal pile runoff and column effluent was limited to S and Fe. As a result of the very good agreement between $SO_4^{2-}$ concentrations via IC and total S via ICPES (<1% deviation), all S was considered as $SO_4^{2-}$ (Table 1).[6,7] The distribution of Fe between the Fe(II) and Fe(III) forms was determined from the o-phenanthroline determined Fe(II), with Fe(III) taken as the difference between ICPES total Fe and Fe(II). For thermodynamic equilibrium speciation calculations, the valence of the remaining elements was assigned assuming forms common to those elements under typical environmental conditions (Table 1), though subsequent discussions will exclude reference to particular valence states except where explicitly determined. Equilibrium

**Table 1**
**Composition of the Coal Pile Runoff Used in the Column Experiments**

| Component | Concentration (mg/L)[a] |
|---|---|
| pH | 2.13 |
| $Al^{3+}$ | 101.0 |
| $Ca^{2+}$ | 83.62 |
| $Co^{2+}$ | 0.461 |
| $Cr^{3+}$ | 0.010 |
| $Cu^{2+}$ | 0.262 |
| $Fe^{2+}$ | 4.70 |
| $Fe^{3+}$ | 119.5 |
| $K^+$ | 1.668 |
| $Li^+$ | 0.232 |
| $Mg^{2+}$ | 62.17 |
| $Na^+$ | 11.52 |
| $Ni^{2+}$ | 0.878 |
| $H_4SiO_4$ | 88.30 |
| $SO_4^{2-}$ | 2024 |
| $Sr^{2+}$ | 0.791 |
| $Zn^{2+}$ | 2.392 |

[a] except pH.

speciation of the coal pile runoff and column effluent was calculated via EPA's MINTEQA2 geochemical equilibrium model.[18]

## RESULTS AND DISCUSSION

### Coal Pile Runoff

The coal pile runoff was very acidic, and possessed high dissolved $SO_4^{2-}$, $Fe^{3+}$, and Al concentrations (Table 1). The runoff also possessed high levels of alkaline earth metals, transition metals, and other elements (Table 1).[6,7]

### Soils

The subsoils were coarse-textured, with low pH, low to modest cation exchange capacities, and low base saturations (Table 2). These characteristics are typical of soils of the highly weathered Coastal Plain physiographic province. The soils also possessed very low organic matter contents (< 0.15%). The bulk density for the first sampling depth (0 to 0.25 m beneath basin surface, or approximately 2 to 2.25 m below natural soil surface) was 1.53 metric ton/m$^3$. A saturated hydraulic conductivity of about 1 cm/hr was estimated from particle size and bulk density data.

The subsoil was found to vary significantly with respect to solid phase composition over the relatively narrow depth to water table. Ox-extractable Al was greatest for the 0.25 to 1.25 interval, whereas Si extracted with ammonium-oxalate was more uniformly distributed with depth. Ox-extractable Fe varied

**Table 2**
**Selected Soil Properties**

| Sampling Depth (m) | pH$_w$ | CEC (cmol$_c$/kg) | Base Sat. (%) | Particle Size (%) | | |
|---|---|---|---|---|---|---|
| | | | | Sand | Silt | Clay |
| 0.00–0.25 | 4.93 | 4.43 | 15.3 | 78 | 2 | 20 |
| 0.25–0.75 | 4.96 | 6.82 | 16.1 | 75 | 5 | 20 |
| 0.75–1.25 | 5.06 | 3.71 | 21.1 | 81 | 6 | 13 |
| 1.25–1.40 | 5.12 | 2.15 | 11.3 | 88 | 2 | 10 |

**Table 3**
**Sequential Extraction Results**

| Sampling Depth (m) | Ox-Al (%) | Ox-Fe (%) | Ox-Si (%) | DCB-Al (%) | DCB-Fe (%) | Ox/DCB-Fe |
|---|---|---|---|---|---|---|
| 0.00–0.25 | 0.022 | 0.028 | 0.005 | 0.036 | 0.477 | 0.06 |
| 0.25–0.75 | 0.033 | 0.054 | 0.007 | 0.062 | 0.826 | 0.06 |
| 0.75–1.25 | 0.033 | 0.085 | 0.006 | 0.075 | 0.910 | 0.09 |
| 1.25–1.40 | 0.025 | 0.029 | 0.006 | 0.020 | 0.295 | 0.09 |

considerably with depth, being highest at the 0.75 to 1.25 m sampling interval, which corresponds to the zone adjacent to the existing (at time of sampling) water table (Table 3). Seasonal fluctuations in the water table likely resulted in an enrichment of relatively poorly crystalline Fe oxide phases at this depth. DCB-extractable Fe was also greatest for this depth interval, which supports the notion of deposition-dissolution of Fe during seasonal water table fluctuations. The presence of Al within the reductant-soluble DCB Fe fraction suggests that Al and Fe may be associated through a coprecipitation reaction, though DCB is known to remove Al from hydroxy-interlayer materials. A relatively constant mole ratio of DCB-extractable Al to Fe (0.15, 0.16, 0.17, and 0.13 for the 0 to 0.25, 0.25 to 0.75, 0.75 to 1.25, and 1.25 to 1.40 m depths, respectively) was noted. The occurrence of natural Al-substituted goethites, often approaching 30%, has been well-documented.[19] Ratios of Ox to DCB-extractable Fe are frequently used as an index of the overall crystallinity of soil Fe phases. Consistent with the proposed role of the fluctuating water table in regulating overall crystallinity of the Fe phases, a significant increase in Ox/DCB Fe was noted for the lower two sampling depths (Table 3).

The mineralogical composition was also found to vary with depth. Quartz was the major component within the subsoil, accounting for all sand and silt-sized material, and its content in the soil increased with depth (Table 4). Kaolinite, a common 1:1 weathering product of 2:1 phyllosilicate minerals and a dominant mineral within the highly weathered Piedmont and Coastal Plain provinces, was the next most abundant mineral for all depths. Smectite was greatest within the first sampling depth and decreased with depth. Mica, vermiculite, hydroxy-interlayered vermiculite, and interstratified mica-vermiculite were also present in

**Table 4**
**Soil Mineralogical Properties (on Whole Soil Basis)**

| Sampling Depth (m) | Q (%) | K (%) | M (%) | S (%) | V (%) | HIV (%) | M-V (%) | G (%) | T (%) |
|---|---|---|---|---|---|---|---|---|---|
| 0.00–0.25 | 80 | 12.8 | 1.0 | 4.8 | Tr | 0.3 | 1.1 | Tr | ND |
| 0.25–0.75 | 80 | 13.2 | 1.2 | 3.4 | Tr | Tr | 2.2 | ND | ND |
| 0.75–1.25 | 87 | 6.3 | 0.9 | 3.3 | 0.2 | Tr | 2.3 | ND | ND |
| 1.25–1.40 | 90 | 5.1 | 0.4 | 1.5 | 0.2 | Tr | 2.0 | ND | 0.9 |

*Note:* Q = quartz, K = kaolinite, M = mica, S = smectite, V = vermiculite, HIV = hydroxy-interlayered vermiculite, M-V = interstratified mica-vermiculite, G = gibbsite, T = talc., Tr = trace, and ND = none detected.

varying quantities within the subsoil (Table 4). A trace of gibbsite was observed for the first sampling interval, indicative of more intense weathering and desilication at the basin surface. Talc was also present within the aquifer material.

## Column Leaching Experiments

Results from the transport experiments are presented using reduced values for time, or equivalently, as the number of pore volumes supplied to the columns ($V/V_0$). Reduced concentrations, $C/C_0$, where C is the effluent concentration relative to that of the influent ($C_0$), are used when different $C_0$ values are plotted on the same figure. Reproducibility of columns and column effluent concentration histories was very good. For example, duplicated columns of the 0.25 to 0.75 m material yielded fitted dispersion coefficients of 2.00 and 2.02 $cm^2$/hr as well as very similar column effluent concentrations histories.[6,7]

### pH

Hydrogen ion activities as a function of eluted pore volumes exhibited similar relative trends for all four soil depths (Figure 2: pH). The pH was essentially constant as the initial volume of equilibrated deionized water was displaced. At varying subsequent pore volumes, a slight rise in pH was observed, immediately followed by a rapid decrease in pH to near 3.5. The pH of the effluent remained at this value for each of the soils for some variable length of time (or volume), ranging from approximately 2 pore volumes for the 1.25 to 1.40 m material to about 6 pore volumes for the 0.25 to 0.75 m material. Buffering reactions within the soils apparently served to maintain solution pH at approximately 3.5. Following this pH plateau, the pH of the column effluents again rapidly decreased, approaching that of the influent pH (Figure 2: pH).

A rapid decrease in pH would be expected to occur at $V/V_0$ near 1 if no reaction occurred. As initial proton breakthrough occurred between 1.3 and 2 pore volumes, some mechanism resulting in reaction or retardation was occurring. One possible mechanism can be gleaned from the sulfate effluent concentrations histories (Figure 2: $SO_4$). Breakthrough of sulfate was coincident with the initial reduction in pH to 3.5 (Figure 2). Given the necessary condition of electroneutrality

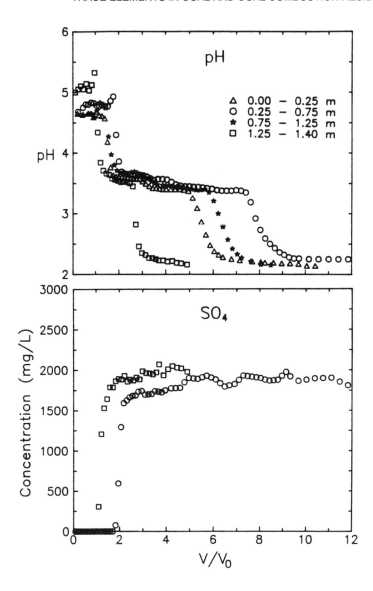

**FIGURE 2.** Column effluent concentration vs. reduced time for the four subsoil materials: pH and sulfate.

in solution, and previous observations for significant adsorption of sulfate to soil and soil components[20-22] and precipitation of basic Al and Fe sulfates,[23,24] sulfate mobility may be important to the mobility of protons. To further evaluate this relationship between anion mobility and breakthrough of protons, equivalent pH solutions containing only sulfate or nitrate were compared with the coal pile runoff. Initial breakthrough of protons from equivalent pH solutions was, in fact, observed to vary significantly as a function of counter ion (Figure 3). Proton

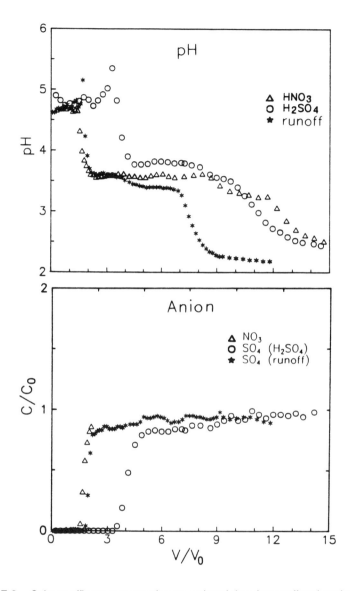

**FIGURE 3.** Column effluent concentration vs. reduced time for runoff and equivalent pH nitric and sulfuric acid influent solutions (0.25 to 0.75 m material): pH and anion.

breakthrough occurred rapidly ($V/V_0$ about 1.7) when the counter ion was nitrate, whereas significant reductions in pH were not noted until about 3.7 pore volumes when sulfate was the counter ion. The differences in pH and sulfate breakthrough behavior for the runoff and $H_2SO_4$ solution appear to be principally due to the four times greater sulfate concentration in the coal pile runoff.

Comparisons between the $H_2SO_4$ solution and the coal pile runoff are not straightforward, however. Metals associated with the runoff would be expected to complex to varying degrees with sulfate, and thus reduce its interactions with the soil surface and hasten transport through the soil. MINTEQA2 calculations indicate that approximately 31% of the sulfate would be complexed with metals at its influent pH. If one allows the remaining sulfate (48.4% as $SO_4^{2-}$ and 20.1% as $HSO_4^-$ to participate in sorption reactions, the free sulfate concentration would remain about three times as great as that for the $H_2SO_4$ solution. This increased sulfate concentration in the coal pile runoff relative to the $H_2SO_4$ solution may result in more rapid filling of available sorption sites. It appears, then, that initial migration of protons through the soils is related through the condition of electroneutrality in solution to the mobility of the accompanying anion.

Another feature of the proton breakthrough curves for these various soil depths, already briefly discussed, is the plateau near pH 3.5 (Figures 2 and 3). A number of reactions can be invoked to account for this buffering, including perhaps dissolution of various 2:1 minerals, oxyhydroxides and/or interlayer materials; protonation of edge sites of phyllosilicates, and proton exchange. An evaluation of the various baseline chemical and mineralogical data for these soils fails to identify a single factor apparently responsible for the observed buffering. Neither Ox nor DCB-extractable Al, Fe, or Si follows observed buffering. Whole-soil kaolinite contents and CEC values qualitatively tend to follow the observed trend, though a direct relationship is lacking. Proton exchange may be in part responsible for the observed buffering, though the high concentration of di- and trivalent metal ions in solution should minimize adsorption of protons except for selective edge sites. Specific adsorption of sulfate may result in pH buffering due to displacement of surface hydroxyls[25] and could account for the slight increase in pH preceding the initial pH reduction (Figures 2 and 3). However, little sulfate removal from solution was observed over this buffer region (Figures 2 and 3). It thus appears that an array of coupled surface reactions, rather than a single, well-defined reaction, is responsible for the observed buffering. Titration of whole soil samples suspended in deionized water underestimated buffer capacity by about 40% (data not shown), likely due in part to the rapid rate of titration relative to the leaching experiments (28 min titration).

Comparison of the runoff- and pH 2.13 $H_2SO_4$-solution breakthrough curves (Figure 3) reveals another reaction occurring within the soil columns. For an equivalent hydrogen ion activity in the influent solution, greater volumes of pH 2.13 $H_2SO_4$ were required to lower the pH to the plateau region near pH 3.5, and also to the influent pH, as compared to the coal pile runoff. The hydrolysis of Fe(III) would impart additional acidity to the solution,[26] and thus the runoff possesses greater overall acidity. MINTEQA2 simulations indicate that at pH 2.13, hydrolysis products are relatively minor components of the total Fe(III) in solution, comprising only 2.8% of the total Fe(III) (as $FeOH^{2+}$). If one allows for hydrolysis to $Fe(OH)_3$, an additional proton loading nearly equivalent to the measured $H^+$ activity would result. That the observed buffer region for the runoff-leached column is two-thirds that of the $H_2SO_4$ leached column under equivalent flux rates suggests that the Fe(III) is not hydrolyzing to $Fe(OH)_3$, but to a lower hydroxyl content phase.

## Major Components

Breakthrough of Al was observed concurrent with the breakthrough of sulfate and the initial reduction in pH for the different soil materials (Figure 4: Al). Peak concentrations in the effluents from the different materials tended to be related to the CEC and native exchangeable Al contents (Table 5). The peak effluent Al concentration for the low Al exchange content-low CEC aquifer material (1.25 to 1.40 m depth) was observed after only 2.5 pore volumes, while the peak Al concentration in the high Al-high CEC 0.25 to 0.75 m subsoil occurred at $V/V_0$ near 6 (Figure 4: Al). Native exchangeable Al (Table 5), and not an amorphous Al solid phase (Table 3), appears to be a primary source for the initial high Al concentrations in solution, and a result of exchange by cations present in the runoff. As a consequence, one can expect Al concentrations several times that of the runoff to follow the sulfate plume in coal pile runoff-contaminated acid soil sites. This is probably a phenomenon analogous to the so-called "hardness halo" noted for groundwater contaminated by municipal landfills in the Midwest.[27]

Iron was retained to a large extent by materials from all depths and did not appear in the effluent (Figure 4: Fe) until the effluent pH approached that of the influent (Figure 2: pH). Low pH would be necessary to keep Fe(III) (the principal form of Fe in the runoff, Table 1) from hydrolyzing and precipitating as a hydroxo- or basic sulfate phase. As discussed in the previous section, it appears that the Fe is precipitating to a basic sulfate phase with an OH/Fe mole ratio near 2. This precipitated phase could also be expected to undergo dissolution as the soil buffer capacity is exhausted and the ambient pH descends toward that of the influent; thus Fe breakthrough behavior is controlled by precipitation-dissolution processes.

Silicon was observed to break through and peak in effluent concentration (Figure 4: Si) immediately preceding the initial drop in pH and the breakthrough of sulfate (Figure 2). A relatively high Si effluent concentration (Figure 4: Si) was maintained during the observed pH buffer regions (Figure 2: pH), which implies a role of phyllosilicate dissolution in the buffering reactions for these soils. Barnhisel and Rotromel[28] subjected kaolinite and illite clay minerals to simulated acid mine drainage solutions and concluded that the mode of attack on both clays occurred at the edges, releasing Al and Si, and also Fe and K in the case of the illite, and suggested that phyllosilicate dissolution is the probable source of Si and also Al in coal spoil solutions. Filipek et al.[29] similarly suggested that the dissolution of phyllosilicate minerals was the principal source of Si found in acid mine drainage.

## Alkali Metals

Breakthrough curves of K, Li, and Na for the four sampling depths are provided in Figure 5. One notices quite disparate behavior for these three alkali metals. Though a considerable amount of scatter in the data was found, peak K concentrations tend to occur later in the leaching process for samples with higher

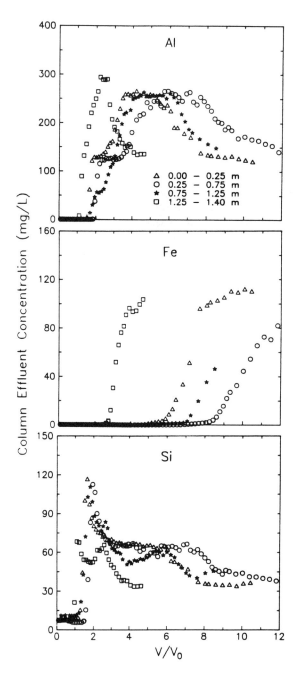

**FIGURE 4.** Column effluent concentration vs. reduced time for the four subsoil materials: Al, Fe, and Si.

**Table 5**
**Exchange Phase Composition: Major Elements (mg/kg)**

| Sampling Depth (m) | Al | Ca | Mg | Na | K |
|---|---|---|---|---|---|
| 0.00–0.25 | 337.4 | 21.6 | 64.0 | 5.3 | 9.4 |
| 0.25–0.75 | 519.5 | 25.6 | 104.8 | 4.1 | 14.9 |
| 0.75–1.25 | 263.4 | 39.9 | 63.2 | 8.3 | 11.7 |
| 1.25–1.40 | 171.5 | ND | 19.6 | 4.6 | 25.0 |

CEC values (Figure 5: K). Unlike that noted for Al, however, no relationship between native exchange phase K contents and effluent concentration profiles was apparent (e.g., the aquifer sample had the highest exchangeable K level (Table 5) though the highest eluted K was noted for the 0.75 to 1.25 m sample) (Figure 5: K). The K-bearing phyllosilicate minerals present within these materials (Table 4) may be an additional source of K to solution.

A general lack of detectable exchange phase Li (Table 6) apparently resulted in the absence of a pronounced effluent peak noted for other elements (Figure 5: Li), with breakthrough again concurrent with that of sulfate (Figure 2: $SO_4$). Rather only subtle peaks near 2 and 4 pore volumes for the 0.25 to 0.75 and 1.25 to 1.40 m samples were noted, with column effluent Li concentrations then corresponding to approximately that of the influent. Native exchangeable Li for Coastal Plain soils is generally very low, on the order of 1 $\mu mol_c/kg$.[30]

Sodium breakthrough was rapid and sharp (Figure 5: Na), quickly returning to a concentration near that of the influent (Table 1), and provides an interesting contrast to that of K. Both Na and K are relatively minor components of the exchange phase of the samples (Table 5), and are monovalent. Yet the K peak for all samples is retarded and also greatly broadened relative to Na (Figure 5). Considerable research has established, however, that a number of 2:1 clay minerals have a high selectivity for K over other common soil cations.[31] Due to its low hydration energy and size, which allows it to fit easily into hexagonal holes within Si tetrahedral layers,[31] K can participate in fixation reactions with certain 2:1 phyllosilicate minerals (e.g., vermiculite) found within these materials (Table 4). The observed retardation of K relative to Na thus may be a result of K sorption to these highly K-selective sites, though precipitation within an alunite or jarosite phase can not be discounted.

## Alkaline Earth Metals

Breakthrough for all the alkaline earth metals (Figure 6) commences coincident with the breakthrough of sulfate and the initial decline in pH (Figure 2). Initial breakthrough was related to the presence of native exchangeable metals, the magnitude of which descends in the order Mg>Ca>Sr.

While exchangeable Ca and Mg are commmon features of soils, reports of native exchangeable Sr is less frequent. Peak effluent concentrations of Mg and Ca (Figure 6) follow the general trend noted for Al (Figure 4: Al), with peak concentration and

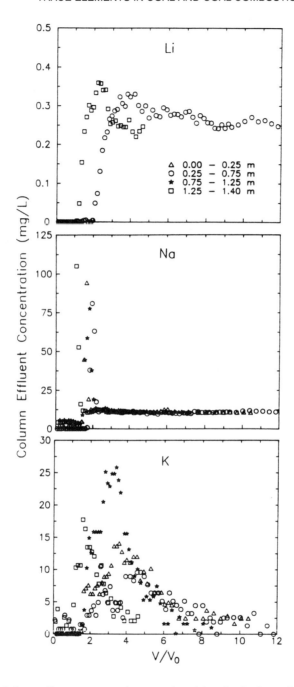

**FIGURE 5.** Column effluent concentration vs. reduced time for the four subsoil materials: Li, Na, and K.

**Table 6**
**Exchange Phase Composition: Minor Elements (µg/g)[a]**

| Sampling Depth (m) | Cu | Li | Ni | Si | Sr | Zn |
|---|---|---|---|---|---|---|
| 0.00–0.25 | 0.05 | ND | ND | 10.3 | 0.75 | 0.02 |
| 0.25–0.75 | 0.34 | 0.04 | ND | 14.1 | 0.79 | ND |
| 0.75–1.25 | 0.21 | ND | ND | 14.1 | 0.79 | ND |
| 1.25–1.40 | 0.02 | ND | 0.05 | 4.6 | ND | 1.77 |

[a] No detectable (ND) Co and Cr at any depth.

retardation being related to CEC (Table 2) and exchange phase content (Table 5). For example, the highest exchange phase Ca content of the 0.75 to 1.25 m sample resulted in the greatest peak concentration and peak intensity, while the peak Ca concentration occurred latest for the highest CEC 0.25 to 0.75 m depth (Figure 6: Ca). Strontium exhibited similar trends to Ca and Mg, though two peaks in the breakthrough curve were noted for the 0.25 to 0.75 m sample (Figure 6: Sr). This depth possessed the highest native exchangeable Sr as well as the highest CEC. Apparently these factors resulted in a chromatographic separation and elution of Sr in two chromatographic waves (Figure 6: Sr).

*Transition Metals*

The breakthrough of Co, Ni, and Zn all tended to follow one another very closely for a given material (Figure 7). As with other elements, initial appearance in the effluent was related to sulfate mobility, while peak concentrations tended to be retarded to a degree related to CEC, with peak concentrations occurring at the earliest pore volumes for the lowest CEC aquifer material and longest for the highest CEC 0.25 to 0.75 m sample (Figure 7). This observation was tested more rigorously by a simple statistical analysis in which CEC, Ox-Fe, and DCB-Fe contents were regressed against the peak $V/V_0$ at which peak effluent concentrations were noted. Notwithstanding the small sample size (only two degrees of freedom), CEC was the only soil property which had a significant correlation with observed peak effluent transition metal concentrations ($r = 0.94$, significant at 0.10). Oxalate and DCB-extractable Fe yielded nonsignificant correlation coefficients of 0.40 and 0.75, respectively, when regressed against $V/V_0$. This suggests that ion exchange is the principal reaction governing transport of these metals for these soils leached with coal pile runoff. Thus, though specific adsorption to native Fe phases is a well-established mechanism for removal of trace metals from solution,[32] neither sorption to native Fe phases nor coprecipitation with freshly precipitating phases wholly regulates Ni, Zn, Co, and Mn activities in solution. Additionally, pH-adsorption edges for these metals on ferric hydroxides tend to be above the native soil pH, and well above the pH range encountered during much of the leaching experiments.[33,34] Considering the 0.25 to 0.75 m sample, one notices that the peak metal concentrations coincide with the second peak in the breakthrough curve for Sr (Figure 6: Sr). Apparently the lack of significant

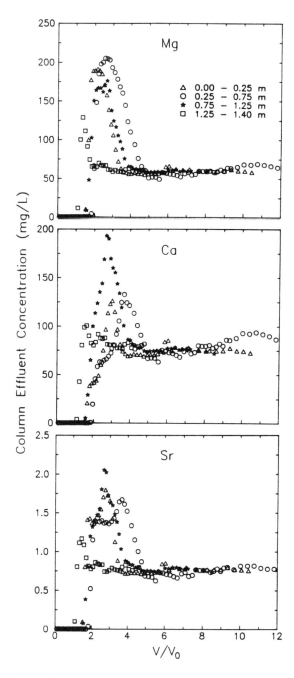

**FIGURE 6.**  Column effluent concentration vs. reduced time for the four subsoil materials: Mg, Ca, and Sr.

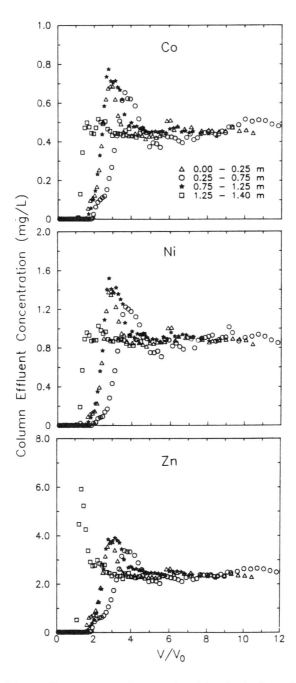

**FIGURE 7.**   Column effluent concentration vs. reduced time for the four subsoil materials: Co, Ni, and Zn.

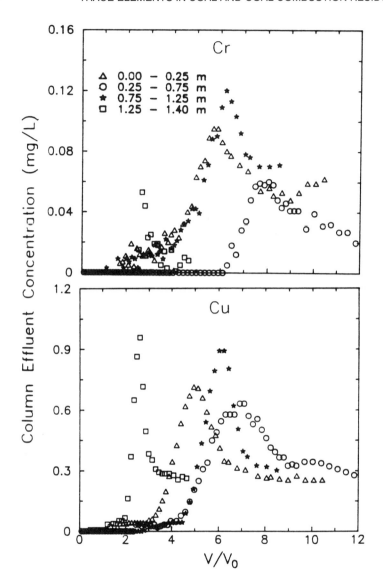

**FIGURE 8.** Column effluent concentration vs. reduced time for the four subsoil materials: Cr and Cu.

exchangeable transition metal contents (Table 5) results in the reduction or absence of the first peak at 2.2 pore volumes most apparent for Sr.

Unlike the previous transition metals, Cr and Cu exhibited quite different break-through behavior (Figure 8). While limited amounts tended to break through with the other transition metals, peak concentrations occurred much later, and approximately coincided with the descent in pH from 3.5 toward the influent pH of 2.13, with peak Cr lagging slightly behind that of Cu (Figure 8). It thus appears that Cr and Cu mobilities are regulated by a mechanism different from that of the other transition metals examined. The dependence on pH implies a precipitation-dissolution, coprecipitation, or specific adsorption-desorption reaction.

## Influence of Flow Rate on Observed Transport

To ascertain the influence of flow rate on the breakthrough behavior of elements, the aquifer (1.25 to 1.40 m) material was also leached under steady saturated flow at a Darcy velocity of 0.2 cm/hr. Selected results are presented in Figures 9 and 10. Hydrogen ion activity as a function of eluted pore volumes exhibited similar behavior to that at q = 1.3 cm/hr at short $V/V_0$, though deviated at higher pore volumes (Figure 9: pH). Decreasing q resulted in longer apparent buffering at pH 3.5 before descending toward influent pH. Effluent Si followed this general trend (Figure 9: Si), which supports the notion of a desilication reaction contributing to the buffer capacity of the soils. Flux of Co through the column, which was suggested to be controlled by ion exchange, was little affected by Darcy velocity (Figure 10: Co). This observation implies local equilibrium with respect to components partitioning between solution and soil exchange sites. Copper, whose mobility was apparently controlled by a more specific interaction with the solid phase, was sensitive to flow rate, however (Figure 10: Cu). Peak Cu concentration again coincided with the decrease in pH from that of the buffer region.

## Solubility Relations

A number of solid phases have been proposed to control element solubilities within acid mine drainage systems.[23,24] Effluent fractions were speciated with the geochemical model MINTEQA2[18] to assess the potential for such phases controlling observed aqueous phase chemistry. Saturation indices (SI) defined as

$$SI = \log\left(\frac{IAP}{K_{sp}}\right) \tag{1}$$

where IAP is the ion activity product of a solution and $K_{sp}$ is the solubility product for a given mineral phase, for selected solid phases as a function of $V/V_0$ for the aquifer material are illustrated, though very similar trends were observed for the other materials as well.

Aluminum solubility most closely followed that of jurbanite, $AlOHSO_4$ (Figure 11: Al Minerals). Both amorphous $Al(OH)_3$ and K-alum were undersaturated over the entire course of the leaching process. Jurbanite and alunite $(KAl_3(SO_4)_2(OH)_6)$ were supersaturated until $V/V_0$ near 2.7, the pore volume in which the soil buffer was exhausted and the pH began to decrease (Figure 2). Van Breeman[35] suggested that $AlOHSO_4$ governs Al solubility for a range of acid

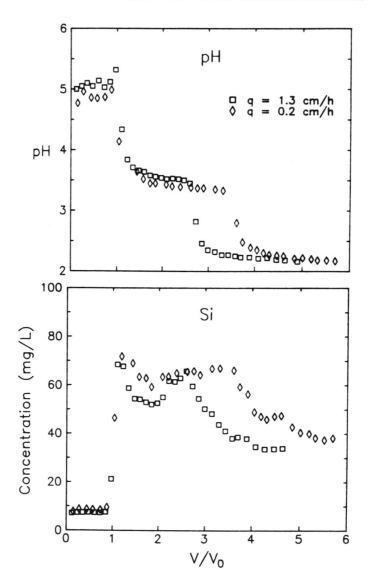

**FIGURE 9.**   Column effluent concentration vs. reduced time as a function of Darcy velocity (1.25 to 1.40 m material): pH and Si.

sulfate soils. Nordstrom[24] outlined the solubility relations for Al in the presence of $SO_4^{2-}$ at low to moderate pH; he reported that jurbanite solubility was in good agreement with the data of van Breeman[35] and proposed that jurbanite controls the solubility of Al in many acid sulfate soils and acid mine drainage systems. Nordstrom and Ball[36] noted more recently, however, that at pH <4.6, Al in surface acid drainage waters was generally transported conservatively and that apparent conformance with solubility limits was coincidental. Similar observations by

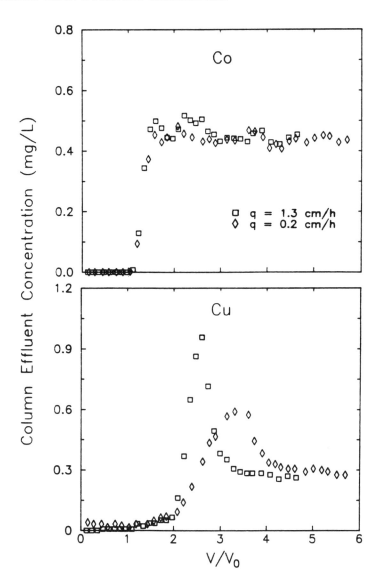

**FIGURE 10.** Column effluent concentration vs. reduced time as a function of Darcy velocity (1.25 to 1.40 m material): Co and Cu.

Filipek et al.[29] suggest that similarities between dissolved Al concentrations and jurbanite solubility may be misleading.

Iron was below detection limits within the effluent at early $V/V_0$ and as a result saturation indices are not available until 2.5 pore volumes. Unlike Al, in which saturation indices generally decreased below 0 at $V/V_0 > 2.5$ to 3, the decrease in pH and appearance of Fe in the effluent yielded positive saturation indices for a number of Fe oxyhydroxide and basic sulfate phases (Figure 11: Fe Minerals).

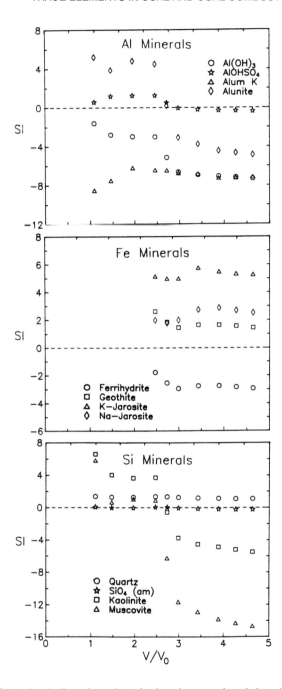

**FIGURE 11.** Saturation indices for selected minerals vs. reduced time (1.25 to 1.40 m material): Al, Fe, and Si minerals.

The column effluent was undersaturated only with respect to the highly soluble ferrihydrite. Results indicate that Fe concentrations in the column effluent most closely follow that predicted by assuming Na-jarosite controls Fe solubility. Variously substituted jarosites have been previously postulated to control Fe solubility within acid sulfate soils and acid mine drainage systems,[23] though significant supersaturation of acid mine waters with respect to jarosite mineral phases has also been reported.[29,37]

A number of solid phases proposed to control trace metal solubilities in soils[38] were also evaluated. No evidence for precipitation-dissolution reactions controlling metal mobility was found. For example, cupric ferrite, which has been suggested to control Cu solubility in soils, was undersaturated over the course of the leaching process, and thus does not explain its breakthrough behavior (Figure 8: Cu). Filipek et al.[29] also reported that Cu and Zn were undersaturated with respect to all potential mineral phases for an acid drainage-surface water confluence.

Solubility calculations for selected Si-bearing minerals demonstrate that aqueous Si concentrations follow very closely that based on an amorphous $SiO_2$ phase (Figure 11: Si Minerals). Transition from supersaturation to undersaturation for the phyllosilicates, as noted for a number of the Al phases, occurs when the effluent pH descends below 3.5 (Figure 11: Si Minerals). Thus, both the kaolinite and mica within the samples (Table 3) can be expected to undergo dissolution under prolonged leaching and, as noted by Barnhisel and Rotromel[28] and Filipek et al.,[29] release Si and other mineral structural components to solution.

As mineral dissolution is generally a slow process, being surface controlled under most conditions,[39] the influence of flow rate on observed saturation indices was also evaluated. Saturation indices as a function of q for the phases which were found to most closely follow observed aqueous chemistries were calculated (data not shown). No discernible influence on the saturation index for the amorphous $SiO_2$ phase was noted. Slight deviations for jurbanite and Na-jarosite at $V/V_0 > 2.5$ apparently resulted from the difference in effluent pH for the two Darcy velocities. It thus appears that deviations from local thermodynamic equilbrium are not significantly influenced by reduction in Darcy velocity from 1.3 to 0.2 cm/hr.

## Simple Modeling of Transport

The classical convection-dispersion equation (CDE), routinely used to model solute transport in column experiments, was evaluated for its applicability to modeling component breakthrough, employed here in the form:

$$R\frac{\partial C}{\partial t} = D\frac{\partial^2 C}{\partial x^2} - v\frac{\partial C}{\partial x} \qquad (2)$$

here C is concentration ($M/L^3$), t is time (T), x is distance (L), D is the hydrodynamic dispersion coefficient ($L^2/T$), v is the pore water velocity (L/T), and R is the retardation coefficient. (M, L, and T refer generically to mass, length, and time, respectively). Central to the CDE is that of linear partitioning between sorbed and solution phases, and on that basis, the simple CDE is theoretically an oversimplification of the chemistry controlling component mobility during this investigation. Widely available codes which expressly accommodate ion exchange, specific adsorption, and precipitation-dissolution reactions are generally lacking, in part due to the tremendous computational effort.[40]

Application of the simple CDE was first evaluated using a pulse of nonreactive solute (Br⁻). Generally good agreement between observed and fitted breakthrough of Br⁻ was noted ($r^2$ = 0.98), suggesting that the simple one-region CDE adequately describes the flow within the columns.[16] Fitted dispersion coefficients were then used as input for simulation of runoff breakthrough curves. Moderate agreement between observed (points) and predicted (lines) $SO_4^2$ breakthrough for the aquifer (1.25 to 1.40 m) and subsoil (0.25 to 0.75 m) materials was obtained (Figure 12: $SO_4$), yielding fitted retardation factors (R) of 1.3 and 2.0, respectively. The model tended to overpredict $SO_4^{2-}$ concentration after the initial rise in concentrations, particularly for the subsoil material, which indicates that in addition to a retardation reaction (i.e., R>1), $SO_4^{2-}$ is also participating in a reaction which is maintaining a solution concentration below the predicted $C/C_0$ value of 1. In general, however, the shape of the predicted breakthrough curve follows that observed. This relative agreement was not noted for the other elements (e.g., Figure 12: Ni). Elements in which some initial reactive mass resides within the soil exhibit pronounced waves in the breakthrough curves, as previously noted. Ion exchange, specific adsorption, and/or precipitation-dissolution reactions which result from the percolation of runoff through the soils tend to mobilize some fraction of that reactive mass and thus serve as sources not explicitly considered in the CDE. Even components in which negligible initial reactive mass exist, as in the case of a number of transition metals (Table 5), exhibit peak concentrations in column effluent in excess of influent concentration (e.g., Figure 7). Thus the transport of reversibly sorbing components also results in this general phenomenon.[41]

## Implications for Groundwater Systems

A general implication of the observed transport of components within coal pile runoff, and more generally acid mine drainage, is that waves of component concentrations well in excess of that of the runoff can be expected. Exchange or other removal processes in which the soil serves as source of components to the aqueous phase can greatly increase component concentrations. Even elements not initially present in the soil to any significant extent exhibit this wave phenomenon and are a result of chromatographic processes within the soil. Thus, even if runoff itself meets applicable water quality standards, groundwater downgradient may well exceed concentration limits as the wave advances with the groundwater flow.

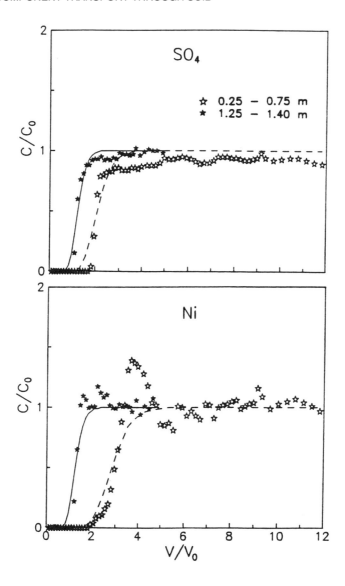

**FIGURE 12.** Observed (points) and fitted (lines) breakthrough curves for the 0.25 to 0.75 m (subsoil) and 1.25 to 1.40 m (aquifer) materials: $SO_4$ and Ni.

## SUMMARY AND CONCLUSIONS

A naturally acidic Coastal Plain subsoil proffered little attenuation of infiltrating acidic, metal-rich coal pile runoff, with peak element concentrations often moving at 0.25 to 1 times the rate of water flow. The appearance of cationic components in the effluent was related to sulfate mobility. Ion exchange controlled peak Ca, Co, Mg, Ni, Sr, and Zn effluent concentrations, whereas precipitation appeared to control Fe(III) mobility, and coprecipitation was important to Cr and Cu mobilities. Reduction in Darcy velocity from 1.3 to 0.2 cm/hr had little influence on relative transport of components affected by ion exchange, but did influence components apparently regulated by precipitation-dissolution reactions. Solubility relations were of limited utility in describing element fluxes, however.

Results from this work indicate that naturally acidic, non-carbonatic soils provide little impedance to the movement of acidic metal-rich runoff through soil. Furthermore, development of chromatographic and precipitation-dissolution waves are associated with the subsurface migration of acidic runoff. These observations suggest that severe degradation of groundwaters may result from coal stockpiles situated on highly weathered soils. As a result, engineering controls are needed to capture and treat coal pile runoff prior to any infiltration and recharge.

## ACKNOWLEDGMENTS

This research was partially supported by contract DE-AC09-76SR00819 between the University of Georgia and the U.S. Department of Energy.

## REFERENCES

1. Davis, E. C. and Boegly, W. J., A review of water quality issues associated with coal storage, *J. Environ. Qual.*, 10, 127, 1981.
2. National Research Council, *Coal Mining and Groundwater Resources in the United States,* National Academy Press, Washington, D.C., 1981, 1.
3. Nichols, C. R., Development Document for the Proposed Effluent Limitations, Guidelines, and New Source Performance Standards for the Steam-Electric Power Generating Point Source Category, U.S. EPA Report-440/1-73-029, 1974, 1.
4. Wangen, L. E. and Williams, J. M., Control by alkaline neutralization of trace elements in acidic coal cleaning waste leachates, *J. Water Pollut. Control Fed.*, 54, 1302, 1982.
5. Wangen, L. E. and Jones, M. M., The attenuation of chemical elements in acidic leachates from coal mineral wastes by soils, *Environ. Geol. Water Sci.*, 6, 161, 1984.
6. Anderson, M. A., A Hydrogeochemical Investigation of Metalliferous Coal Pile Runoff with Soil and Groundwater, Ph.D. thesis, Virginia Polytechnic Institute and State University, Blacksburg, 1990, 1.
7. Anderson, M. A., Bertsch, P. M., Feldman, S. B., and Zelazny, L. W., Interactions of acidic metal-rich coal pile runoff with a subsoil, *Environ. Sci. Technol.*, 25, 2038, 1991.
8. Blake, G. R. and Hartge, K. H., Bulk density, in *Methods of Soil Analysis, Part 1,* Klute, A., Ed., American Society of Agronomy, Madison, WI, 1986, 363.
9. Danielson, R. E. and Sutherland, P. L., Porosity, in *Methods of Soil Analysis, Part 1,* Klute, A., Ed., American Society of Agronomy, Madison, WI, 1986, 443.
10. Miller, W. P. and Miller, D. M., A micropipette method for soil mechanical analysis, *Commun. Soil Sci. Plant Anal.*, 18, 1, 1987.

11. McLean, E. O., Soil pH and lime requirement, in *Methods of Soil Analysis, Part 1*, Page, A. L., Miller, R. H., and Keeney, D. R., Eds., American Society of Agronomy, Madison, WI, 1986, 199.

12. Jackson, M. L., Lim, C. H., and Zelazny, L. W., Oxides, in *Methods of Soil Analysis, Part 1*, Klute, A., Ed., American Society of Agronomy, Madison, WI, 1986, 101.

13. Allison, L. E., Organic carbon, in *Methods of Soil Analysis, Part 1*, Klute, A., Ed., American Society of Agronomy, Madison, WI, 1986, 1367.

14. Whittig, L. D. and Allardice, W. R., X-ray diffraction techniques, in *Methods of Soil Analysis, Part 1*, Klute, A., Ed., American Society of Agronomy, Madison, WI, 1986, 331.

15. Rich, C. I. and Barnhisel, R. I., Preparation of clay samples for X-ray diffraction analysis, in *Minerals in Soil Environments*, Dixon, J. B. and Weed, S. B., Eds., Soil Science Society of America, Madison, WI, 1977, 797.

16. Parker, J. C. and van Genucthen, M. Th., Determining Transport Parameters from Laboratory and Field Tracer Experiments, Virginia Agricultural Experimental Station Bulletin 84-3, 1984, 1.

17. *Standard Methods for the Examination of Water and Wastewater*, 14th Ed., American Public Health Association, Washington, D.C., 1976, 3.

18. Brown, D. S. and Allison, J. D., MINTEQA2, an Equilibrium Metal Speciation Model, U.S. EPA Report-600/3-87/012, 1987, 1.

19. Norrish, K. and Taylor, R. M., The isomorphous replacement of iron by aluminum in soil goethites, *J. Soil Sci.*, 12, 294, 1961.

20. Singh, B. R., Sulfate sorption by acid forest soils: 3. desorption of sulfate from adsorbed surfaces as a function of time, desorbing ion, pH, and amount of adsorption, *Soil Sci.*, 138, 346, 1984.

21. Sigg, L. and Stumm, W., The interaction of anions and weak acids with the hydrous goethite surface, *Coll. Surf.*, 2, 101, 1980.

22. Aylmore, L. A., Karim, M., and Quirk, J. P., Adsorption and desorption of sulfate ions by soil constituents, *Soil Sci.*, 103, 10, 1967.

23. Nordstrom, D. K., Aqueous pyrite oxidation and the consequent formation of secondary iron minerals, in *Acid Sulfate Weathering*, Kittrick, J. A., Fanning, D. S., and Hossner, L. R., Eds., Soil Science Society of America, Madison, WI, 1982, 37.

24. Nordstrom, D. K., The effect of sulfate on aluminum concentrations in natural waters: some stability relations in the system $Al_2O_3$-$SO_3$-$H_2O$ at 298 K, *Geochim. Cosmochim. Acta*, 46, 681, 1982.

25. Rajan, S. S. S., Sulfate adsorbed on hydrous alumina, ligands displaced, and changes in surface charge, *Soil Sci. Soc. Am. J.*, 42, 39, 1978.

26. Baes, C. R., Jr. and Mesmer, R. E., *The Hydrolysis of Cations*, John Wiley & Sons, New York, 1976, 226.

27. Cartwright, K., Griffin, R. A., and Gilkeson, R. H., Migration of landfill leachate through glacial tills, *Ground Water*, 15, 294, 1977.

28. Barnhisel, R. I. and Rotromel, A. L., Weathering of clay minerals by simulated acid coal spoil-bank solutions, *Soil Sci.*, 118, 22, 1974.

29. Filipek, L. H., Nordstrom, D. K., and Ficklin, W. H., Interaction of acid mine drainage with waters and sediments of west Squaw Creek in the West Shasta mining district, California, *Environ. Sci. Technol.*, 21, 388, 1987.

30. Anderson, M. A., Bertsch, P. M., and Miller, W. P., Distribution of lithium in selected soils and surface waters of the southeastern U.S.A., *Appl. Geochem.*, 3, 205, 1988.

31. Sposito, G., *The Surface Chemistry of Soils*, Oxford University Press, New York, 1984, 15.
32. Jenne, E. A., Controls on Mn, Fe, Co, Ni, Cu, and Zn concentrations in soils and waters: the significant role of hydrous Mn and Fe oxides, in *Trace Inorganics in Water*, Gould, R. F., Ed., American Chemical Society, Washington, D.C., 1968, 337.
33. Benjamin, M. M. and Leckie, J. O., Multiple-site adsorption of Cd, Cu, Zn, and Pb on amorphous iron oxyhydroxide, *J. Colloid. Interface Sci.*, 79, 209, 1981.
34. Forbes, E. A., Posner, A. M., and Quirk, J. P., Specific adsorption of divalent Cd, Co, Cu, Pb, and Zn on goethite, *J. Soil Sci.*, 27, 1654, 1976.
35. van Breemen, N., Dissolved aluminum in acid sulfate soils and in acid mine waters, *Soil Sci. Soc. Am. Proc.*, 37, 694, 1973.
36. Nordstrom, D. K., and Ball, J. W., The geochemical behavior of aluminum in acidified surface waters, *Science,* 232, 54, 1986.
37. Chapman, B. M., Jones, D. R., and Jung, R. F., Processes controlling metal ion attenuation in acid mine drainage streams, *Geochim. Cosmochin. Acta*, 47, 1957, 1983.
38. Lindsay, W. L., *Chemical Equilibria in Soils*, John Wiley & Sons, New York, 1976, 211.
39. Berner, R. A., Rate control of mineral dissolution under earth surface conditions, *Am. J. Sci.*, 278, 1235, 1978.
40. Yeh, G. T. and Tripathi, V. S., A critical evaluation of recent developments in hydrogeochemical transport models of reactive multichemical components, *Water Resour. Res.*, 25, 93, 1989.
41. Clancy, K. M., and Jennings, A. A., Experimental verification of multicomponent groundwater contamination predictions, *Water Res.*, 24, 307, 1988.

# 8

# Leachability of Ni, Cd, Cr, and As from Coal Ash Impoundments of Different Ages on the Savannah River Site

**S. S. Sandhu,[1] G. L. Mills,[2] and K. S. Sajwan[3]**

[1]Claflin College, Orangeburg, SC
[2]Savannah River Ecology Laboratory, Aiken, SC
[3]Savannah State College, Savannah, GA

## ABSTRACT

This study was undertaken to evaluate and compare the leachability of Cd, Ni, Cr, and As in ash disposal basins of different ages located at the U.S. Department of Energy (DOE) Savannah River Site. Intact ash cores were removed from the basins and leached in two successive pulses with solutions of varying pH and redox status. The leaching under all of the experimental conditions resulted in the slow release of metals from the ashes. Nickel was the most mobile and Cd was the least mobile. All of the metals examined were released in the highest amounts by leaching under reducing conditions in cores from both the new and old ash disposal sites. The amount of metal released was kinetically controlled under the leaching used. Although the amount of metal released from cores from the new basins was typically greater than those from the old, measurable amounts of metal continued to be released from the old basin-ash cores even after weathering for over 10 years.

## INTRODUCTION

The utility industry produces more than $7.5 \times 10^7$ metric tons of solid waste annually, of which only $2.0 \times 10^7$ metric tons (26.67%) is put to practical use.[1] Available estimates indicate that solid wastes generated by the utility industry will double by the turn of the century. The electrical utilities reported that 70% of the current ash production is sluiced to disposal ponds in an aquatic environment, posing contamination potential for ground and surface waters.

Several coal-burning power plants have been in operation at the Savannah River Site, the largest of which is located in the D-Area facility. Fly ash is generated at the D-Area Site by routine operations of a four-unit, $4 \times 10^8$ Btu/hr, coal-burning power plant that has been operating since 1952.[2] In 1987, the D-Area burned $2 \times 10^5$ metric tons of coal, averaging 1.9% sulfur. The ash content of the coal is currently 10.4%, a decrease of 14% from the mid-1970s. Prior to 1976 about 67% of the ash was collected by mechanical cyclones. The remaining ash (33%) went up the stacks. Since 1976, electrostatic precipitators have been in operation to remove fly ash from stack emissions with greater than 99% efficiency. The ash particles collected by the electrostatic precipitators are smaller in size than those trapped by the cyclone collectors and contain high concentrations of many trace metals.

Several studies have examined the chemical composition and water chemistry of the ash basin water at the SRS as well as the stream system which receives overflow discharged from the basins.[3-9] One of these studies[9] also examined the changes in water chemistry within the ash basin water system during normal seasonal cycles and compared these changes to those occurring in nonimpacted impoundments.[9] However, the mobility and fate of metals in the ash basin sediments formed by the accumulation of ash in the impoundments has not been reported. The older ash basins were abandoned when filled and presently have been colonized by vegetation which has accumulated high levels of several toxic and nontoxic metals.[10] The abandoned ash basins have aged by weathering for several years; thus, the partitioning of the trace metals among solid phase components and their leachability from the ash sediments may have changed. Quantitative rates for long-term leaching of fly ash contaminants from utility waste disposal sites are necessary to accurately predict the potential for ground water contamination. Several coal ash leaching studies have previously been reported.[3,8,11-14] Generally, these studies have focused on fresh ashes collected directly from the power plants prior to sluicing into disposal ponds. Most of these studies used either strong acids, bases, or chelating agents as extractants which would not simulate field conditions. One of these studies conducted *in vitro* leaching of heavy metals from coal ashes using biological ligands. The objective of the present study was to evaluate the leachability of trace elements from impounded ashes of different ages. Weak extracting solutions were employed using intact ash cores to simulate the amount of leaching that might be expected to occur under field conditions at the SRS disposal area.

## MATERIALS AND METHODS

### Study Area

The basins received ash sluiced from power production facilities in the D-Area at SRS (in service since plant start-up in 1952). D-Area (Figure 1) has three adjacent basins. The primary ash basin and secondary ash basin are active disposal sites and receive $5.07 \times 10^4$ m$^3$ of ash generated by the power plant. About $4.9 \times 10^9$ L/year

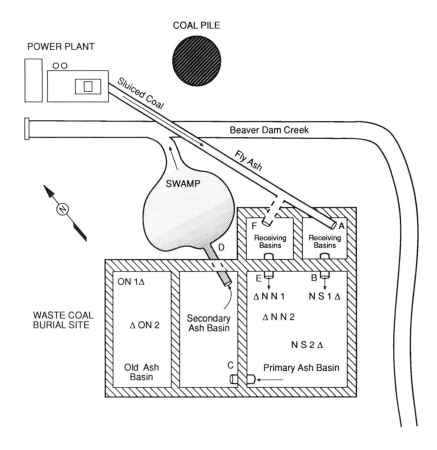

**FIGURE 1.** Sampling location in the D-area ash basins at the Savannah River Site.

of water is used to transport the ash. The sluiced water flows by gravity alternately into the small receiving basins at sites A or F and subsequently moves into the primary basins for sedimentation. The turbid supernatant water flows into the secondary ash basin. Overflow water from the secondary disposal basin, along with small-sized or less dense suspended ash particles, drains into a small swamp eventually discharging into Beaver Dam Creek, a tributary of the Savannah River. The dimensions of the primary and secondary ash disposal basins are 100m × 100m × 3.0 and 100m × 50m × 3.0m, respectively. Calculated retention time for the primary and the secondary ash disposal basins are 39 and 22 days,[9] respectively.

## Sampling Sites

Selection of sampling sites at each ash disposal basin was based on its representativeness of the study area and accessibility for sampling throughout the project. Two sampling sites were selected in the old ash basin and designated ON-1 and ON-2. Four sites were selected in the primary ash basin and named NN-1, NN-2, NS-1, and NS-2 (Figure 1).

The sampling sites NS-1 and NS-2 had a thick and healthy growth of wax myrtle *(Myrica Cerifera* L.), and an undergrowth of dog fennel *(Eupatorium compositifolium* walter) and cattails *(Typha latifolia* L.). The sampling sites NN-1 and NN-2 were dominated by several varieties of aster *(Aster paternus* Cronquist), cattails, dog fennel, rabbit tobacco *(Guaphallium obtusifolium* L.) and dandelion *(Taraxacum wiggens* L.). Sea myrtle *(Baccharis halimifolia* L.) and wax myrtle were also present. Swamp cottonwood *(Populus heterophylla* L.), black willow *(Salix nigra* Marshall) and red maple *(Acer rubrum* L.) were scattered in the primary ash disposal basin. Half of the primary ash basin was under several feet of water; no vegetation was observed in that area during the time of the study.

The old ash basin was located to the northwest of the secondary ash disposal basin (Figure 1) and adjoined a waste coal disposal basin (burial site). It had a thick growth of common swamp vegetation, including swamp cottonwood, sycamore *(Platanus occidentalis* L.), black willow, wax myrtle, pokeweed *(Phytolacca americana* L.), and elderberry *(Sambucus canadensis* L.), and its surface was covered with a thick layer of organic matter, in various stages of decomposition. All the ash basin sites were surrounded by a containment soil berm over 2m high.

## Sample Collection

Several intact ash cores were removed from each site (Figure 1) using a Wildco soil sampling apparatus with a PVC core liner. Each core was 5.12 cm in diameter and about 90 cm long, providing approximately 850 g of ash. Six ash columns were removed at one time.

## Leaching Studies

The leaching apparatus of six cores has been diagrammed (Figure 2). Reservoirs (1 L) containing the leaching solutions were arranged to allow gravity flow into a thistle tube with microscopic pores installed in the center of the column (Figure 3). The flow rate of solution dripping into the thistle tube was controlled by a pinch clamp (Figure 2).

The effect of solution pH on the mobilization of trace metals was studied using $H_2O$ (pH 6.2), dilute $HNO_3$ (pH 3.0), and dilute $Ca(OH)_2$ (pH 8.0). The effect of oxidizing and reducing conditions was determined, using solutions containing $1M$ $H_2O_2$ and $0.04M$ $NH_2OH\text{-}HCL$, respectively. A separate set of columns was used for each solution treatment. Five aliquots, each containing 100 mL of leaching solution, were collected from each ash core. The movement of leaching solution through the ash columns was slow and usually required several days to collect 100 mL of leachate. The experiments were designed to simulate the nonequilibrium condition between the ash material and leaching solutions to simulate field conditions. There was continuous movement of solution through the ash columns during leaching

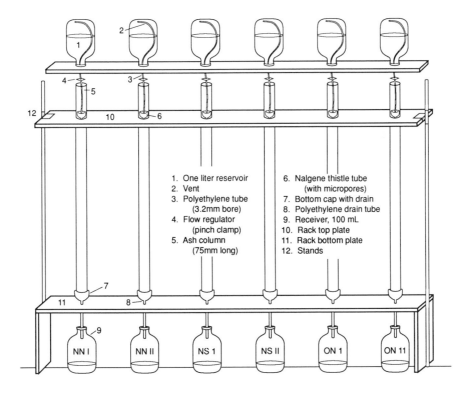

1. One liter reservoir
2. Vent
3. Polyethylene tube
   (3.2mm bore)
4. Flow regulator
   (pinch clamp)
5. Ash column
   (75mm long)

6. Nalgene thistle tube
   (with micropores)
7. Bottom cap with drain
8. Polyethylene drain tube
9. Receiver, 100 mL
10. Rack top plate
11. Rack bottom plate
12. Stands

**FIGURE 2.** Arrangement showing leaching of ash columns.

periods. After the collection of the first 500 mL of leachate, the columns were capped, stored for 5 weeks, and leached a second time under the same solution conditions. This pulse-like manner of leaching is similar to what can be expected at the field sites with successive rain events.

## Sample Analysis

The nutritional levels, mechanical composition, and porosity of ash sample were also determined using standard procedures of the University of Georgia's Soils Testing Laboratory.[15] The mechanical analysis[16] and porosity determination[17] were carried out with composite samples for each basin prepared by mixing equal amounts of ash from the corresponding depth of the respective cores.

The elemental analyses were conducted with a Hitachi Model 180-80 Zeeman Effect Atomic Absorption Spectrophotometer using an air-acetylene flame or furnace which was equipped with a HGA-3 graphite atomizer.[18] Argon gas flow and standard conditions were used for all determinations, except As. Zirconium acetate treated, pyrolytically coated cuvett[19] and $NiNO_3$ additions were used to improve As analysis. Matrix effects for all elements were corrected using the method of standard additions.

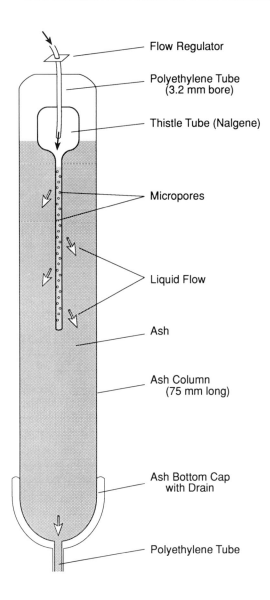

**FIGURE 3.**　Cross section of ash column undergoing leaching.

## RESULTS

### Physical and Chemical Characteristics of the Ash

No significant differences in porosity were found between the old (ON) and new (NS-NN) sites, indicating that with aging no further compaction occurred (Table 1). There was a significant ($r^2$=0.98) decreasing gradient in the amount of organic matter found in the ash basins with the increasing depth. The thick layer

**Table 1**
**Physical Characteristics of Ash (Composite Samples)**

| Sampling Site | Depth (cm) | Density (mg/m³ × 10⁻³) | | Porosity (%) | Organic Matter | Material Composition (%) | | | | Textural Classes |
|---|---|---|---|---|---|---|---|---|---|---|
| | | Bulk | Particle | | | Gravel | Sand | Silt | Clay | |
| NS-NN | 0-7 | | | | 2.95 | 0.11 | 13.58 | 72.05 | 9.20 | Silt Loam[a] |
| NS-NN | 7-15 | 0.78 | 1.86 | 58.1 | 2.53 | 2.84 | 20.28 | 63.20 | 8.03 | Silt Loam |
| NS-NN | 15-30 | | | | 1.78 | 2.57 | 47.50 | 34.56 | 7.80 | Sandy Loam |
| NS-NN | 30-60 | 0.81 | 2.19 | 65.3 | 1.99 | 3.29 | 23.12 | 60.53 | 6.35 | Silt Loam |
| ON | 0-7 | | | | 5.15 | 2.61 | 40.39 | 38.72 | 14.16 | Loam |
| ON | 7-15 | 0.7 | 1.81 | 61.4 | 4.25 | 0.13 | 33.89 | 52.57 | 9.97 | Silt Loam |
| ON | 5-30 | | | | 3.78 | 1.20 | 29.20 | 54.82 | 8.76 | Silt Loam |
| ON | 30-60 | 0.68 | 1.79 | 62.1 | 2.60 | 7.91 | 31.13 | 51.15 | 7.64 | Silt Loam |
| ON | 60-77 | | | | 1.25 | 17.80 | 28.23 | 47.44 | 9.56 | Loam |

[a] Using the nomenclature for soils.

**Table 2**
**Chemical Characteristics of Ash (Composite Samples)**

| Sampling Site | Depth (cm) | pH | Concentration (mg/kg) | | | | | | | | | | | | |
|---|---|---|---|---|---|---|---|---|---|---|---|---|---|---|---|
| | | | P | K | Na | Ca | Mg | Al | Pb | Mn | Fe | B | Cu | Zn | Mo |
| NS | 0–7 | 6.7 | 33.2 | 82.0 | 63.2 | 485 | 48.4 | 545 | 0.8 | 8.5 | 143 | 4.7 | 7.9 | 1.8 | 0.4 |
| NS | 7–15 | 6.4 | 41.1 | 44.0 | 46.0 | 287 | 26.9 | 456 | 0.5 | 3.8 | 153 | 3.4 | 8.6 | 1.7 | 0.4 |
| NS | 15–30 | 6.5 | 52.6 | 52.7 | 39.8 | 247 | 23.1 | 377 | 0.5 | 2.9 | 126 | 2.7 | 7.9 | 1.6 | 0.2 |
| NS | 30–60 | 7.2 | 139 | 64.8 | 43.7 | 416 | — | 454 | 0.5 | 4.8 | 196 | 2.4 | 7.9 | 2.2 | 0.3 |
| NN | 0–7 | 6.0 | 67.8 | 86.0 | 249.9 | 497 | 67.9 | 577 | 0.7 | 5.0 | 156 | 5.9 | 4.5 | 1.1 | 0.5 |
| NN | 7–15 | 6.4 | 35.9 | 74.6 | 155.4 | 155 | 56.8 | 630 | 0.7 | 3.1 | 165 | 5.5 | 5.0 | 1.6 | 0.4 |
| NN | 15–30 | 6.5 | 34.3 | 66.0 | 126.3 | 350 | 46.4 | 466 | 0.5 | 2.9 | 165 | 4.4 | 4.1 | 1.5 | 0.3 |
| NN | 30–60 | 7.3 | 71.4 | 67.8 | 195.9 | 601 | 68.9 | 779 | 0.7 | 4.6 | 264 | 9.6 | 5.3 | 2.7 | .4 |
| ON | 0–7 | 5.8 | 62.2 | 106.8 | 34.9 | 861 | 46.4 | 468 | 0.4 | 13.7 | 117 | 1.3 | 0.7 | 3.2 | 0.2 |
| ON | 7–15 | 5.6 | 15.9 | 95.8 | 22.2 | 487 | 38.5 | 422 | 0.3 | 3.1 | 101 | 1.3 | 1.4 | 0.5 | 0.2 |
| ON | 15–30 | 5.6 | 46.7 | 119.7 | 68.4 | 721 | 54.4 | 504 | 0.7 | 4.2 | 89 | 2.8 | 2.1 | 0.9 | 0.3 |
| ON | 30–60 | 6.3 | 54.0 | 145.1 | 95.8 | 1847 | 171.3 | 63 | 0.8 | 8.9 | 84 | 4.5 | 1.6 | 1.5 | 0.3 |
| ON | 60–77 | 7.3 | 29.6 | 162.4 | 108.3 | 640 | 240.0 | 313 | 0.7 | 8.1 | 85 | 4.3 | 1.5 | 1.2 | 0.2 |

of organic matter in the ON site was in various stages of decomposition and was mobilized and deposited through the ash horizons (Table 1). The decomposition of organic matter produced both inorganic and organic acids (e.g., acetic, citric) that contributed to the acidification of the sediments. Thus, the acidity of the ash cores was inversely correlated with organic matter concentration. The pH was generally lowest near the surface and increased with depth and was lower in the old basin (ON) than in the new basin (NS and NN) (Table 2). A detailed discussion of the chemical data (Table 2) is presented elsewhere.[21]

## Effects of Solution pH

### Leaching with Water

The cumulative masses of metals Ni, As, Cr, and Cd mobilized by water from the intact ash columns were plotted against the cumulative volume of leachate from the NN and ON sites (Figure 4). The initial leachates for all metals showed a slight but steady increase in amount released with progressive leaching. However, there was a sharp increase after 500 mL. This increase was particularly pronounced for Ni and Cd at the NN site and for Cr and As at the ON site. The amounts of Cr and As extracted by water from both sites were about the same; however, the amount of Ni leached from the NN core was generally an order of magnitude greater than that obtained from the ON core. The results for Cd were similar to the NN site, yielding about five times that of the ON site. Thus, Ni and Cd were easily solubilized in the early period of ash deposition, whereas the release of Cr and As proceeded more slowly.

### Leaching with Acid

The concentration of $HNO_3$ (.001 $M$, pH 3) was deliberately kept low to simulate the field conditions that may be expected to prevail in ash deposits. With the exception of Cr, there was a slight but steady increase in the release of all metals during the initial 500 mL of leaching for NN and ON (Figure 5) basins; however, there was a sharp increase in the release of all metals after 500 mL. The amounts of As and Cr extracted under acid conditions from the NN core was slightly higher than extracted from the ON core, whereas the reverse was true for Cd. The amount of Ni extracted from the NN and ON cores was about the same. The amount of Cd extracted under mildly acidic conditions from both the ON and NN cores was nearly exhausted after about 900 mL of leachate.

### Leaching with Base

The sluiced fly ash water as it was discharged at location A (Figure 1) in the sediment pond contained 16.5 mg/L of calcium and had a pH of 8.98. The initial 500 mL of leachate for both the NN and ON (Figure 6) cores showed a small but continuous release of Cd and As, but negligible amounts of Ni and Cr. However,

## Water Leachate

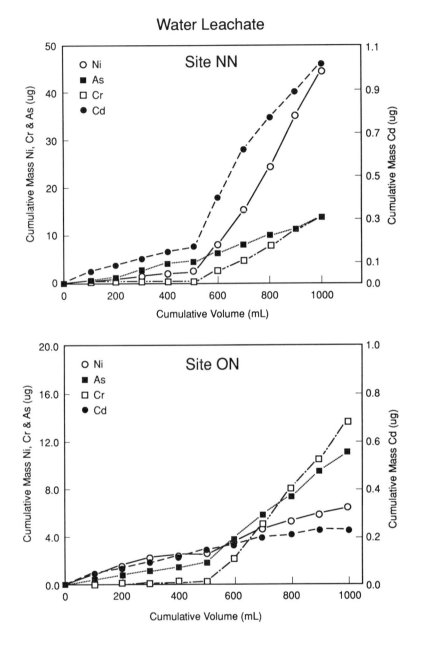

**FIGURE 4.**    Relationship between the cumulative volume and cumulative mass of Ni, Cr, As, and Cd in water leachate of site NN and site ON.

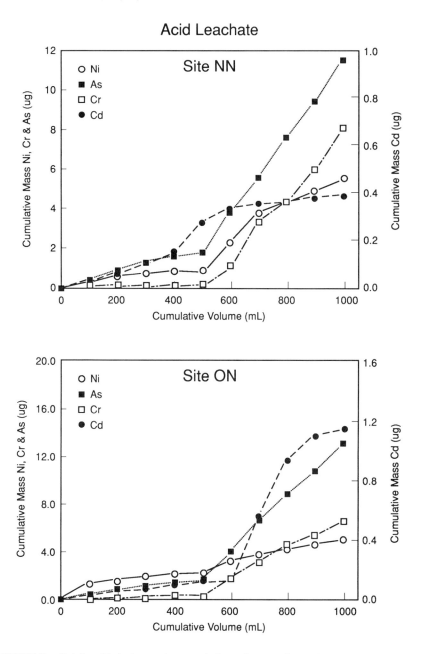

**FIGURE 5.** Relationship between the cumulative volume and cumulative mass of Ni, Cr, As, and Cd in acid leachate of site NN and site ON.

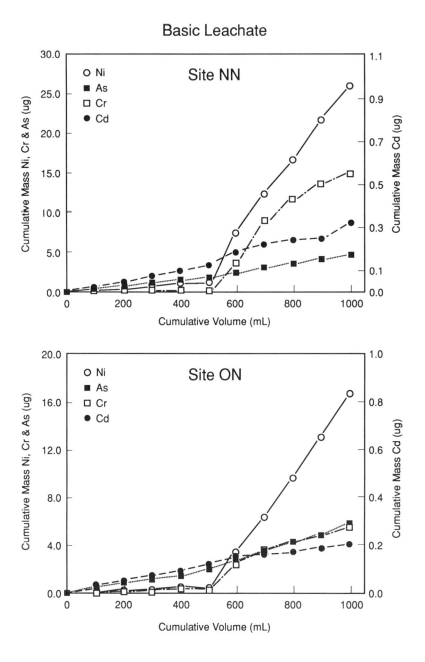

**FIGURE 6.** Relationship between the cumulative volume and cumulative mass of Ni, Cr, As, and Cd in basic leachate of site NN and site ON.

in the second leaching period (500 to 1000 mL) there was a marked increase in the release of Ni for both NN and ON sites and a less pronounced increase in the release of Cr for NN site only. Cadmium and As continued to be leached at about the same rate. The total mass of metals released was small, with highest and lowest values found for Ni and Cd, respectively.

## Effect of Redox Conditions

### Peroxide Leachate

The amounts of As extracted by the oxidizing solution ($H_2O_2$) from NN and ON (Figure 7) ash samples were comparable. A continuous release of all metals were observed in both cores. The lowest amounts of metal extracted from the NN and ON ash sites (Figure 7) were Cd and Ni, respectively. The ashes from the NN site yielded considerably higher quantity of Ni then the ON site. The oxidizing solution was relatively ineffective in mobilizing trace elements, probably because the metals were present in a stable oxidized form.

### Reducing Leachate

A progressive release by reducing leachate of most metals was observed for both NN and ON sites (Figure 8) cores throughout the leaching period. However, there was marked increase in the rate of release of Ni from the NN core during the second leaching period (500 to 1000 mL). This was also true for As, but to a lesser extent. For most metals, the greatest amount of total metal released was observed under reducing conditions. This was particularly true for Ni and Cd at the ON site (Figure 8). The total mass of metal released was greater in the new basin core (NN) than the old basin core (ON) for all of the measured metals.

## DISCUSSION

The study was undertaken to evaluate and compare the leachability of selected metals in ash disposal basins of different ages. The four metals (Cd, Ni, Cr, and As) selected for study are on the priority list of pollutants of the U.S. Environmental Protection Agency.[26] These metals are highly toxic, and Ni, Cr, and As are suspected carcinogens. Each of these metals exhibits different geochemical behavior.

The leaching under the experimental conditions were characterized by a very slow initial release of the metals from the ashes. The elements Cd, Ni, Cr, and As studied are expected to be present in the ashes associated with the solid phase oxides of Fe, Al, and Mn or bound to organic ligands.[21] Attenuation of the metals, initially released, by precipitation and adsorption reactions in the lower horizons of the ash columns may have also been responsible for the low levels of these elements in the leachates. The greatest amounts of metal were released during the second leaching period under reducing conditions after 5 weeks of equilibration

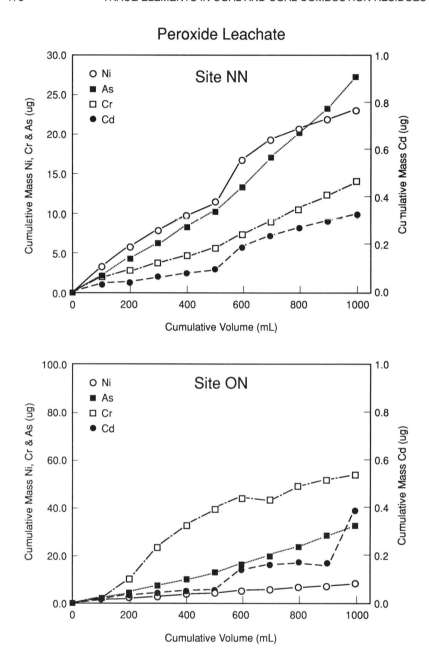

**FIGURE 7.** Relationship between the cumulative volume and cumulative mass of Ni, Cr, As, and Cd in peroxide leachate of site NN and site ON.

# Reducing Leachate

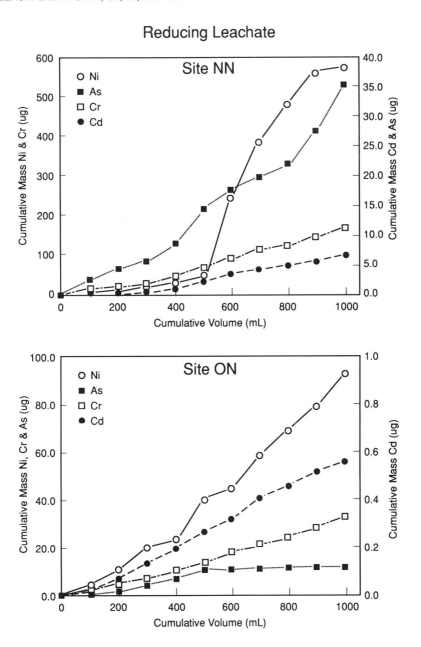

**FIGURE 8.** Relationship between the cumulative volume and cumulative mass of Ni, Cr, As, and Cd in reducing leachate of site NN and site ON.

in storage. Thus, most of the leachable fraction of these metals requires a period of equilibration in a reductive stage and are then released upon leaching.

In all of the leaching solutions examined, there was a measurable release of metals in both the new and old ash basin sediments. The ash deposits in the old basin have been weathered and leached for over 10 years, yet still may provide a source of metal contamination to infiltrating water. Thus, ash disposal basins may be potential sources of ground water contamination for many years after ash deposition has ceased.

## CONCLUSIONS

The mild leaching solutions employed in this study resulted in the slow release of Cd, Ni, Cr, and As from intact ash cores obtained from both new and old ash disposal basins. For the leaching solutions used in this study, Ni was found to be the most mobile and Cd the least mobile. The highest concentrations of all metals were obtained by leaching under reducing conditions in cores from both the old and new disposal sites. The release of these metals to infiltrating water appears to be kinetically controlled. The cumulative mass of metal released was generally greater in the cores from the new basin compared with the old. However, measurable amounts of metal continue to be released from ash cores from the old basin even after the deposits have weathered for over 10 years.

## ACKNOWLEDGMENTS

This research was supported by a grant from the U.S. Department of Energy, HBCU Program (DE-FG09-865R.15170). Manuscript preparation was supported by contract DE-AC09-76SROO-819 between the University of Georgia and the U.S. Department of Energy. The authors express their gratitude to Dr. Robert F. Keefer for his helpful suggestions, review, and comments in the preparation of this manuscript and to Mr. Larry Adriano for assistance in sample collection and analysis.

## REFERENCES

1. Muraka, I. P. and McIntosh, D. A., Solid-waste Environmental Studies: Description, Status and Available Results, Report EA-5322-SR, Electric Power Research Institute, Palo Alto, CA, 1987.
2. Horton, J. H., Dorsett, R. S., and Cooper, R. E., Trace elements in the terrestrial environmental of coal-fired power plant, U.S. DOE Report DP-1475, E. I. DuPont de Nemours & Co., Savannah River Laboratory, Aiken, SC, 1977, 49.
3. Dressen, D. R., Gladney, E. S., Owens, J. W., Perkins, B. L., Wienke, C. L., and Wangen, L. E., Comparison of levels of trace elements extracted from fly ash and levels found in effluent waters from a coal-fired power plant, *Environ. Sci. Technol.*, 11, 1017, 1977.

4. Cherry, D. S. and Guthrie, R. K., Mode of elemental dissipation from ash basin effluent, *Water Air Soil Poll.,* 9, 403, 1978.

5. Chu, T. Y., Ruane, R. J., and Krenkel, P. A., Characterization and reuse of ash pond effluents in coal-fired power plants, *J. Water Pollut. Control Fed.,* 50, 2494, 1978.

6. Evans, D. W. and Giesy, J. P., Jr., Ecology and coal resources development, in *Proc. Int. Cong. Energy Ecosystem,* Walie, M. K., Ed., Grand Fork, ND, 1978, 782.

7. Evans, D. W., Alberts, J. J., and Clark, R. A., III, Reversible ion-exchange fixation of Cesium[137] leading to mobilization from reservoir sediments, *Geochim. Cosmochim. Acta,* 47, 1041, 1983.

8. Theis, T. L. and Wirth, J. L., Sorptive behavior of trace metals on fly ash in aqueous systems, *Environ. Sci. Technol.,* 11, 1096, 1977.

9. Alberts, J. J., Newman, M. C., and Evans, D. W., Seasonal variations of trace elements in dissolved and suspended loads for coal ash ponds and pond effluents, *Water Air Soil Poll.,* 26, 111, 1985.

10. Babcock, M. F., Evans, D. W., and Alberts, J. J., Comparative uptake and translocation of trace elements from coal ash by *Typha latifolia* L., *Sci. Total Environ.,* 28, 203, 1983.

11. Harris, W. R. and Silberman, S., Time dependent leaching of coal fly ash by chelating agents, *Environ. Sci. Technol.,* 17, 139, 1983.

12. Grisafe, D. A., Angino, E. E., and Smith, S. M., Leaching characteristics of a high-calcium fly ash as a function of pH: a potential source of selenium toxicity, *Appl. Geochem.,* 3, 601, 1988.

13. Talbot, R. W., Anderson, M. A., and Andren, A. W., Qualitative model of heterogeneous equilibria in a fly ash pond, *Environ. Sci. Technol.,* 12, 1056, 1978.

14. Kuryk, B. A., Bodek, L., and Santhanam, C. J., Leaching studies on utility solid wastes. Feasibility experiments, Report EZ 4215 A-1F1, Electric Power Research Institute, Palo Alto, CA, 1985.

15. University of Georgia, Soils Testing Laboratory, Personal communication, 1990.

16. Day, P. R., Particle fractionation and particle size analysis, in *Methods of Soil Analysis, Part I,* Black, C. A., Ed., American Society of Agronomy, Madison, WI, 1965, 545.

17. Blake, G. R., Particle density and bulk density, in *Methods of Soil Analysis, Part I,* Black, C. A., Ed., American Society of Agronomy, Madison, WI, 1965, 371.

18. Hitachi Applications and Software Manual, Zeeman Atomic Adsorption Spectrometer, NSI Hitachi Scientific Instruments, Mountainview, CA, 1982, 20.

19. Vickery, T. M., Harrison, G. V., and Ramelow, G. J., Treated graphite surfaces for determination of tin by graphite furnace atomic adsorption spectometry, *Anal. Chem.,* 53, 12573, 1981.

20. American Society for Testing and Materials, Fly ash and raw or calcined natural pozzolan for use as a mineral admixture in Portland Cement concrete, Part 14, Annual Book of ASTM Standards, 1982, 381.

21. Sandhu, S. S. and Mills, G. L., Mechanisms of mobilization and attenuation of inorganic contaminants in coal ash basins, in *Emerging Technologies of Hazardous Waste Management, Part II,* Tedder, D. W. and Pohland, F. G., Eds., 1991, 342.

22. Nebergall, W. H., Schmidt, F. C., and Holtzclaw, H. F., Jr., *College Chemistry with Qualitative Analysis,* 4th ed., D. C. Heath & Company, Lexington, MA, 1982, 814.

23. Weast, R. C., *Handbook of Chemistry and Physics,* CRC Press, Cleveland, 1977, Table F 90.

24. Douglas, G. S., Mills, G. L., and Quinn, J. G., Organic copper and chromiun complexes in the interstial waters of Narragansett Bay sediments, *Marine Chem.*, 19, 161, 1986.
25. Theis, T. L. and Richter, R. O., Chemical speciation of heavy metals in power plant ash pond leachate, *Environ. Sci. Techol.*, 13, 219, 1979.
26. U.S. Environmental Protection Agency, Second Annual Report on Carcinogens, NTP 81-43, 1981, 31.
27. Windholz, M., *The Merck Index*, 10th ed., Merck and Co., Rahway, NJ, 1983.

*Use of Coal Ash for*
*Plant Growth*

# 9

# Extractable and Plant Concentrations of Metals in Amended Coal Ash

A.P. Schwab

Kansas State University, Manhattan, KS

## ABSTRACT

Plants growing directly in coal ash, without a cover layer of soil, might contain excessive concentrations of potentially toxic metals. A field study was initiated to evaluate metal contents of revegetated plants in coal ash and to determine the capabilities of the $NH_4HCO_3$-DTPA soil test to predict plant-availability of metals. A revegetation site was established in eastern Kansas near a coal-fired power plant which uses high-S eastern coal and low-S western coal. Plots were constructed with 90 cm of ash mixtures containing eastern scrubber sludge plus eastern bottom ash, western fly ash plus eastern scrubber sludge, or western bottom ash. Sorghum and soybean were seeded directly into the ash. Selected plots were amended with 5% composted manure or 5% composted manure plus 1% soil. Unamended ash consistently supported very poor yields. Addition of manure greatly increased plant dry matter, but there was no advantage to adding soil to the amended plots. Soybean plants in all treatments showed symptoms of severe B toxicity, but sorghum growing in amended ash exhibited no symptoms of stress. High concentrations of B, Mo, Zn, and Cd were observed in at least some treatments. All other metals were within normal concentration ranges. Metal concentrations extracted by the $NH_4HCO_3$-DTPA soil test were correlated with the concentrations of metals in plants ($p < 0.05$) for nine elements, including Zn, Cu, Mo, Cd, B, and Pb.

## INTRODUCTION

Many states rely exclusively on the combustion of coal to generate electricity, and the utilities in these states are faced with disposing of tremendous quantities of waste. Approximately 80 million metric tons of fly ash, bottom ash, and flue gas desulfurization (FGD) sludge were generated in 1990 from the burning of

0-87371-890-9/93/$0.00+$.50
© 1993 by Lewis Publishers

coal.[1] Depending upon the nature of the coal and the scrubbing processes, the solid wastes will contain high concentrations of potentially hazardous elements such as As, B, Ba, Cr, Cu, Mo, Ni, Se, and Zn. These elements are often volatilized during the combustion process and may condense and precipitate in a thin film on the surfaces of the ash.[2] The surface deposition and the solubility of the supporting matrix make these elements soluble, potentially mobile, sometimes toxic, and potentially an environmental hazard. Other physical and chemical properties of the ash, including high pH, fine particle size, and highly saline leachates, can have a negative impact on the surrounding environment that can be damaging to aquatic life.[3]

The establishment of permanent vegetative cover can reduce the risk of water pollution by decreasing runoff and leaching from the ash piles. Many studies have found that unweathered coal combustion ash is often high in pH, B content, and total salts[4–6] that make revegetation extremely difficult. Also, cementation of fine particles[7] can impede drainage, reduce germination, prevent emergence, and restrict root growth.[3] Regulations and guidelines for the reclamation and revegetation of coal combustion products often require covering the ash with soil before plant establishment. This can be an expensive procedure and is damaging to the land from which the soil was obtained. An alternative to burying the ash is establishing vegetative cover directly in the ash after adding small amounts of organic material and/or fertilizer.[4,5,8] The properties of the ash which are hazardous to plants (high salinity, high B, and high pH) are usually less severe after two or three years,[6,9] so that it becomes possible to establish acceptable cover and obtain permanent stands.

State environmental regulatory agencies often require that revegetated waste products contain less than maximum permissible concentrations of plant-available metals as determined by chemical extractants. The standards were established to guarantee that plant concentrations of certain metals would not reach levels which are toxic to the plants or grazing fauna. Unfortunately, the chemical extractants used in these standards, such as DTPA[10] or $NH_4HCO_3$-DTPA,[11] were developed to predict nutrient deficiencies and have not been adequately calibrated to predict metal toxicities. Lewis[12] correlated plant concentrations with DTPA-extractable concentrations. He was able to obtain highly significant correlations for Zn, Mn, and Cu, but not Fe. Similarly, Keefer and Singh[13] studied the vegetation growing in soil amended with coal ash and found significant correlations between plant uptake of Zn and DTPA-Zn and between plant Cd and DTPA-Cd, but correlations were poor for Cu, Pb, and Cr. Adams et al.[14] found significant correlations ($r^2 \geq 0.64$) between water soluble, $CaCl_2$, or DTPA extractable Zn, Ni, and Cu and uptake of these elements by plants growing in sewage sludge. Pierzynski and Jacobs[15,16] evaluated the $NH_4HCO_3$-DTPA soil test for Mo in soils amended with inorganic Mo and sewage sludge. Extractable Mo was highly correlated with plant concentrations in both greenhouse and field trials. Many researchers have demonstrated the efficacy of the extraction of B in predicting uptake of B by plants growing in coal ash or ash-amended soils.[17–19] Although plant uptake and plant concentrations of other metals from coal ash or polluted soils have been examined in several publications,[20-25] very few studies have examined the use of DTPA or similar extractants for correlation of plant uptake in coal ash.

Ion activities of the soluble constituents in the ash can be useful in the evaluation of environmental impacts of ash leachates. If the measured activities correspond to the solubility of a given mineral, then it may be assumed that the mineral controls the solubility of the ion.[26,27] Therefore, changes in solubility in response to changes in pH, ionic strength, or solution composition could be determined. The plant uptake of certain elements is proportional to activities in solution.[14,28,29]

The coal-burning power plant at LaCygne, KS, is jointly owned and operated by the Kansas City Power and Light Company (KCPL) and Kansas Gas and Electric (KGE) and currently has two coal-fired boiler and generating units. LaCygne generating unit No. 1 is designed to burn eastern coal, which is high in S and ash. Unit 1 uses a fluid limestone suspension as the scrubbing system. Approximately 120,000 m³ (96 ac-ft) of black, glassy bottom ash and 420,000 m³ (335 ac-ft) of FGD sludge are produced annually. Generating unit No. 2 burns low-S western coal and is not equipped with a desulfurizing system. Particulate matter (fly ash) is removed from flue gases by an electrostatic precipitator. This unit produces 16,000 m³ (13 ac-ft) of a light-brown, alkaline bottom ash and 45,000 m³ (36 ac-ft) of highly alkaline fly ash per year. KCPL is attempting to revegetate the ash without the usual addition of a covering layer of topsoil. It was determined that satisfactory plant covers could be established by mixing the ash with small quantities of soil and manure, adequate fertilization, and proper selection of plant species. After three years, excellent stands and aerial cover were attained.

As part of this effort, the impacts of ash revegetation on metal uptake by plants and metal leaching were investigated. The objectives of this study were to: (1) measure the uptake and concentrations of metals in plants growing in unamended ash and ash that had received manure and soil (2) determine the concentrations of metals extractable with $NH_4HCO_3$-DTPA, and (3) correlate the plant concentrations with $NH_4HCO_3$-DTPA extractable concentrations. This information will be useful in the calibration of the $NH_4HCO_3$-DTPA soil test for uptake of metals and the prediction of potentially toxic concentrations in plants.

## EXPERIMENTAL

### Site Preparation

Revegetation plots were established in the spring of 1989 and consisted of two replications each of 90% eastern bottom ash plus 10% fly ash (EB+fly); 100% western bottom ash (WB); and 75% eastern bottom ash plus 25% FGD sludge (EB+sludge). A 90-cm layer of gravel-sized eastern bottom ash was used as a base and to provide drainage. Ninety centimeters of the ash mixtures were added on top of the base layer. A chisel plow and a four-bottom plow were used to combine the materials. A typical mixing operation included four passes with the chisel plow and eight passes with the bottom plow.

Within 7 days after the first rainfall, the plots were deep-ripped with a chisel plow and further mixed with a bottom plow to break up any surface crusting and cemented layers that may have formed. After 2 weeks, the plots received 2.5 cm of irrigation water and 7.5 cm of rainfall.

Selected reclamation plots were amended with beef feedlot manure which had been composted over a 12-month period. Each $10 \times 20$ m plot was divided into six subplots ($3.3 \times 10$ m) in which amendments were incorporated. The amended plots received 100 metric tons/ha manure or 100 metric tons/ha manure plus 20 metric tons/ha soil. Control plots were not amended. The amendments were added to the plots with a manure spreader and incorporated with a tractor-pulled rototiller. The manure was incorporated into the top 12 cm, after which the soil was incorporated into the top 2.5 cm. The EB+sludge and EB+fly required numerous passes before mixing was uniform.

A Truax grass drill was modified to side-band liquid 10-34-0 fertilizer at a rate of 33 kg N/ha and 112 kg P/ha while simultaneously planting sorghum sudan *(Sorghum vulgare sudanese),* soybean *(Glycine max),* and a mixture of soybean plus sorghum sudan. At the time of planting, each plot was fertilized with 80 kg P/ha and 20 kg N/ha.

The experimental design was a split-split plot. The whole plots were ash mixtures, the first split was amendment (none, manure, manure plus soil), and the second split was crop species (sorghum-sudan, soybean, sorghum-sudan plus soybean). The experiment was replicated twice.

## Sampling and Analysis

Original, unmixed ash materials were analyzed for moisture content, particle size distribution, electrical conductivity, mineralogy, and total elemental content. Total contents were determined by HF digestion followed by ICP-AES analysis.

Ash materials were sampled in June 1989, prior to seeding the crops. Duplicate cores to a depth of 15 cm were removed from each plot, air dried, ground to pass a 2 mm sieve, and analyzed at the Colorado State University Soil Testing Laboratory. The samples were extracted using the $NH_4HCO_3$-DTPA soil test[11] and analyzed for a large number of metals by inductively coupled plasma atomic emission spectroscopy.

The plots were planted on June 6, 1989, and harvested after 90 days. Plant leaves were sampled from each plot in September 1989. They were dried at 70°C for 48 hours, ground to pass a 0.5 mm sieve, and sent to Colorado State University Soil Testing Laboratory for digestion and analysis. The samples were digested in 3:1 nitric:perchloric acid and analyzed by inductively coupled plasma atomic emission spectroscopy.

The experimental design and data analysis were established under consultation with the Kansas State University Statistics Department. Data were analyzed using the SAS statistical package.

## RESULTS AND DISCUSSION

### Properties of the Ash

The eastern bottom ash has very large particle size, with only 4% of the particles being smaller than sand (Table 1). This analysis is consistent with the gravelly appearance and the very low moisture contents at all water potentials except saturation. The pH of 8.2 is relatively moderate, and the electrical conductivity is typical for soil materials. The eastern scrubber sludge is a powdery material when dry and is dominated by silt-sized particles. The material has a very high moisture content at saturation (0 MPa moisture tension) and at field moisture (0.03 MPa) but low moisture at the wilting point (1.5 MPa). The high pH (8.9) reflects the high content of calcium oxides, and the sludge is fairly saline with an electrical conductivity of 2.8 dS/m. Western bottom ash has a fairly coarse texture, although not as coarse as eastern bottom ash. Some of the coarse particles are the result of cementing action of the free oxides. The western bottom ash has relatively high moisture contents at all potentials and is extremely alkaline (pH 11.2) with moderate salinity (E.C. = 2.9 dS/m). Of all the unmixed ash materials, the western bottom ash is the only one which has the appearance and texture of soil. The western fly ash is a fine powder and subject to blowing when dry. The typical physical measurements are meaningless for this material because of its extreme tendency to form cement when wet. Thus, measurement of moisture contents was impossible because the ash formed hardened disks while on the moisture plates. During the particle size analysis, the fly ash again cemented to form large particles which were incorrectly determined to be sand sized. The pH is unusually high (12.3) and the electrical conductivity is high enough to be detrimental to most plants.

As demonstrated by the total analysis (Table 2), the ash materials had high concentrations of B, Cd, Cu, Na, Pb, Sr, and Zn. Total analyses of soils are generally poor indicators of plant-availability, and the same is expected of these ash materials. Physical, thermodynamic, and kinetic considerations will often assume a greater role in determining plant availability than total analysis. For example, the 500 mg P/kg in the fly ash and 250 mg P/kg in the western bottom ash do not ensure that these materials will have greater P availability than the eastern ash products with less than 40 mg P/kg. In fact, the very high pH and Ca oxide content of the western ashes greatly reduce the availability of P, and the eastern ashes can support plant growth with minimal P fertilization. Some trace elements are volatilized during the combustion of the coal only to be condensed and precipitated in a thin film on the surfaces of the ash.[2] These elements have higher solubility and plant-availability than their total contents might indicate because a significant fraction of these elements is physically situated for ready dissolution.

### Vegetative Yield

The vegetative yields (dry-weight basis) of the three crops (soybean, sorghum, sorghum+soybean) for the three ash mixtures were quite different, but all yields

**Table 1**
**Selected Properties of the Ash Materials Used in This Study**

| | | Moisture Content | | | Particle Size (%) | | | | |
| | | 0 mPa | 0.03 mPa | 1.5 mPa | Sand | Silt | Clay | E.C.[a] (dS/m) | Predominant Minerals (X-ray) |
| Ash Type | pH | | | | | | | | |
|---|---|---|---|---|---|---|---|---|---|
| Eastern bottom | 8.2 | 39 | <1 | <1 | 96 | 2 | 2 | 0.9 | X-ray amorphous |
| Western bottom | 11.2 | 35 | 21 | 8 | 64 | 18 | 18 | 2.9 | Gypsum, calcite, quartz |
| Eastern sludge | 8.9 | 31 | 19 | 6 | 8 | 28 | 64 | 2.8 | Gypsum, quartz, hannebachite, calcite |
| Western fly ash | 12.3 | 55 | ND | ND | 68 | 16 | 16 | 8.3 | Quartz, lime |

*Note:* ND = not determined.

[a] E.C. = electrical conductivity.

**Table 2**
**Total Concentrations of Selected Elements in the Unweathered, Unmixed Ash Materials**

| Ash Type | HF Digest (%) | | | | | | | | | |
|---|---|---|---|---|---|---|---|---|---|---|
| | C | Ca | Mg | Na | K | P | Al | Fe | Mn | Ti |
| Eastern bottom | 0.0 | 8 | 0.66 | 0.43 | 1.67 | 0.04 | 6.1 | 11 | 0.064 | 0.37 |
| Western fly ash | 0.3 | 18 | 2.21 | 1.25 | 0.29 | 0.50 | 7.3 | 3 | 0.048 | 0.82 |
| Western bottom | 0.7 | 12 | 1.70 | 0.72 | 0.52 | 0.25 | 6.3 | 5 | 0.056 | 0.66 |
| Eastern sludge | 1.1 | 23 | 0.23 | 0.10 | 0.37 | 0.03 | 0.9 | 2 | 0.036 | 0.07 |

| | HF Digest (mg/kg) | | | | | | | | | | |
|---|---|---|---|---|---|---|---|---|---|---|---|
| | Cu | Zn | Ni | Mo | Cd | Cr | Sr | B | Ba | Pb | V |
| Eastern bottom | 66 | 714 | 81 | 14 | 5 | 91 | 331 | 985 | 300 | 70 | 128 |
| Western fly ash | 276 | 362 | 52 | 17 | 8 | 74 | 3480 | 2143 | 49 | 109 | 251 |
| Western bottom | 132 | 302 | 61 | 8 | 6 | 107 | 2092 | 1330 | 132 | 63 | 154 |
| Eastern sludge | 47 | 3010 | 34 | 11 | 35 | 24 | 587 | 48 | 48 | 341 | 32 |

increased with manure applications (Figure 1). Within cropping system, the sorghum and sorghum+soybean had the highest yield and soybean had the lowest. The soybean plants were quite sensitive to the concentrations of B in all the ash materials and tended to lose more than 90% of their leaves. Despite the difference in yield, the response to treatment (manure or manure+soil) was similar for all cropping systems and all ash mixtures. The yields in the unamended control plots were approximately ten times less than in the amended plots. Within the control plots, the yields were not significantly different, but EB+sludge had slightly greater dry matter than the other materials. The addition of manure and manure+soil significantly increased yields for all ash materials and cropping systems.

The ash materials were initially devoid of organics. (The C content listed in Table 2 was inorganic carbon as carbonate or uncombusted C.) Without organic matter in the ash, it was difficult for the plants to establish a normal rhizosphere due to low microbial activity and limited nutrient cycling. Root growth was limited which restricted water and nutrient uptake. The addition of manure improved the rooting environment by adding the critical organic matter. In addition, the microbial decomposition of the manure also radically decreased the pH of the western materials (Figure 2). This was accomplished largely by the generation of $CO_2$ which reacted with the free oxides in the ash and formed $CaCO_3$. The $CaCO_3$ buffered the ash at a much lower pH than was found in its absence. Changing pH from above 9 to below 9 has the potential of increasing the uptake of P[30] and decreasing B uptake,[31] but the response of other elements will depend upon the controlling chemical reactions, and these will be discussed in subsequent sections.

## Availability Indices and Plant Uptake

The paramount objective of this study was to determine the relationship between soil test level and plant concentration of a wide array of elements in the ash mixtures. The only extractant used in this study was $NH_4HCO_3$-DTPA

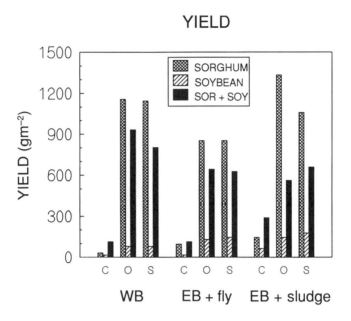

**FIGURE 1.** Dry-matter yields for sorghum, soybean, and the mixture sorghum+soybean as influenced by ash mixture and amendments. Legend for amendments: C = unamended control, O = composted manure, S = composted manure+soil.

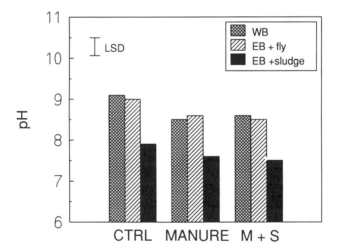

**FIGURE 2.** The effects of amendment on the pH of the ash materials. Legend for amendments: CTRL = unamended control, MANURE = amended with 5% composted manure, M + S = manure plus 1% soil; LSD = least significant difference ($p<0.05$).

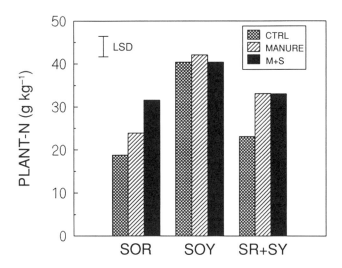

**FIGURE 3.** Plant concentrations of N as affected by amendments. Legend for plant species: SOR = sorghum; SOY = soybean; SR+SY = sorghum+soybean; LSD = least significant difference *(p*<0.05).

because it was developed for high-pH media. It has been tested in soils as an index of availability for several nutrients, but there is limited information about its ability to predict plant-availability of potential toxins. The availability indices could be used as predictors of concentrations in the plant material because ash was sampled in June and plants were sampled in September.

*Nitrogen*

The most significant effect for N concentration in the plants was the interaction between crop and amendment (Figure 3). Soybean in all treatments had the highest N concentrations with little response to amendment. Despite the chronic B toxicity and low yields, the soybeans were nodulated and apparently were fixing considerable amounts of $N_2$. Sorghum plants in the unamended plots had the lowest N concentrations but had a significant, positive response to manure and manure+soil. The N contents in the plants from the sorghum+soybean plots were midway between the individual crops and showed a response to manure. The response to manure applications was undoubtedly due to the N content in the manure and the increased cycling of nutrients as a result of increased microbial activity. The combination of greater yield and greater N concentrations in amended plots resulted in ten-times-greater uptake of N in the amended plots (22 g N/m²) compared to the control plots (2.6 g N/m²). Availability indices were not determined for N, and correlations were not possible.

Table 3
Concentrations (mg/kg) of Elements Extractable with NH₄HCO₃DTPA for Which There
Was No Statistical Interaction Between Ash Mixture and Amendment

| Amendment | K | Na | Cu | B | Ni |
|---|---|---|---|---|---|
| None | 87 | 11 | 4.6 | 22 | 0.9 |
| Manure | 334 | 21 | 5.2 | 10 | 1.5 |
| Manure+soil | 356 | 19 | 4.5 | 9 | 1.4 |
| LSD[a] | 58 | 3 | 0.6 | 4 | 0.7 |

*Note:* Concentrations were averaged over plant species and ash mixtures.

[a] LSD = least significant difference (*p* <0.05).

Table 4
Plant Concentrations of Elements for Which There Was No Statistical Interaction
Between Ash Mixture and Amendment

| | | | Concentration (mg/kg) | | |
|---|---|---|---|---|---|
| Amendment | K (%) | Na (%) | Cu | B | Ni |
| None | 2.23 | 0.05 | 10.44 | 382 | 4.33 |
| Manure | 3.79 | 0.02 | 7.66 | 265 | 3.32 |
| Manure+soil | 3.67 | 0.02 | 9.36 | 274 | 3.21 |
| LSD[a] | 1.40 | 0.03 | 2.18 | 23 | 1.12 |

*Note:* Concentrations were averaged over plant species and ash mixtures.

[a] LSD = least significant difference (*p* <0.05).

## *Potassium*

There were no interactions between concentrations (in either ash or plant) and
ash materials, so K concentrations were averaged over ash materials and only the
effects of amendments are presented. There was a highly significant response of
the extractable K to amendment (Table 3). In the unamended plots, the extractable
K was less than the "critical level" (the extractable concentration below which a
response is expected to fertilization) of 80 mg K/kg, and the addition of a source
of available K should have resulted in an increase in soil-test K. Indeed, the
extractable concentrations increased more than fivefold following manure applications, and plant concentrations reflected the increased availability (Table 4).
The plants in the unamended plots had an average concentration of 22 g/kg and
the mean plant concentrations in the amended plots were nearly 40 g/kg. All plant
concentrations were within the range considered sufficient for plants. Thus,
despite the fact that the soil-test concentrations were below the critical level and
the plants responded to additions of K, the plants were not K-deficient.

The correlation of extractable K with plant concentrations was highly significant (Table 5, Figure 4). When all cropping systems were considered, the r² was
0.50 (n=18, P<0.01). However, the correlations for the individual systems were

**Table 5**
**Coefficients of Determination (r²) for Plant Concentrations Against NH₄HCO₃-DTPA Extractable Concentrations**

| Element | Total | Sorghum | Soybean | Sor+Soy |
|---------|-------|---------|---------|---------|
| P | 0.61[c] | 0.94[c] | 0.37[a] | 0.62[b] |
| K | 0.50[c] | 0.68[c] | 0.87[c] | 0.32NS |
| Mg | 0.20[b] | 0.31NS | 0.07NS | 0.60[b] |
| Zn | 0.71[c] | 0.78[c] | 0.81[c] | 0.56[b] |
| Cu | 0.37[c] | 0.40[a] | 0.42[a] | 0.74[c] |
| Mo | 0.15[b] | 0.56[b] | 0.17NS | 0.08NS |
| Cd | 0.52[c] | 0.68[c] | 0.87[c] | 0.82[c] |
| B | 0.48[c] | 0.88[c] | 0.22NS | 0.87[c] |
| Pb | 0.44[c] | 0.99[b] | 0.81[b] | — |

*Note:* NS = not significant.

[a]  $p < 0.10$
[b]  $p < 0.05$
[c]  $p < 0.01$

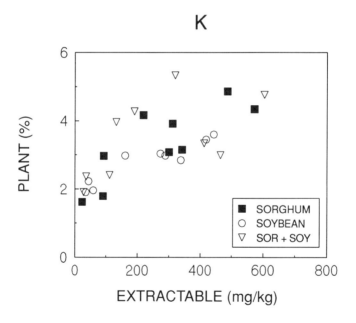

**FIGURE 4.**  The relationships between plant concentrations of K and concentrations of K extracted from the ash. (Correlation coefficients are given in Table 5.)

different. The correlations for sorghum ($r^2 = 0.68$) and soybean ($r^2 = 0.87$) were highly significant but were not for the sorghum+soybean intercropping. Apparently, the relationship between uptake by plants and extractability from the ash was different enough that, when the plants were combined in the sorghum+soybean plots, the relationship was no longer significant. However, it seems likely that the

$NH_4HCO_3$-DTPA extractant can be used as a predictor of plant-available K in coal ash.

## Phosphorus

Due to the significant interaction between ash mixture and amendments, the means of P concentrations as influenced by both factors will be discussed. The extractable P in all control plots was relatively low, averaging 15 to 28 mg P/kg (Table 6). These concentrations were greater than the critical level (8 mg/kg) for agronomic crops,[11] and that is not surprising considering the rate at which these plots were fertilized. None of the plants displayed P stress, but severe P deficiency was observed previously when other ash plots were unfertilized. Amending the ash mixtures with manure or manure+soil did not produce a significant increase in extractable P except for the EB+sludge. The pH in this mixture was low enough to prevent the rapid fixation of the fertilizer P observed in the mixtures with higher pH.

Concentrations of P in plants were significantly greater in amended plots than in control plots only for EB+sludge (Table 7). The plant-P concentrations in the amended WB and EB+fly were statistically equivalent to concentrations in the control plots. Assuming that the $NH_4HCO_3$-DTPA concentrations are an indication of plant availability in coal ash, this was the trend expected. The extractable P and plant P were highly correlated (Table 5, Figure 5), especially for the sorghum ($r^2 = 0.94$). The correlations were significant for the sorghum+soybean and combined systems as well. Although the $NH_4HCO_3$-DTPA is useful for predicting uptake of P by these plants, we did not test the ability of the extractant to predict P deficiency. This would require a different approach than that used in this study.

## Magnesium

There was no effect of amendment on the extractable concentrations of Mg within ash mixtures, but there were large differences between mixtures (Table 6). The WB had extractable concentrations in excess of 200 mg Mg/kg, EB+fly averaged approximately 150 mg/kg, and EB+sludge averaged less than 50 mg/kg. Plant concentrations were well within normal ranges for these plants[32] and did not show a significant response to amendment. Similarly, the plant concentrations of Mg were only weakly correlated with extractable Mg (Table 5, Figure 6). This lack of correlation is not of great concern because Mg deficiencies of plants growing in revegetated materials are rare, and the plants in this study did not exhibit symptoms indicating Mg stress.

## Sodium

Sodium concentrations in ash or plant material are important to examine because of the potential of Na toxicity in plants and the detrimental effect that

**Table 6**
**Concentrations NH$_4$HCO$_3$-DTPA Extractable Elements for Those Which Show a Significant Statistical Interaction Between Ash Mixture and Amendment**

| Ash Mixture | Amendment | P (%) | Mg (%) | Concentration (mg/kg) | | | | | | |
|---|---|---|---|---|---|---|---|---|---|---|
| | | | | Mn | Pb | Cd | Mo | Zn | Cr | Ba |
| Western bottom | None | 15 | 231 | 5.5 | 1.78 | 0.31 | 0.24 | 4.1 | 0.59 | 0.66 |
| | Manure | 43 | 247 | 8.2 | 1.29 | 0.24 | 0.16 | 7.7 | 0.22 | 0.85 |
| | Manure+soil | 39 | 246 | 8.8 | 1.37 | 0.27 | 0.17 | 7.7 | 0.24 | 0.78 |
| EB+fly | None | 24 | 104 | 2.9 | 0.93 | 0.06 | 0.11 | 2.3 | 0.18 | 0.95 |
| | Manure | 61 | 165 | 5.9 | 0.85 | 0.08 | 0.10 | 6.6 | 0.14 | 0.56 |
| | Manure+soil | 70 | 133 | 7.9 | 0.77 | 0.08 | 0.09 | 7.6 | 0.14 | 0.49 |
| EB+sludge | None | 26 | 18 | 7.0 | 3.58 | 0.71 | 0.12 | 13.9 | 0.08 | 0.33 |
| | Manure | 154 | 58 | 19.4 | 4.29 | 0.99 | 0.24 | 24.4 | 0.13 | 0.33 |
| | Manure+soil | 126 | 55 | 18.5 | 3.89 | 1.00 | 0.19 | 18.2 | 0.12 | 0.32 |
| LSD[a] | | 47 | 42 | 4.4 | 1.40 | 0.21 | 0.08 | 3.3 | 0.08 | 0.19 |

*Note:* Values were averaged over plant species.

[a] LSD = least significant difference ($p < 0.05$).

**Table 7**
**Plant Concentrations of Elements Which Statistically Showed an Interaction Between Ash Mixture and Amendment**

| Ash Mixture | Amendment | P (%) | Mg (%) | Zn | Mn | Mo | Cd | Cr | Ba | As | Se | Pb |
|---|---|---|---|---|---|---|---|---|---|---|---|---|
| | | | | | | | | | Concentration (mg/kg) | | | |
| Western bottom | None | 0.29 | 0.39 | 20.0 | 19.4 | 9.17 | 0.33 | 3.36 | 37.58 | 0.30 | 0.12 | 1.25 |
| | Manure | 0.25 | 0.46 | 43.8 | 28.4 | 7.17 | 0.67 | 2.38 | 22.89 | 0.23 | 0.06 | 1.25 |
| | Manure+soil | 0.23 | 0.45 | 36.1 | 26.8 | 7.75 | 0.67 | 2.58 | 34.67 | 0.28 | 0.07 | 1.00 |
| EB+fly | None | 0.26 | 0.39 | 27.3 | 29.3 | 6.67 | 0.17 | 2.53 | 19.92 | 0.20 | 1.37 | 0.83 |
| | Manure | 0.29 | 0.47 | 39.2 | 19.5 | 7.17 | 0.58 | 2.93 | 41.58 | 0.17 | 0.43 | 1.83 |
| | Manure+soil | 0.30 | 0.44 | 45.3 | 22.3 | 6.50 | 0.42 | 2.82 | 45.17 | 0.19 | 0.36 | 3.50 |
| EB+sludge | None | 0.21 | 0.31 | 54.5 | 30.8 | 5.92 | 4.92 | 3.01 | 16.83 | 2.44 | 0.73 | 0.25 |
| | Manure | 0.47 | 0.41 | 63.5 | 28.3 | 7.84 | 3.69 | 2.86 | 9.64 | 2.67 | 0.32 | 0.67 |
| | Manure+soil | 0.44 | 0.39 | 69.7 | 22.8 | 9.21 | 3.58 | 3.18 | 4.33 | 2.17 | 0.60 | 0.67 |
| LSD[a] | | 0.11 | 0.17 | 14.3 | 13.1 | 2.25 | 1.22 | 0.77 | 12.0 | 0.45 | 0.36 | 1.33 |

*Note:* Concentrations were averaged over plant species.

[a] LSD = least significant difference (*p* <0.05).

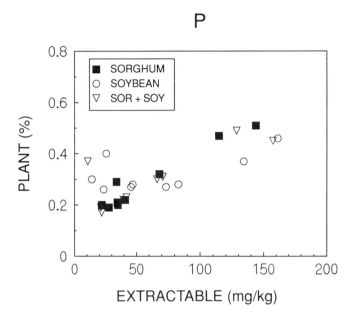

**FIGURE 5.** Uptake of P by plants as a function of P extracted from ash. (Correlation coefficients are given in Table 5.)

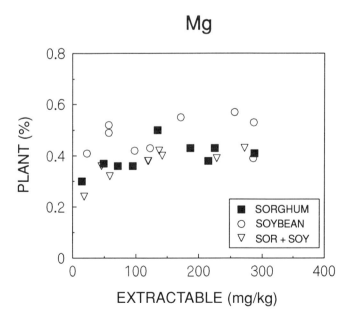

**FIGURE 6.** The relationships between plant concentrations of Mg and concentrations of Mg extracted from the ash. (Correlation coefficients are given in Table 5.)

excess Na can have on the physical properties of soils. If Na on the exchange sites exceeds 10 to 15% of the cation exchange capacity, then dispersion of the colloids is possible and water infiltration can be severely restricted. The total sodium contents of the western ash materials were high (>0.7%), indicating that Na could be a problem in these materials (Table 2).

The Na concentrations in the coal extracts ranged from 10 mg Na/kg in the controls to 21 mg/kg in the plots amended with manure. As with Mg, the $NH_4HCO_3$-DTPA extractant has not been tested for Na so it is not possible to compare these concentrations to other values. Nevertheless, the Na concentrations increased upon addition of manure. This probably occurred because $CaCO_3$ precipitated as the pH dropped in response to manure amendment, and competition by Ca with Na in the extractant decreased.

The concentrations of Na in the plant material were greater in the control plots than in the amended plots (Table 3). Although the mechanism by which manure additions decreased Na availability is not readily apparent, manure additions had a positive effect on the Na content of the plants. The $NH_4HCO_3$-DTPA extractant was not useful for Na as evidenced by the lack of correlation of extractable Na with plant Na for all species (data not shown).

## Boron

Perhaps the most important element in the revegetation of coal ash is boron. The very high total concentrations (Table 2) combined with the soluble nature of the element presents great risk for B toxicity in the ash. Boron toxicity is recognized as one of the greatest impediments to revegetation of many waste materials.[6,13,17,19] Boron availability to plants is usually evaluated by solubility in hot water,[18,19] but $NH_4HCO_3$-DTPA may be a suitable extractant because the $HCO_3^-$ is an effective agent in anion or ligand exchange.

The $NH_4HCO_3$-DTPA extractable B concentrations in the ash mixtures ranged from 9 mg/kg in the amended plots to 22 mg/kg in the unamended plots (Table 3). Generally, hot-water-soluble B concentrations are less than 2 mg B/kg, with concentrations of 1 mg/kg corresponding to 60 to 100 mg B/kg in plant tissue. If the $NH_4HCO_3$-DTPA extracts B from the same pool as hot water, it would be expected that the plant concentrations would be extremely high. Indeed, the mean B concentrations in the plants ranged from 270 mg/kg in the amended plots to nearly 400 mg/kg in the controls (Table 4). The soybean plants in the control plots were highly stressed and showed classic B toxicity symptoms. The soybean in the amended plots were also stressed, but not to the degree of the control plots. In the most stressed plants, the margins of immature trifoliates were necrotic, and the tissue continued to die until the entire leaf was dead within 2 weeks after trifoliate development. At the end of the growing season, there were few leaves on the stalks of the soybean plants. In contrast, the sorghum plants showed very little stress, and the leaves did not display toxicity symptoms any time during the growing season.

Boron concentrations in all cropping systems except soybean were highly correlated with extractable concentrations. The high correlation (Table 5) for all

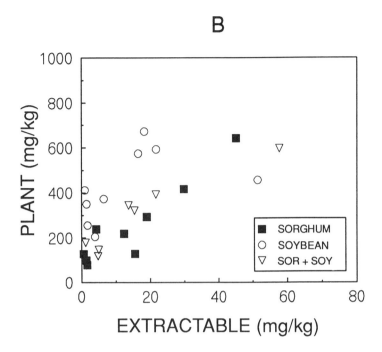

**FIGURE 7.**  Uptake of B by plants as a function of B extracted from ash. (Correlation coefficients are given in Table 5.)

plants ($r^2 = 0.48$) was largely the result of the high correlations for sorghum ($r^2 = 0.87$) and sorghum+soybean ($r^2 = 0.87$) overcoming the nonsignificant correlation of the soybean ($r^2 = 0.22$). All plots with sorghum or sorghum+soybean (Figure 7) displayed a similar relationship between extractable B and plant concentrations, but some outlying data points markedly reduced the correlation for soybean. The $NH_4HCO_3$-DTPA extractant is highly suited for predicting B concentrations in the sorghum plants and might be useful in healthy soybean. Also, the sorghum and soybean have much different tolerances for B because the sorghum appeared healthy with the same B concentrations that induced severe necrosis in soybean.

Addition amendments to the ash reduced the concentration of B in the plant tissue. The combination of amendments, leaching, and passage of time will ameliorate the B toxicity which was so prevalent during the first year. Previous studies indicate that the B toxicity will decrease 2 to 3 years after establishment.[17]

## Copper and Mo

The total concentration of Cu in the ash mixtures (Table 2) was much greater than normal background levels of 10 to 40 mg Cu/kg.[24] The total concentrations in the western ashes were especially high with concentrations of 132 mg Cu/kg for western bottom ash and 276 mg Cu/kg for fly ash. The high total contents were

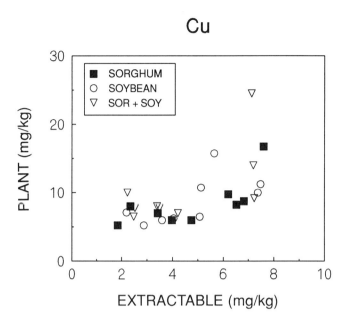

**FIGURE 8.** The relationship between concentrations of Cu extractable from the ash and plant concentrations of Cu. (Linear correlations are given in Table 5, but notice the nonlinear trends for all plants.)

reflected to a small degree by the $NH_4HCO_3$-DTPA extractable Cu concentrations (4 to 5 mg/kg) (Table 3) that were about twice the concentrations seen in soils.[11,12] However, the concentration of Cu in the plants (Table 4) were within the normal range[24,32] and typical of plants growing in soil amended with fly ash.[13,20] Copper concentrations in neither the ash nor in the plants were significantly affected by amendment or ash material.

Linear coefficients of determination for extractable Cu vs. Cu concentrations in the plants (Table 5) were significant for the combination of all plants and the sorghum+soybean. The individual sorghum and soybean correlations were significant only at $p<0.10$. The relationship is not linear but is curvilinear (Figure 8). The Cu concentrations in the plants remained fairly constant over the $NH_4HCO_3$-DTPA extractable range from 2 to 5 mg/kg, beyond which the plant concentrations increase dramatically. Both higher[12,14] and lower[25] correlations have been observed between DTPA extractable and plant Cu, so the usefulness of the soil test in predicting the behavior of Cu in coal ash is unclear.

Molybdenum in ash and plants is important not just to the growth of the vegetation but also to grazing ruminants. High concentrations of Mo can induce Cu deficiencies in the animals. Total Mo in the ash materials was ten times the normal range in soils.[24] The solubility of Mo usually decreases with decreasing pH; thus, we expected that the additions of amendments would decrease the

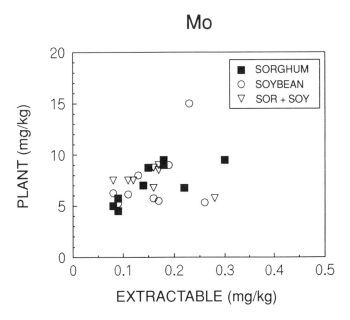

**FIGURE 9.**  The relationships between plant concentrations of Mo and concentrations of Mo extracted from the ash.

extractable Mo levels. However, there were no significant decreases in extractable Mo resulting from manure additions (Table 6). The observed concentrations were higher than average soil levels,[24] but much larger $NH_4HCO_3$-DTPA concentrations were observed in retorted oil shale[33] and in soil amended with extreme concentrations of inorganic Mo or Mo in sludge.[15,16] Typical Mo concentrations are less than 1 mg/kg.

Plant concentrations of Mo (Table 7) were ten times higher than the 0.4 mg Mo/kg reported to be typical.[32] In order to avoid molybdenosis in grazing ruminants, concentrations in the plant should be lower than 5 mg/kg.[34–36] This critical concentration was exceeded in all but one plant sample in this study. A more sensitive parameter in predicting potential molybdenosis for ruminants is the Cu to Mo ratio.[37] A ratio of 6:1 is considered to be safe, and 2:1 will likely induce molybdenosis.[38] This 2:1 ratio was exceeded only in one sample, further indicating that coal ash at this time was unsuitable for supporting forages safe for grazing ruminants. However, the high availability of Mo may decrease with time and make the forage safer after several years.

The $NH_4HCO_3$-DTPA extractant proved to be a poor indicator of Mo availability in these ash materials (Figure 9). Although the coefficients of determination for the sorghum and the combination of all species were significant $(p<0.05)$, the $r^2$ values were not large and were not significant for the soybean and sorghum+soybean plots (Table 5). Pierzynski and Jacobs[15,16] found a strong correlation between

$NH_4HCO_3$-DTPA extractable Mo and plant concentrations. In their study, very high amounts of Mo were added as treatments, and the concentrations of Mo in plants and soil were very high. In untreated plots, soil and plant Mo concentrations were low; thus, the correlation included very high and very low values which strongly influenced the magnitude of $r^2$. The $NH_4HCO_3$-DTPA extractant needs to be tested on more materials with reasonable ranges of Mo before it is possible to judge whether or not it is suitable to evaluate the status of Mo.

### Zinc

The total concentrations of Zn in the ash mixtures were high, especially for the EB+sludge. Soil materials are usually ten times less than these concentrations.[24] The $NH_4HCO_3$-DTPA extractable Zn was higher than for soils but comparable to other waste materials.[12,20,29,30] In this study, the greatest concentrations of extractable Zn were observed in the sludge samples (Table 6). This would be expected from the very high total content and the low pH of the mixture. The solubility of Zn increases with decreasing pH, so greater extractable Zn in the sludge was anticipated. For EB+fly and EB+sludge, the addition of manure caused a significant increase in extractable Zn. This was likely due to two effects of the manure: (1) the complexation of Zn by organic constituents in the manure that made Zn more soluble and more available for extraction by $NH_4HCO_3$-DTPA, and (2) the manure invoked a decrease in pH for all mixtures causing Zn to be more soluble. The extractable concentrations were considerably higher than those observed in a similar study.[12]

The plant concentrations (Table 7) followed trends similar to the ash, but the differences between the mixtures were not as great and there was more variability within replicates. The EB+sludge again had the highest concentrations. Normal concentrations of plant Zn are 6 to 40 mg/kg,[32] and concentrations greater than 200 mg/kg are usually toxic to the plants. Zinc concentrations in all plants in the EB+sludge exceeded 40 mg/kg but were less than 80 mg/kg. In amended plots, Zn concentrations in plants were near 40 mg/kg. Similar concentrations for plants growing in waste materials were observed in other published studies.[12,20,39,41]

The correlation between extractable and plant Zn was significant for both plant species and their combinations (Table 5, Figure 10). Despite the fact that the soil test was developed for alkaline soils (not ash) to determine deficiency levels (not potential toxicity), the correlations are very encouraging, and this has been observed by others.[12,14,40] The correlation is strong for all plants, and the relationship between extractable and plant concentrations is similar for all plants; therefore, the $NH_4HCO_3$-DTPA extractant could be used in a predictive sense for Zn in waste products.

### Lead and Cd

For both elements, there was a statistical interaction between ash mixture and amendment. Amendment had no effect on the $NH_4HCO_3$-DTPA extractable Pb, but there were large differences between ash mixtures (Table 6). The EB+sludge

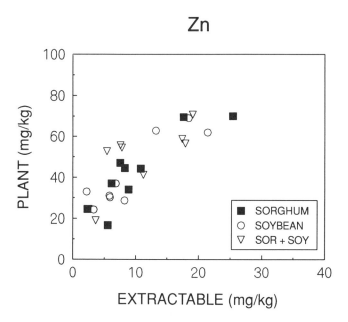

**FIGURE 10.** Uptake of Zn by plants as a function of Zn extracted from ash. (Correlation coefficients are given in Table 5.)

had the highest Pb (ca. 4 mg/kg), and WB and EB+fly had lower concentrations (ca. 1 mg/kg). There was no apparent relationship between the concentrations of Pb in the plants and the amendment (Table 7). The highest concentration was observed in the EB+sludge amended with manure+sludge, but the other treatments had statistically equivalent concentrations. There was a significant negative relationship between plant Pb and extractable Pb (Table 5, Figure 11). The reason for this trend is not known, but poor reliability of soil tests in predicting plant Pb has been observed previously.[42] More of the aspects of the Pb and Cd behavior in the ash were published previously.[42]

Although there was not a clear effect of amendment on extractable and plant Cd, there were marked differences between ash mixtures. Extractable Cd was greatest in EB+sludge with a slight increase in the amended treatments (Table 6). The lowest concentrations were observed in the WB and EB+fly and there was little change in concentrations from amendment. The trends for plant Cd (Table 7) were similar to those for extractable Cd. Plant and extractable Cd were closely correlated for the individual crops and in combination (Table 5, Figure 12). It is unclear why the correlation of Cd would be so much better than the correlation of Pb. The chemistry of the elements is similar in the ash materials with a strong relationship between metal activity and $CO_3^{2-}$ activity.[42] Therefore, the action of the $NH_4HCO_3$-DTPA extractant on the elements should be similar. The difference must lie with the uptake (or exclusion) mechanism for the elements.[43] The uptake of Cd by plants has been shown to be passive,[28] and Cd is readily accumulated.

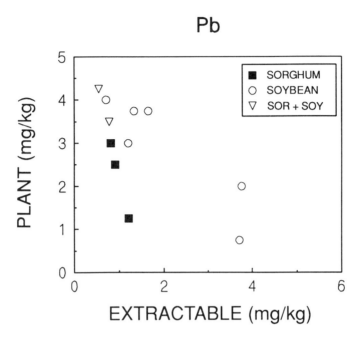

**FIGURE 11.** The negative relationship between plant concentrations of Pb and Pb concentrations extracted from the ash.

The accumulation of Pb through the roots is not well defined. Extremely large increases in soluble Pb had little effect on Pb in plant.[44] High concentrations of Fe or $PO_4$ can reduce Pb uptake.

## Manganese

The total Mn concentrations in the ash materials (Table 2) were similar to those in soil.[24] The extractable concentrations in the unamended control plots (Table 6) are marginally low, but typical for ash materials.[12,39,40] Although all ash mixtures showed at least some increase in extractable Mn in response to amendment addition, the response was significant only in the EB+sludge. As with some of the other metals showing this response, the increase could be due to the decreased pH or organic complexation of Mn. However, seeking a mechanism for the increase in the case of Mn is not a fruitful exercise because the plant concentrations did not show a similar increase with manure additions (Table 7). All plant concentrations are low[32] and are lower than those normally observed in fly ash-amended soils.[13,30]

Correlation between $NH_4HCO_3$-DTPA extractable Mn and plant concentrations was nonsignificant with $r^2 < 0.1$. This is probably due to the narrow range in plant Mn concentrations, and the poor relation between the soil test plant uptake of Mn. Lewis[12] found a strong relationship between DTPA-extractable and plant concentrations of Mn (r=0.97), but the high correlation is misleading. In his experiments, most of the data clustered near 0 mg/kg DTPA-Mn with plant Mn

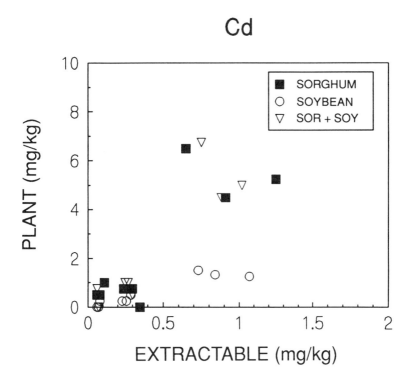

**FIGURE 12.** Uptake of Cd by plants as a function of Cd extracted from ash.

ranging from 50 to 125 mg Mn/kg. There was another small cluster of points with DTPA-Mn of 8 to 10 mg/kg and plant Mn 250 to 300 mg/kg. The result, therefore, was essentially a two-point correlation. Therefore, the $NH_4HCO_3$-DTPA extractant is not suitable for predicting availability of Mn in coal ash.

*Barium, Ni, and Cr*

These three elements have not been studied in the context of correlation between extractable and plant concentrations in coal ash. $NH_4HCO_3$-DTPA extractable Ba showed little response to amendment in the EB+sludge and WB (Table 6), but the concentrations in the EB+fly significantly decreased with amendment. Plant concentrations did not display a consistent response to amendment (Table 7). The Ba concentrations in plants in the EB+fly increased with amendment, decreased in EB+sludge, and did not respond to amendment in WB. The observed concentrations appear to be typical for plants growing in amended soil.[24] When plant Ba was regressed against extractable Ba, the overall regression was not significant for all samples, nor was it significant for the individual cropping systems. There was a significant, negative correlation for plants in WB and EB+fly (r = –0.88) and no correlation for the plants in EB-SLUDGE. It is not clear if the significant correlation for WB and EB+sludge was due to chance; therefore, the $NH_4HCO_3$-DTPA extractant needs further testing as a predictor of Ba availability.

Extractable Cr significantly decreased with the addition of manure to WB, but otherwise there was no response to amendments (Table 6). Likewise, plant concentrations did not respond to amendment and were essentially constant across all treatments (Table 7). The Cr contents of the plants are well within safe limits.[24,25] Correlation between plant Cr and extractable Cr was nonsignificant.

For Ni, there were no significant interactions of amendment with ash mixture. Therefore, Ni concentrations were averaged over ash mixture. There were no significant effects of amendment on extractable Ni (Table 3) or plant concentrations (Table 4). Although the Ni concentrations are fairly high,[39] they were far less than the toxic level of 26 mg Ni/kg.[25] DTPA-Ni was found by Adams et al.[14] to correlate with plant Ni, but there was no correlation in this study.

## Arsenic and Se

Although plant As and Se were measured in the plants, their extractable concentrations were not determined in the ash materials. Arsenic concentrations in the plants were highly dependent upon ash mixture but were unaffected by amendment (Table 7). The highest As concentrations were observed in the EB+sludge (2.67 mg/kg), exceeding the concentrations in the WB and EB+fly by five times. The concentration of As in plants generally ranges from 0.1 to 1 mg/kg,[25] but the toxicity level is approximately 20 mg/kg. The reason for the high availability of As in the EB+sludge is not clear. If it was strictly because the pH of the EB+sludge was lower than the other ash mixtures, then amendment should have changed the uptake because the manure additions significantly dropped the pH. Obviously, there was no effect of organic complexation. It could be that the EB+sludge contained greater quantities of labile As adsorbed on the surface of the solids.

Selenium in plants was affected by both ash mixture and amendment (Table 7). Plants from unamended EB+fly had the greatest concentrations of Se, but these concentrations were cut in half after manure amendment. The Se contents of plants in the other ash mixtures also decreased upon amending with manure or manure+soil, but the changes were not statistically significant. The Se content of these plants was higher than the normal range of 0.05 to 0.2 mg/kg, but far below the phytotoxicity limit of 30 mg/kg and the 5 mg/kg necessary to produce toxicity in grazing animals. For many anions, solubility and availability increases with increasing pH. This appears to be the case for Se because the availability decreases in each ash mixture upon amendment. It is expected, therefore, that the Se concentrations would continue to decrease as time passes and the pH of the ash mixtures continues to decline.

## SUMMARY AND CONCLUSIONS

The plant concentrations were well within the acceptable range for all elements except B and Mo. Although K was predicted to be deficient by the

$NH_4HCO_3$-DTPA soil test and the plants responded to K additions, the concentrations in the plants were not in the deficient range. The B concentrations, however, were quite high and were toxic to the soybean plants. Although the sorghum did not show visible signs of toxicity, the very high concentrations (>100 mg B/kg) were enough to suppress yields. Additions of manure and manure+soil decreased the mean concentrations of B in the plants. Copper and Mo concentrations in range plants are often important because high Mo or a low Cu to Mo ratio can induce copper deficiencies in grazing ruminants. The Cu contents of the plants were normal, but the Mo concentrations exceeded the critical level of 5 mg/kg. Also, the critical Cu to Mo ratio of 2:1 was exceeded in only one sample, indicating potential danger for grazing ruminants. The Cu to Mo ratio did not show any trends with amendments. Molybdenosis can be corrected by dietary supplement, but caution needs to be observed in the use of these ashes for revegetation, and Cu to Mo needs to be monitored. The addition of a cover of topsoil would probably not alleviate this problem, as indicated in previous studies.[33]

The $NH_4HCO_3$-DTPA extractable concentrations of P, K, Mg, Zn, Cu, Mo, Cd, B, and Pb were significantly correlated with plant concentrations. This is especially encouraging for the elements B, Cu, and Mo which, as mentioned above, were the only elements with potentially toxic concentrations. Therefore, if one merely monitors the $NH_4HCO_3$-DTPA extractable concentrations of these elements, the trends in plant concentrations may be predicted. The $NH_4HCO_3$-DTPA extractant was originally designed for P, K, Zn, and Cu, so a strong correlation with these elements was anticipated. However, the high coefficient of determination for Pb ($r^2$ = 0.44) was a negative correlation (r = –0.66), and the $NH_4HCO_3$-DTPA extractant will not be useful for this element. The extractant was not tested for As and Se.

## ACKNOWLEDGMENT

This research was supported in part by a grant from Kansas City Power and Light Corporation, contribution No. 91-354-B of the Kansas Agricultural Experiment Station.

## REFERENCES

1. Rehage, J. A. and Holcombe, L. J., EPRI EN-6532 Project 2796-1, Electric Power Research Institute, Palo Alto, CA, 1990.
2. Linton, R. W., Williams, P., Evans, C. A., Jr., and Natusch, D. F. S., Determination of the surface predominance of toxic elements in airborne particles by ion microprobe mass spectrometry and auger electron spectrometry, *Anal. Chem.*, 49, 1515, 1977.
3. Smith, M. A. and Harris, M. R., The environmental implications of the disposal and utilisation of fly ash, *M & Q Environment*, 1, 10, 1983.
4. Wysocki, W. and Hill, R. D., Reclamation of alkaline ash piles, *EPA/600/S2-86/049*, 1986.

5. Schivley, W. W., Fly ash revegetation - a case history in reclaiming Eastern and Western fly ash, in *Coal Technol. '83*, Houston, 1983

6. Adriano, D. C., Page, A. L., Elseewi, A. A., Chang, A. C., and Straughan, I., Utilization and disposal of fly ash and other coal residues in terrestrial ecosystems: a review, *J. Environ. Qual.*, 9, 333, 1980.

7. Howard, G. S., Schuman, G. E., and Rauzi, F., Growth of selected plants on Wyoming surface-mined soils and fly ash, *J. Range Manage.*, 30, 306, 1977.

8. Petrikova, V., The effect of semi-liquid manure on electricity power station ash-dump reclamation, *Agric. Waste*, 2, 37, 1980.

9. Hodgson, D. R. and Townsend, W. N., The amelioration and revegetation of pulverized fuel ash, in *The Ecology and Reclamation of Devastated Land, Part II*, Jutnik, R. J. and Davis, G., Eds., Gordon and Breach, London, 1980.

10. Lindsay, W. L. and Norvell, W. A., Development of a DTPA test for zinc, iron, manganese, and copper, *Soil Sci. Soc. Am. J.*, 42, 421, 1978.

11. Soltanpour, P. N. and Schwab, A. P., A new soil test for the extraction of micro- and macronutrients in alkaline soils, *Commun. Soil Sci. Plant Anal.*, 8, 195, 1977.

12. Lewis, B. G., Extractable trace elements and sodium in Illinois coal-cleaning wastes: correlation with concentrations in tall fescue, *Reclam. Reveg. Res.*, 2, 55, 1983.

13. Keefer, R. F. and Singh, R. N., Fly ash/soil mixtures plus additives for field corn production, in *Proc. 7th Int. Ash Util. Symp. Expo., Vol. 1*, DOE/METC-85/6018, 1985, 94.

14. Adams, T. McM., McGrath, S. P., and Sanders J. R., The effect of soil pH on solubilities and uptake into ryegrass of zinc, copper, nickel added to soils in sewage sludges, in *Proc. Int. Conf. Heavy Metals Environ.*, Lekkas, T. D., Ed., Athens, Greece, 1985, 484.

15. Pierzynski, G. M. and Jacobs, L. W., Extractability and plant availability of molybdenum from inorganic and sewage sludge sources. *J. Environ. Qual.*, 15, 323, 1986.

16. Pierzynski, G. M. and Jacobs, L. W., Molybdenum accumulation by corn and soybeans from a molybdenum-rich sewage sludge, *J. Environ. Qual.*, 15, 394, 1986.

17. Ransome, L. S. and Dowdy, R. H., Soybean growth and boron distribution in a sandy soil amended with scrubber sludge, *J. Environ. Qual.*, 16, 171, 1987.

18. Aitken, R. L., Jeffrey, A. J., and Compton, B. L, Evaluation of selected extractants for boron in some Queensland soils, *Aust. J. Soil Res.*, 25, 263, 1987.

19. Plank, C. O. and Martens, D. C., Boron availability as influenced by application of fly ash to soil, *Soil Sci. Soc. Am. Proc.*, 38, 974, 1974.

20. Mishra, L. C. and Shukla, K. N., Elemental composition of corn and soybean grown on fly ash amended soil, *Environ. Pollut. (Ser. B)*, 12, 313, 1986.

21. Xian, X., Response of kidney bean to concentration and chemical form of cadmium, zinc, and lead in polluted soils, *Environ. Pollut.*, 57, 127, 1989.

22. Bollag, J. M. and Czaban, J., Effect of microorganisms on extractability of Cd from soil with NaOH and DTPA, *J. Soil Sci.*, 40, 451, 1989.

23. van Erp, P. J. and van Lune, P., A new method for determining the relation between soil- and plant-cadmium, *Plant Soil*, 116, 119, 1989.

24. Roffman, H. K., Kary, R. E., and Hudgins, T., Ecological distribution of trace elements emitted by coal-burning power generating units employing scrubbers and electrostatic precipitators, in *Proc. 4th Symp. Coal Util.*, 1977, 192.

25. Jastrow, J. D., Zimmerman, C. A., Dvorak, A. J., and Hinchman, R. R., Plant growth and trace-element uptake on acidic coal refuse amended with lime or fly ash, *J. Environ. Qual.*, 10, 154, 1981.

26. Mattigod, S. V., Rai, D., Eary, L. E., and Ainsworth, C. C., Geochemical factors controlling the mobilization of inorganic constituents from fossil fuel combustion residues: I. Review of the major elements, *J. Environ. Qual.*, 19, 188, 1990.

27. Eary, L. E., Rai, D., Mattigod, S. V., and Ainsworth, C. C., Geochemical factors controlling the mobilization of inorganic constituents from fossil fuel combustion residues: II. Review of the minor elements, *J. Environ. Qual.*, 19, 202, 1990.

28. Baker, D. E., Rasmussen, D. S., and Kotuby, J., Trace metal interactions affecting soil loading capacities for cadmium, in *Hazardous and Industrial Waste Management and Testing: Third Symposium, ASTM STP 8851*, Jackson, L. P., Rohlik, A. R., and Conway, R. A., Eds., American Society for Testing and Materials, Philadelphia, 1984, 118.

29. Schwab, A. P. and Lindsay, W. L., Computer simulation of Fe(III) and Fe(II) complexation in limited nutrient solution. II. Application to growing plants, *Soil Sci. Soc. Am. J.*, 53, 34, 1989.

30. Lindsay, W. L., *Chemical Equilibria in Soils*, Wiley Interscience, New York, 1979.

31. Sposito, G., *The Chemistry of Soils*, Oxford University Press, New York, 1989, 160.

32. Chapman, H. D., Ed., *Diagnostic Criteria for Plants and Soils*, University of California Division of Agricultural Science, Riverside, CA, 1965.

33. Schwab, A. P., Lindsay, W. L., and Smith, P. J., Elemental contents of plants growing on soil-covered retorted shale, *J. Environ. Qual.*, 12, 301, 1983.

34. Webb, J. S. and Aktinson, W. J., Regional geochemical reconnaissance applied to some agricultural problems in Co. Limerick, Eire, *Nature*, 208, 1056, 1965.

35. Alloway, B. J., Copper and molybdenum in swayback pastures, *J. Agric. Sci.*, 80, 521, 1973.

36. Thornton, I., Biogeochemical studies on molybdenum in United Kingdom, in *Proc. Symp. Molybdenum in the Environment*, Vol. 2, Chappell, W. R. and Peterson, K., Eds., Marcell Dekker, New York, 1977, 341

37. Erdman, J. A., Ebens, R. J., and Case, A. A., Molybdenosis: a potential problem in ruminants grazing on coal mine spoils, *J. Range Manage.*, 31, 34, 1978.

38. Dollahite, J. W., Rowe, L. D., Cook, M., Hightower, D., Bailey, E. M., and Kyzar, J. R., Copper deficiency and molybdenosis intoxication associated with grazing near a uranium mine, *Southwest. Vet.*, 19, 47, 1972.

39. Severson, R. C. and Gough, L. P., Rehabilitation materials from surface-coal mines in Western U.S.A. III. Relations between elements in mine soil and uptake by plants, *Reclam. Reveg. Res.*, 3, 185, 1984.

40. Gajbhiye, K. S., Goswami, N. N., Banerjee, N. K., Rajat, D., and Singh, R. K., Evaluation of a common extractant for estimating available Fe, Mn, Zn and Cu in soil, *J. Indian Soc. Soil Sci.*, 32, 309, 1984.

41. Keefer, R. F., Singh, R. N., Bennett, O. L., and Horvath, D. J., Chemical composition of plants and soils from revegetated mine soils, in *Symp. Surface Mining, Hydrol., Sedimentol., Reclam.*, 1983, 155.

42. Schwab, A. P., Tomecek, M. B., and Ohlenbusch, P. D., Plant availability of lead, cadmium, and boron in amended coal ash, *Water Air Soil Poll.*, 57-58, 297, 1990.

43. Eriksson, J. E., The influence of pH, soil type and time on adsorption and uptake by plants of Cd added to the soil, *Water Air Soil Poll.*, 48, 317, 1989.

44. Lagerwerff, J. V., Uptake of cadmium, lead, and zinc by radish from soil and air, *Soil Sci.*, 111, 129, 1971.

# 10

# Uptake of Chemical Elements by Terrestrial Plants Growing on a Coal Fly Ash Landfill

L. H. Weinstein,[1] M. A. Arthur,[1] R. E. Schneider,[1] P. B. Woodbury,[1] J. A. Laurence,[1] A. O. Beers,[2] and G. Rubin[1]

[1]Cornell University, Ithaca, NY
[2]N.Y. State Electric and Gas Corp., Binghamton, NY

## ABSTRACT

Disposal of coal fly ash in soil-capped landfills has raised questions regarding the presence of potentially toxic elements in the ash and their future disposition in the local ecosystem. Experiments were carried out on the uptake of chemical elements by cultivated and indigenous plants grown on a coal fly ash landfill and on control sites. Selenium was found to concentrate in amounts more than 50 times higher in landfill-grown plants than in control plants, and was the element of greatest interest and concern. Other elements, such as Mo and B, were often higher in the tissues of control plants, but the degree of uptake was usually less than two-fold.

Rutabaga *(Brassica napus* L., Cruciferae), birdsfoot trefoil *(Lotus corniculatus* L., Leguminoseae), and alfalfa *(Medicago sativa* L., Leguminoseae) absorbed more Se than other species such as carrot *(Daucus carota* L., Umbelliferae), corn *(Zea mays* L., Graminae), timothy *(Phleum pratense* L., Graminae), bromegrass *(Bromus inermis* Leyss, Graminae), red clover *(Trifolium pratense* L., Leguminoseae), or milkweed *(Asclepias syriaca* L., Asclepiadaceae). Cauline leaves of wild carrot and bitterweed *(Picris hieracioides L.,* Compositeae) contained more Se than older rosette leaves. In sweet and field corn cultivars, greater amounts of Se were found in leaves and kernels than in stems, cobs, or roots.

Coal flue gas desulfurization will result in the production of $CaSO_3$ as a by-product, from which gypsum $(CaSO_4)$ can be made easily. Research by others indicated that application of gypsum to Se-containing soils often resulted in reduced Se uptake in plants. In the studies described here, application of gypsum at the rate of 2 metric tons/ha reduced the uptake of Se in both rutabaga and carrot shoots and roots. We conclude that gypsum application offers a possible management tool to control uptake and cycling of Se through plants to other biota.

## INTRODUCTION

In 1988, electric utilities generated about 80 million metric tons of coal ash,[1] of which about 62% was fly ash, 23% was bottom ash and boiler slag, and 15% was flue gas desulfurization (FGD) sludge. Annual production of coal ash is projected to increase to about 120 million metric tons by the year 2000. The proportion of ash in coal ranges from 3 to 30%, with an industrywide mean for electric utility power generating plants of 10.1%.[2] Of the total amount of ash produced, 20 to 25% is used in cement products, road bases, and asphalt mixes; some of the bottom ash and slag is used for blasting grit and roofing granules.[1,2] The remaining 75 to 80% is disposed of in landfills,[3] creating a potential waste disposal problem, both in the volume of the waste, but also in its unknown effects on plants and the local ecosystem. Coal contains every naturally occurring chemical element, so it is not surprising that many of these elements are found also in fly ash.[4] Fly ash produced from coal in central New York State contained significant amounts of Se, As, B, Cu, Cr, Pb, Mo, and Zn.[5–12]

Many of the elements found in coal ash are essential to plants and animals, including some macroelements (e.g., K, P, Mg, Ca) and trace elements (e.g., Fe, Mn, B, Mo, Cu, Zn). Selenium is commonly found in coal ash and is essential to animals. However, the range between beneficiality and toxicity is relatively narrow. Much of the literature on the effects of coal ash on plant and animal life and on Se uptake and accumulation in plants has been reviewed recently.[13]

Little information is available concerning the uptake and movement of potentially toxic elements from coal ash into the ecosystem, or concerning methods of amelioration. Plants grown in soils amended with fly ashes from various sources are known to accumulate a number of elements, of which As, Mo, and Se have been reported to reach levels potentially toxic to grazing animals in some areas.[9–16] Hurd-Karrer[17] reported more than 50 years ago that plants belonging to the Cruciferae (e.g., rutabaga) were accumulators of Se, while members of the Umbelliferae (e.g., carrot) were not. Hurd-Karrer,[18] Mikkelsen et al.,[19] Elrashidi et al.,[20] and others have reported that there is an antagonistic relationship between S and Se in plant nutrition, and that under many, but not all, conditions, the presence of S reduces uptake of Se by plants. The lack of information on some aspects of Se in plants provided the impetus for studies by the Boyce Thompson Institute for Plant Research, the Ecosystems Research Center, and the Biometrics Unit, all located at Cornell University, at a soil-capped landfill in central New York State. This report summarizes a part of our field research program. The purposes of the research were (1) to study uptake of Se and other elements present in the landfill in several species of indigenous and cultivated plants representing several plant families, and (2) to determine the efficacy and value of added S from gypsum applications to the landfill in reducing Se uptake by plants. A parallel study was carried out by the Department of Natural Resources which investigated the uptake of Se and other elements in insects, fish, frogs, birds, rooted aquatic vegetation, and small mammals found on the landfill. Results of these studies were reported elsewhere.[14]

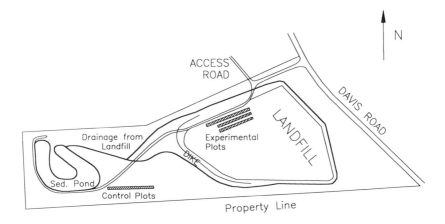

DAVIS ROAD FLY ASH LANDFILL

**FIGURE 1.**   General plan of the Davis Road fly ash landfill in Lansing, NY.

## MATERIALS AND METHODS

Preliminary research during the first year of study (1987) included a survey of four soil-capped fly ash landfill sites in central New York state.[16] During the second and third years, research was confined to the Davis Road landfill site in Lansing, NY (Figure 1). The landfilled part of the site covers an area of about 2 ha. A typical soil-capped fly ash landfill used in the northeastern U.S. is diagrammed in Figure 2. It consists of a layer of about 10 to 12 m of fly ash, underlain by a plastic sheet or a layer of compacted clay, surrounded and held in place by a dike of soil, and capped with about 0.7 m of compacted soil. A pasture mix is then sown over the area. Drainage from the landfill is conducted to a sedimentation pond. When completed, it has not been the policy to use it for agricultural activities.

Ash used in the construction of the landfill came from the Milliken Station in Lansing, NY and is relatively high in Se, As, Cu, and Cr content.[6,13] Construction of the landfill began in 1976 and was completed in 1978; thus, the landfill was about 9 to 11 years old during the 3-year period of research.

Field plots — arranged in three parallel rows, with the most northerly row containing 15 plots, and the other two each having 9 plots — were established on the landfill and the Lansing control areas. The $2 \times 5$ m plots were oriented in an east-west direction (Figure 1). Plants grown in plots received fertilizer additions and general care as in normal agronomic practices. Indigenous plots were selected randomly on the landfill, although differences in distribution of plant species tended to concentrate most of the plots toward the west end of the landfill. Gypsum was applied to plots containing rutabaga and carrot plants at a rate of 2

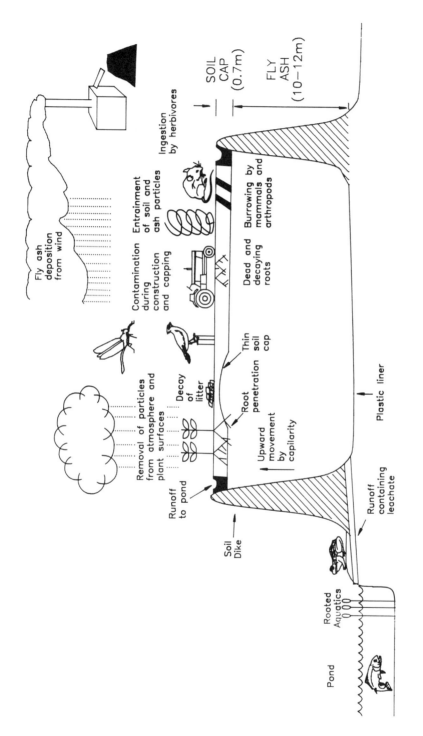

**FIGURE 2.**   Schematic of typical soil-capped fly ash landfill in northeast U.S.

metric tons/ha (ca. 5 tons/acre) as a side dressing, after which it was watered in thoroughly.

Aboveground tissues of alfalfa, clover, and timothy for the first cutting, of alfalfa and clover for the second cutting, and of alfalfa for the third cutting were analyzed for Se and other elements. There was inadequate regrowth of timothy for the second cutting, and of clover and timothy for the third cutting. Soil samples were removed with a soil auger at depths of 0 to 15 cm and 30 to 45 cm from three randomly selected locations in each plot. Samples of indigenous and cultivated plants were harvested from the landfill and control areas. One control was located adjacent to the landfill (Lansing control); a second was established at the Boyce Thompson Institute farm in Ithaca, NY (Ithaca control). Plant parts such as leaves, roots, kernels, cobs, and stems were dried at 313°K (40°C) to reduce loss of Se from plant tissues by volatilization, and ground to pass a 40-mesh sieve. Soils were dried in the same manner. Aliquots of plant tissues and of soils were analyzed chemically for Se by the diaminonaphthalene fluorometric method,[21] and plant tissues were analyzed also for other elements by use of an inductively coupled argon plasma spectrometer (Jarrell Ash Model 975) equipped with a fixed cross-flow nebulizer. The whole organism or selected tissues such as leaves, stems, or roots of plants were used. Samples were dry ashed at 973°K (450°C) for 6-hr, then digested with hydrogen peroxide and ashed at 973°K (450°C) for 2-hr.

Cultivated crops were established on landfill and control sites in equal-sized plots, arranged in a block design, with three blocks per site. Statistical analyses, therefore, were based on a split-unit experimental design in which the block was the experimental unit for the site, while the individual plot was the experimental unit for the crop. Effects of site were tested using the variation in mean response among the blocks at each site as the error term (i.e., block-within-site). Effects of crop, and interactions between site and crops, were tested with the variation among the blocks in the difference in response between crops (i.e., error term). In a split-unit design, whole plot means, and thus effects, are based on averages over the subunit factor. Therefore, site effects were based on differences between blocks in the mean response of the crops.

The indigenous vegetation study had a factorial design where the experimental unit for both site and species (the factors) was a sample of a given species at a particular location on the landfill or on a control site. All effects were tested with the same error term, since the experimental unit was the same size for both factors.

Generally, interactions were evaluated at a conservative $\alpha$ level (0.1) while main effects and contrasts were evaluated at an $\alpha$ level of 0.05. For some elements, the differences between the mean values for a crop grown on the landfill or nonlandfill sites were so large that the values would normally be significantly different. These differences were due to the great plot-to-plot variability in elemental content in replicate samples of plants grown on the landfill. Data manipulation and statistical analyses were carried out using the SAS system, Version 5.

We determined the Se, K, P, Ca, Mg, Mn, Fe, Cu, B, Zn, Mo, and Al concentrations in plant tissues. Other elements, such as Co, Cd, Cr, Ni, Pb, S, Si, V, and As were below the level of detection or of accurate analysis.

## RESULTS

### Soils

On the landfill, the mean content of Se in soils from the 0 to 15 cm profile was significantly greater (0.7431±0.033 mg/kg; $p = 0.0012$) than the content of Se at 30 to 45 cm (0.560±0.033 mg/kg). The first five plots at the easterly end of the cropped area had the lowest Se contents, but produced crops that contained the highest concentration of Se in their tissues. Statistical analysis of soil from all plots showed that there was no significant systematic relationship in the Se content in the plot areas as a whole (data not shown). Soil in all plots had more than 0.4 mg/kg Se; four plots had more than 0.9 mg/kg.

At the Lansing control site adjacent to the Davis Road landfill, the Se concentration in soils at both depths was similar (0.459±0.012 mg/kg at 0 to 15 cm; 0.464±0.012 mg/kg at 30 to 45 cm). These values were significantly higher at this control site than at the Ithaca control site at both depths (0.357±0.12 mg/kg at 0 to 15 cm; 0.285±0.012 mg/kg at 30 to 45 cm). At the Lansing control site there was a systematic gradient of decreasing Se concentrations in soils extending away from the fly ash landfill (Figure 3), and there was a small, but statistically significant, correlation between soil Se and tissue Se in alfalfa. At the Ithaca control site, the Se concentration was greater in the 0 to 15 cm than in the 30 to 45 cm layer $(p = 0.0001)$, but there were no significant differences between plots throughout the cropped area (not shown).

### Forage Crops

#### First Cutting (Alfalfa, Clover, Timothy)

The crop (alfalfa, clover, timothy) and site (control, landfill) interactions significantly affected the following elements: Mg, Mn, Fe, B, Zn, and Mo. The crop affected K, Ca, Cu, and Al concentrations. Timothy and clover differed significantly in their concentrations of K, Ca, Cu, and Al, while alfalfa and clover differed significantly in their concentrations of Cu. Statistically significant differences resulted between the sites in forage tissue concentrations of P, Mg, Fe, B, Cu, and Mo (Table 1), but not for Se because of the high plot-to-plot variability. At the control site, the partial correlation between the soil and tissue concentrations of Se ($r = 0.84$; $p = 0.0002$) was positive and significant. After accounting for the variation among crops and blocks, a positive relationship between the Se concentration in the soil and in the plant tissue was found.

#### Second Cutting (Alfalfa, Clover)

The crop and site interaction significantly affected the B concentration; the concentration of B was greater in alfalfa grown on the landfill, whereas it was nearly equal in both crops on the control site. Clover had significantly greater concentrations than alfalfa for K, P, Mg, Mn, Cu, and Zn. The concentrations of

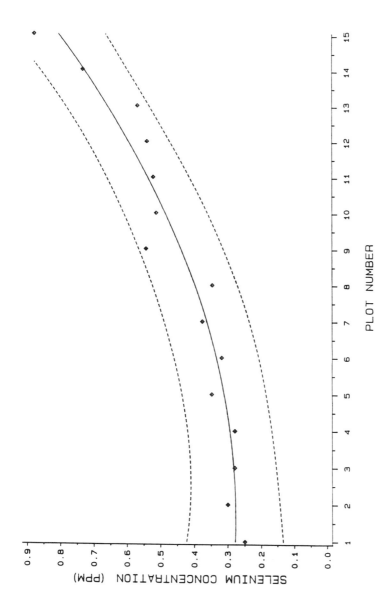

**FIGURE 3.** Concentration of total Se in samples of soil obtained from contiguous plots along a transect in the control area toward the fly ash landfill site. The solid line represents the regression of concentration of total Se in soil on distance along a transect using a second-degree polynomial function. Dashed lines represent 95% confidence intervals.

**Table 1**
**Mean Concentrations of Elements in Aboveground Tissues of Alfalfa, Clover and Timothy from the Fly Ash Landfill and Lansing Control Sites for the First Cutting**

| Treatment | Dry Weight (mg/kg) | | | | | | | |
|---|---|---|---|---|---|---|---|---|
| | Se | Mn | Fe | Cu | B | Zn | Mo | Al |
| **Control** | | | | | | | | |
| Alfalfa | 0.05 (0.89) | 23.48 (2.29) | 62.00 (2.85) | 8.52 (0.35) | 30.77 (0.98) | 21.67 (1.02) | 3.33 (0.88) | 25.02 (3.07) |
| Clover | 0.04 (0.68) | 30.22 (1.73) | 54.93 (2.16) | 10.91 (0.26) | 28.61 (0.74) | 22.93 (0.77) | 3.37 (0.66) | 24.80 (2.32) |
| Timothy | 0.03 (0.68) | 49.22 (1.73) | 28.58 (2.16) | 4.93 (0.26) | 4.89 (0.74) | 17.24 (0.77) | 1.14 (0.66) | 7.50 (2.32) |
| **Landfill** | | | | | | | | |
| Alfalfa | 2.33 (0.68) | 23.23 (1.73) | 75.37 (2.16) | 9.62 (0.26) | 39.75 (0.74) | 23.73 (0.73) | 7.71 (0.66) | 33.56 (2.32) |
| Clover | 0.27 (0.68) | 34.98 (1.73) | 63.05 (2.16) | 11.96 (0.26) | 31.14 (0.74) | 21.98 (0.77) | 4.53 (0.66) | 23.80 (2.32) |
| Timothy | 0.13 (0.68) | 36.47 (1.73) | 27.24 (2.16) | 5.02 (0.26) | 5.84 (0.74) | 20.53 (0.77) | 1.56 (0.66) | 8.27 (2.32) |

| | Dry Weight (%) | | | |
|---|---|---|---|---|
| | K | P | Ca | Mg |
| **Control** | | | | |
| Alfalfa | 2.12 (0.10) | 0.21 (0.01) | 1.71 (0.06) | 0.24 (0.01) |
| Clover | 2.20 (0.08) | 0.17 (0.01) | 1.47 (0.04) | 0.36 (0.01) |
| Timothy | 2.58 (0.08) | 0.20 (0.01) | 0.30 (0.04) | 0.09 (0.01) |
| **Landfill** | | | | |
| Alfalfa | 2.19 (0.08) | 0.27 (0.01) | 1.55 (0.04) | 0.25 (0.01) |
| Clover | 2.32 (0.08) | 0.25 (0.01) | 1.57 (0.04) | 0.42 (0.01) |
| Timothy | 2.62 (0.08) | 0.24 (0.01) | 0.30 (0.04) | 0.12 (0.01) |

*Note:* Standard errors are given in parentheses.

P, Mn, B, and Mo were greater in tissues grown on the landfill, whereas the concentrations of K and Zn were greater in tissues from plants grown on the control site. There were no statistically significant effects for Se, Fe, and Al concentrations, although the Se concentration was marginally greater statistically in the plants grown on the landfill than in those grown on the control site $(p = 0.14)$ (Table 2).

## Third Cutting (Alfalfa)

Plants grown on the landfill contained more P, Mn, and Mo in the alfalfa tissue than from the Lansing control. The plot-to-plot variability in Se content precluded statistical significance $(p = 0.14)$ (Table 3).

# Root Crops

## Early Summer Harvest

*Shoots:* The concentrations of K and P were nearly equal in the shoots of carrot and rutabaga on the landfill site; on the Ithaca control site, the concentration of K was significantly greater in carrot than in rutabaga, and the concentration of P was significantly greater in rutabaga than in carrot. The concentration of Mg was nearly equal in the shoots of both crops on the Ithaca control site; on the landfill, the concentration of Mg was greater in carrot than in rutabaga. Copper concentrations in rutabaga were nearly equal for both sites and were much lower than in carrots; carrot shoots from the landfill were higher than from the control site in Cu concentration, but carrot shoots are Cu accumulators. Carrot had significantly higher concentrations of B than rutabaga, but not for Ca, Se, K, Mg, Cu, and Mo.

The concentrations of Se, Ca, Mg, Cu, and Mo were significantly higher in shoots of plants grown on the landfill, whereas the concentration of K was significantly higher in shoots of plants grown on the Ithaca control site. Manganese, Fe, Al, and Zn were not significantly different between sites (Table 4).

*Roots:* The concentrations of K were nearly equal in the roots of both crops on the Ithaca control and landfill sites. The concentrations of Se, B, Mg, Mo, and Zn were nearly equal in the roots of both crops at the Ithaca control site; on the landfill, however, the concentrations of Mg, B, and Zn were higher in carrot than in rutabaga and the concentrations of Se and Mo were higher in rutabaga than in carrot. The concentration of P was marginally higher in rutabaga roots than in carrot roots for the Ithaca control site, but was higher in carrot roots than in rutabaga roots for the landfill. Carrot roots had significantly greater concentrations of Fe and Al than rutabaga roots; rutabaga roots had significantly higher concentrations of Ca than carrot roots (Table 5).

The concentrations of Se, Mg, Cu, B, and Mo were significantly greater in roots of plants from the landfill; the opposite was true for K and P in roots of plants grown at the control site. The pattern of differences between sites in concentrations of Se, K, Mg, Cu, and Mo (Table 4) in roots was similar to that for the shoots (Table 5).

**Table 2**
**Mean Concentrations of Elements in Aboveground Tissues of Alfalfa and Clover from the Fly Ash Landfill and Lansing Control Sites for the Second Cutting**

| Treatment | Dry Weight (mg/kg) | | | | | | | |
| --- | --- | --- | --- | --- | --- | --- | --- | --- |
| | Se | Mn | Fe | Cu | B | Zn | Mo | Al |
| Lansing control | | | | | | | | |
| Alfalfa | 0.04 (1.03) | 21.25 (1.86) | 56.74 (4.08) | 8.82 (0.43) | 24.37 (1.01) | 23.48 (1.71) | 1.96 (1.58) | 18.37 (2.41) |
| Clover | 0.03 (0.73) | 22.6 (1.31) | 49.96 (2.88) | 12.61 (0.31) | 23.73 (0.71) | 26.24 (1.21) | 2.02 (1.12) | 15.28 (1.71) |
| Landfill | | | | | | | | |
| Alfalfa | 2.81 (0.73) | 23.7 (1.31) | 49.06 (2.88) | 8.81 (0.31) | 34.34 (0.71) | 17.37 (1.21) | 7.08 (1.12) | 16.49 (1.71) |
| Clover | 0.37 (0.73) | 31.65 (1.31) | 53.32 (2.88) | 12.02 (0.31) | 27.98 (0.71) | 23.55 (1.21) | 4.33 (1.12) | 14.49 (1.71) |

| Treatment | Dry Weight (%) | | | |
| --- | --- | --- | --- | --- |
| | K | P | Ca | Mg |
| Lansing control | | | | |
| Alfalfa | 1.79 (0.04) | 0.17 (0.01) | 1.71 (0.08) | 0.27 (0.02) |
| Clover | 1.99 (0.03) | 0.19 (0.01) | 1.38 (0.06) | 0.39 (0.01) |
| Landfill | | | | |
| Alfalfa | 1.52 (0.03) | 0.22 (0.01) | 1.63 (0.06) | 0.22 (0.01) |
| Clover | 1.81 (0.03) | 0.25 (0.01) | 1.39 (0.06) | 0.39 (0.01) |

*Note*: Standard errors are given in parentheses.

**Table 3**
**Mean Concentrations of Elements in Aboveground Tissues of Alfalfa from the Fly Ash Landfill and Lansing Control Sites for the Third Cutting**

| Treatment | Dry Weight (mg/kg) | | | | | | | |
|---|---|---|---|---|---|---|---|---|
| | Se | Mn | Fe | Cu | B | Zn | Mo | Al |
| Lansing control | 0.07 (0.00) | 24.75 (1.54) | 72.23 (3.81) | 10.41 (0.67) | 28.93 (4.86) | 20.63 (1.05) | 1.36 (0.04) | 28.31 (1.72) |
| Landfill | 3.93 (2.30)[a] | 37.66 (2.48) | 64.92 (2.16) | 10.65 (0.42) | 51.09 (5.53) | 18.33 (0.67) | 8.90 (1.72) | 20.72 (1.08) |

| | Dry Weight (%) | | | |
|---|---|---|---|---|
| | K | P | Ca | Mg |
| Lansing control | 1.90 (0.03) | 0.21 (0.01) | 2.35 (0.07) | 0.32 (0.01) |
| Landfill | 2.04 (0.04) | 0.26 (0.01) | 2.12 (0.05) | 0.30 (0.02) |

*Note*: Standard errors are given in parentheses.

[a] Not significantly different from control because of high plot-to-plot variability ($p=0.14$).

**Table 4**
**Mean Concentrations of Elements in Shoots of Carrot and Rutabaga (Early Summer Harvest) from the Fly Ash Landfill and Ithaca Control Sites**

| Treatment | Dry Weight (mg/kg) | | | | | | | |
|---|---|---|---|---|---|---|---|---|
| | Se | Mn | Fe | Cu | B | Zn | Mo | Al |
| Ithaca control | | | | | | | | |
| Carrot | 0.04 (0.21) | 109.66 (11.01) | 382.46 (161.51) | 11.70 (0.36) | 31.83 (1.17) | 30.68 (1.49) | 0.16 (0.28) | 263.41 (87.71) |
| Rutabaga | 0.04 (0.21) | 80.02 (11.01) | 254.00 (161.51) | 3.82 (0.36) | 25.12 (1.17) | 33.51 (1.49) | 0.35 (0.28) | 161.12 (87.71) |
| Landfill | | | | | | | | |
| Carrot | 1.00 (0.21) | 76.03 (11.01) | 764.74 (161.51) | 14.92 (0.36) | 35.16 (1.17) | 30.62 (1.49) | 0.53 (0.28) | 455.92 (87.71) |
| Rutabaga | 1.72 (0.73) | 65.04 (1.31) | 210.59 (2.88) | 3.70 (0.31) | 28.35 (0.71) | 30.60 (1.21) | 0.64 (1.12) | 126.16 (1.71) |

| Treatment | Dry Weight (%) | | | |
|---|---|---|---|---|
| | K | P | Ca | Mg |
| Ithaca control | | | | |
| Carrot | 6.46 (0.22) | 0.25 (0.01) | 1.84 (0.07) | 0.39 (0.03) |
| Rutabaga | 5.18 (0.22) | 0.38 (0.01) | 3.07 (0.07) | 0.42 (0.03) |
| Landfill | | | | |
| Carrot | 2.87 (0.22) | 0.29 (0.01) | 2.39 (0.07) | 0.88 (0.03) |
| Rutabaga | 3.17 (0.03) | 0.27 (0.01) | 3.51 (0.06) | 0.61 (0.01) |

*Note:* Standard errors are given in parentheses.

**Table 5**
**Mean Concentrations of Elements in Roots of Carrot and Rutabaga (Early Summer Harvest) from the Fly Ash Landfill and Ithaca Control Sites**

| Treatment | Dry Weight (mg/kg) | | | | | | | |
|---|---|---|---|---|---|---|---|---|
| | Se | Mn | Fe | Cu | B | Zn | Mo | Al |
| Ithaca control | | | | | | | | |
| Carrot | 0.02 (0.14) | 25.45 (2.50) | 66.95 (4.49) | 8.42 (0.92) | 26.64 (0.79) | 28.08 (1.49) | 0.21 (0.07) | 43.49 (4.72) |
| Rutabaga | 0.02 (0.14) | 18.88 (2.50) | 38.13 (4.49) | 4.08 (0.92) | 24.15 (0.79) | 23.26 (1.49) | 0.25 (0.07) | 16.06 (4.72) |
| Landfill | | | | | | | | |
| Carrot | 0.42 (0.14) | 17.81 (2.50) | 93.31 (4.49) | 15.79 (0.92) | 32.77 (0.79) | 31.48 (1.49) | 0.39 (0.07) | 53.45 (4.72) |
| Rutabaga | 1.05 (0.14) | 13.19 (2.50) | 46.70 (4.49) | 3.26 (0.92) | 23.56 (0.79) | 17.86 (1.49) | 1.32 (0.07) | 21.94 (4.72) |

| Treatment | Dry Weight (%) | | | |
|---|---|---|---|---|
| | K | P | Ca | Mg |
| Ithaca control | | | | |
| Carrot | 3.86 (0.11) | 0.33 (0.02) | 0.32 (0.02) | 0.18 (0.01) |
| Rutabaga | 3.58 (0.11) | 0.39 (0.02) | 0.40 (0.02) | 0.16 (0.01) |
| Landfill | | | | |
| Carrot | 2.14 (0.11) | 0.37 (0.02) | 0.36 (0.02) | 0.27 (0.01) |
| Rutabaga | 2.48 (0.11) | 0.26 (0.02) | 0.40 (0.02) | 0.18 (0.01) |

*Note:* Standard errors are given in parentheses.

## Late Summer Harvest — Gypsum Experiment

*Shoots:* Concentrations of Se were similar in shoots of rutabaga and carrot from plants grown on the Ithaca control site but the concentration in shoots of rutabaga was much higher than in carrot for plants grown on the landfill (Table 6). The Ca concentration was similar in carrot shoots regardless of the gypsum treatment; whereas, in rutabaga, the Ca concentration was lower when grown without gypsum than with gypsum (Table 7).

*Roots:* The Se concentrations were similar in roots of both species grown on the Ithaca control site, but the roots of rutabaga were higher than those of carrot on the landfill (Table 8). The Al concentrations were similar for rutabaga regardless of the gypsum treatment, whereas the concentration in carrot was higher when grown without gypsum than with gypsum ($11.87 \pm 0.60$ mg/kg vs. $8.87 \pm 0.60$ mg/kg, respectively).

## Gypsum Effects

*Shoots:* The Se concentrations were similar in shoots of plants grown on the Ithaca control site, regardless of the gypsum treatment (Table 7). However, the Se concentration in shoots on the landfill was significantly lower (by about 60%) in plants grown in soil with gypsum. Calcium concentrations in shoots of plants grown on the landfill were similar regardless of the gypsum treatment, but the concentration was higher in shoots of plants with gypsum on the control site.

*Roots:* Although the concentration of Se in roots was lower for plants grown in soil with gypsum than in soil without gypsum by about 65% (Table 7), the statistical evidence for a gypsum effect was relatively weak $(p = 0.1111)$, due to high plot-to-plot variability among plants in Se content.

## Sweet Corn and Field Corn

Elemental contents were determined in leaves, roots, stalks, cobs, and kernels of Golden Cross Bantam sweet corn from a late September harvest and Agway 135 and Halsey field corn from a late October harvest.

## Sweet Corn

In Golden Cross Bantam, there were differences in tissue concentrations of Se, K, P, Ca, Mg, Zn, and Mo in the various tissues studied between the landfill and control sites (data not shown). The concentration of Se was significantly different in leaves and in kernels of Golden Cross Bantam grown on the two sites. There was also evidence that roots, stalks, and cobs of plants grown on the landfill were higher than the control ($p$-values of site comparisons were 0.0731, 0.0845, and 0.0820 for roots, stalks, and cobs, respectively) (Table 9).

**Table 6**
**Mean Concentrations of Elements in Shoots of Carrot and Rutabaga (Late Summer Harvest) from the Fly Ash Landfill and Ithaca Control Sites**

| Treatment | Dry Weight (mg/kg) | | | | | | | |
|---|---|---|---|---|---|---|---|---|
| | Se | Mn | Fe | Cu | B | Zn | Mo | Al |
| Ithaca control | | | | | | | | |
| Carrot | 0.03 (0.72) | 130.03 (6.99) | 162.37 (17.51) | 7.25 (0.11) | 35.91 (1.05) | 24.77 (1.60) | 0.16 (0.28) | 112.94 (8.33) |
| Rutabaga | 0.04 (0.72) | 85.84 (6.99) | 262.56 (17.51) | 2.84 (0.11) | 39.53 (1.05) | 27.20 (1.60) | 0.35 (0.28) | 152.69 (8.33) |
| Landfill | | | | | | | | |
| Carrot | 0.82 (0.72) | 57.71 (6.99) | 267.66 (17.51) | 9.22 (0.11) | 40.14 (1.05) | 21.05 (1.60) | 0.53 (0.28) | 187.06 (8.33) |
| Rutabaga | 4.48 (0.72) | 61.96 (6.99) | 232.34 (17.51) | 2.44 (0.11) | 37.30 (1.05) | 22.41 (1.60) | 0.64 (0.28) | 115.17 (8.33) |

| Treatment | Dry Weight (%) | | | |
|---|---|---|---|---|
| | K | P | Ca | Mg |
| Ithaca control | | | | |
| Carrot | 6.77 (0.17) | 0.21 (0.01) | 2.31 (0.07) | 0.38 (0.03) |
| Rutabaga | 4.44 (0.17) | 0.41 (0.01) | 3.56 (0.07) | 0.36 (0.03) |
| Landfill | | | | |
| Carrot | 3.24 (0.17) | 0.21 (0.01) | 2.75 (0.07) | 0.76 (0.03) |
| Rutabaga | 2.30 (0.17) | 0.29 (0.01) | 3.82 (0.07) | 0.46 (0.03) |

*Note:* Standard errors are given in parentheses.

**Table 7**
**Mean Concentrations of Elements in Shoots and Roots of Root Crops from the Fly Ash
Landfill and Ithaca Control Sites, Grown with and without Application of Gypsum to the
Soil**

| Treatment | Se (Dry Weight, mg/kg) | | | |
|---|---|---|---|---|
| | Shoot | | Root | |
| Ithaca control | | | | |
| Without gypsum | 0.04 | (0.28) | 0.02 | (0.24) |
| With gypsum | 0.04 | (0.08) | 0.01 | (0.24) |
| Landfill | | | | |
| Without gypsum | 3.77 | (0.28) | 1.52 | (0.24) |
| With gypsum | 1.53 | (0.28) | 0.54 | (0.24) |
| | Ca (Dry Weight, %) | | | |
| | Shoot | | Root | |
| Ithaca control | | | | |
| Without gypsum | 2.71 | (0.05) | 0.40 | (0.01) |
| With gypsum | 3.16 | (0.05) | 0.41 | (0.01) |
| Landfill | | | | |
| Without gypsum | 3.29 | (0.05) | 0.35 | (0.01) |
| With gypsum | 3.28 | (0.05) | 0.34 | (0.01) |

*Note:* Standard errors are given in parentheses.

## Field Corn

The Se concentrations in leaves, stalks, cobs, and kernels of field corn varieties and site were statistically significant. The concentrations in these tissues were similar for Halsey and Agway 135 when grown on the control site, but were significantly higher in Agway 135 than Halsey on the landfill. The Se concentrations in roots for the two varieties at either site were not significantly different. In Agway 135, the concentration of Se in all tissues was significantly greater in plants grown on the landfill than in controls (Table 9).

## Indigenous Species

### Bromegrass, Milkweed, and Birdsfoot Trefoil

The Se concentrations in plants on the Lansing control site were similar for all three species. Selenium was higher in each species on the landfill, but differences were statistically significant only for bromegrass and birdsfoot trefoil. On the landfill, Se in birdsfoot trefoil was significantly greater than in bromegrass or milkweed.

The Mo concentration in tissue was significantly greater on the landfill than on the control site for bromegrass and birdsfoot trefoil; the concentration in milkweed was similar for the sites. Birdsfoot trefoil had greater Mo concentrations than the other species when grown on either site.

**Table 8**
**Mean Concentrations of Elements in Roots of Carrots and Rutabaga (Late Summer Harvest) from the Fly Ash Landfill and Ithaca Control Sites**

| | Dry Weight (mg/kg) | | | | | | | | | | | | |
|---|---|---|---|---|---|---|---|---|---|---|---|---|---|
| Treatment | Se | | Mn | | Fe | | Cu | | B | | Zn | | Al | |
| Ithaca control | | | | | | | | | | | | | |
| Carrot | 0.01 | (0.34) | 26.47 | (2.15) | 30.64 | (1.80) | 6.43 | (0.20) | 28.58 | (0.66) | 23.05 | (1.17) | 10.91 | (0.56) |
| Rutabaga | 0.02 | (0.34) | 12.80 | (2.15) | 27.31 | (1.80) | 2.04 | (0.20) | 25.53 | (0.66) | 17.03 | (1.17) | 6.31 | (0.56) |
| Landfill | | | | | | | | | | | | | |
| Carrot | 0.30 | (0.34) | 10.83 | (2.15) | 27.94 | (1.80) | 8.90 | (0.20) | 26.90 | (0.66) | 21.43 | (1.17) | 9.82 | (0.56) |
| Rutabaga | 1.75 | (0.34) | 9.46 | (2.15) | 20.58 | (1.80) | 1.57 | (0.20) | 23.13 | (0.66) | 12.00 | (1.17) | 7.94 | (0.56) |

| | Dry Weight (%) | | | | | | | |
|---|---|---|---|---|---|---|---|---|
| | K | | P | | Ca | | Mg | |
| Ithaca control | | | | | | | | |
| Carrot | 3.93 | (0.12) | 0.30 | (0.01) | 0.36 | (0.01) | 0.14 | (0.00) |
| Rutabaga | 3.18 | (0.12) | 0.37 | (0.01) | 0.45 | (0.01) | 0.15 | (0.00) |
| Landfill | | | | | | | | |
| Carrot | 2.26 | (0.12) | 0.28 | (0.01) | 0.32 | (0.01) | 0.18 | (0.00) |
| Rutabaga | 1.88 | (0.12) | 0.24 | (0.01) | 0.37 | (0.01) | 0.14 | (0.00) |

*Note*: Standard errors are given in parentheses.

**Table 9**
**Mean Concentrations of Selenium (mg/kg) in Tissues of Golden Cross Bantam Sweet Corn (Late September Harvest), and Agway 135 and Halsey Field Corn (Late October Harvest) from the Fly Ash Landfill and Ithaca Control Sites**

| | Ithaca Control | | | | | | Landfill | | | | | |
|---|---|---|---|---|---|---|---|---|---|---|---|---|
| Tissue | Halsey | | Agway 135 | | Golden Cross Bantam | | Halsey | | Agway 135 | | Golden Cross Bantam | |
| Leaves | 0.06 | (0.27) | 0.06 | (0.27) | 0.06 | (0.01) | 0.40 | (0.27) | 1.64 | (0.27) | 0.56 | (0.20) |
| Roots | 0.06 | (0.05) | 0.07 | (0.05) | 0.08 | (0.01) | 0.20 | (0.05) | 0.27 | (0.05) | 0.21 | (0.04) |
| Stalks | 0.01 | (0.10) | 0.01 | (0.10) | 0.01 | (0.00) | 0.11 | (0.10) | 0.57 | (0.10) | 0.17 | (0.09) |
| Cobs | 0.01 | (0.06) | 0.01 | (0.06) | 0.01 | (0.00) | 0.07 | (0.06) | 0.33 | (0.06) | 0.13 | (0.04) |
| Kernels | 0.02 | (0.18) | 0.03 | (0.18) | 0.01 | (0.00) | 0.23 | (0.18) | 1.11 | (0.18) | 0.37 | (0.15) |

*Note*: Standard errors are given in parentheses.

The concentrations of B and Cu were significantly different in the three species from the landfill: birdsfoot trefoil had the highest concentration of B, whereas milkweed had the highest concentration of Cu. Bromegrass had significantly lower B and Cu concentrations than those of the other two species when grown on the control site.

Aluminum was significantly higher in milkweed than in birdsfoot trefoil or in bromegrass grown at either site. The Al concentration in tissue was significantly lower on the landfill than on the control site for milkweed and birdsfoot trefoil; the concentration in bromegrass was similar for both sites.

The Zn concentration in tissue was significantly greater on the landfill for milkweed and was significantly lower on the landfill than on the control site for birdsfoot trefoil; the concentration in bromegrass was similar for both sites. Zinc in milkweed from the landfill was significantly greater than that of the other two species; in bromegrass from the control site, Zn was lower than in the other two species.

Potassium, Ca, Mg, Mn, and Fe concentrations differed among the three species regardless of the site. For milkweed, K, Mn, and Fe were significantly greater on the landfill than on the control site, whereas those of Ca and Mg were significantly lower on the landfill. Potassium, Ca, Mg, Mn, and Fe in bromegrass were not significantly different between sites. Calcium, Mg, Mn, and Fe were not significantly different for birdsfoot trefoil, but K was significantly lower on the landfill than on the control site.

The tissue concentration of P differed significantly between species grown on the landfill, and was greatest in milkweed. On the control site, P was significantly greater in milkweed than in birdsfoot trefoil. Phosphorus was significantly greater on the landfill than on the control site for milkweed and for birdsfoot trefoil, and was significantly lower on the landfill than on the control site for bromegrass (Table 10).

## Bitterweed and Wild Carrot Grown on the Landfill

The wild carrot on the control site lacked cauline leaves, so only the data from the landfill site could be used to test for effects of species and leaf type.

Regardless of the site on which they were grown, there were significant differences between rosette leaves of bitterweed and wild carrot in the concentrations of Se, K, P, Mg, Cu, B, and Zn. The concentrations of Se, K, P, Mg, Cu, and Zn were higher in bitterweed than in wild carrot, whereas the concentration of B was higher in wild carrot than in bitterweed (Table 11). The concentrations of Se, K, P, Mg, Cu, B, and Zn were higher in bitterweed, whereas the concentrations of Ca, Mn, and Mo were higher in wild carrot (Table 12).

## DISCUSSION

Although an area adjacent to the fly ash landfill was expected to serve as a control site, it exhibited a strong gradient in soil Se concentration, with the highest

**Table 10**
**Mean Concentrations of Elements in Aboveground Tissues of Bromegrass, Milkweed, and Birdsfoot Trefoil from the Fly Ash Landfill and Lansing Control Sites (Plants harvested in early September 1989)**

| Treatment | Dry Weight (mg/kg) | | | | | | | |
|---|---|---|---|---|---|---|---|---|
| | Se | Mn | Fe | Cu | B | Zn | Mo | Al |
| Lansing control | | | | | | | | |
| Bromegrass | 0.08 (0.32) | 122.37 (4.75) | 45.31 (2.69) | 2.95 (0.29) | 7.18 (1.89) | 12.52 (2.51) | 1.48 (0.64) | 20.50 (1.54) |
| Milkweed | 0.09 (0.32) | 65.57 (4.75) | 93.37 (2.69) | 9.31 (0.29) | 34.35 (1.89) | 27.69 (2.51) | 0.48 (0.64) | 57.67 (1.54) |
| Trefoil | 0.13 (0.32) | 39.01 (4.75) | 70.89 (2.69) | 8.81 (0.29) | 37.11 (1.89) | 32.50 (2.51) | 6.37 (0.64) | 32.97 (1.54) |
| Landfill | | | | | | | | |
| Bromegrass | 1.70 (0.32) | 113.53 (4.75) | 42.01 (2.69) | 2.24 (0.29) | 7.79 (1.89) | 11.25 (2.51) | 5.21 (0.64) | 17.48 (1.54) |
| Milkweed | 0.91 (0.32) | 139.60 (4.75) | 107.63 (2.69) | 12.80 (0.29) | 27.50 (1.89) | 49.04 (2.51) | 2.48 (0.64) | 47.50 (1.54) |
| Trefoil | 4.67 (0.32) | 36.64 (4.75) | 64.97 (2.69) | 7.02 (0.29) | 48.26 (1.89) | 21.89 (2.51) | 15.56 (0.64) | 21.63 (1.54) |

| Treatment | Dry Weight (%) | | | |
|---|---|---|---|---|
| | K | P | Ca | Mg |
| Lansing control | | | | |
| Bromegrass | 0.95 (0.05) | 0.15 (0.01) | 0.60 (0.06) | 0.15 (0.01) |
| Milkweed | 1.36 (0.05) | 0.17 (0.01) | 4.12 (0.06) | 0.65 (0.01) |
| Trefoil | 1.88 (0.05) | 0.14 (0.01) | 1.35 (0.06) | 0.23 (0.01) |
| Landfill | | | | |
| Bromegrass | 1.19 (0.05) | 0.12 (0.01) | 0.69 (0.06) | 0.18 (0.01) |
| Milkweed | 2.15 (0.05) | 0.25 (0.01) | 3.41 (0.06) | 0.53 (0.01) |
| Trefoil | 1.55 (0.05) | 0.20 (0.01) | 1.36 (0.06) | 0.25 (0.01) |

*Note:* Standard errors are given in parentheses.

**Table 11**
**Mean Concentrations of Elements in Rosette Leaves of Bitterweed and Wild Carrot (Early September Harvest) from the Fly Ash Landfill**

| Species | Se | Mn | Fe | Cu | B | Zn | Mo | Al |
|---|---|---|---|---|---|---|---|---|
| | | | | | Dry Weight (mg/kg) | | | |
| Wild Carrot | 0.93 (0.05) | 47.45 (2.21) | 70.67 (5.80) | 8.00 (0.21) | 41.41 (1.88) | 31.95 (0.30) | 4.24 (0.31) | 37.59 (4.15) |
| Bitterweed | 1.60 (0.06) | 41.37 (1.08) | 73.83 (1.50) | 14.05 (0.81) | 32.21 (0.27) | 60.83 (1.14) | 3.82 (0.49) | 36.81 (0.91) |

| | K | P | Ca | Mg |
|---|---|---|---|---|
| | | Dry Weight (%) | | |
| Wild Carrot | 3.75 (0.06) | 0.31 (0.01) | 1.78 (0.07) | 0.34 (0.01) |
| Bitterweed | 4.92 (0.09) | 0.38 (0.01) | 1.77 (0.09) | 0.43 (0.01) |

*Note:* Standard errors are given in parentheses.

**Table 12**
**Mean Concentrations of Elements in Cauline Leaves of Wild Carrot and Bitterweed (Early September Harvest) from the Fly Ash Landfill**

| Species | Dry Weight (mg/kg) | | | | | | |
| --- | --- | --- | --- | --- | --- | --- | --- |
| | Se | Mn | Fe | Cu | B | Zn | Mo | Al |
| Wild carrot | 1.73 (0.29) | 61.43 (2.58) | 66.88 (3.55) | 6.30 (0.32) | 34.13 (0.92) | 46.90 (1.36) | 23.10 (0.86) | 41.57 (3.99) |
| Bitterweed | 4.70 (0.40) | 36.20 (1.10) | 71.31 (2.59) | 14.51 (1.06) | 54.70 (0.99) | 65.01 (3.35) | 7.71 (0.31) | 40.04 (1.91) |

| | Dry Weight (%) | | | |
| --- | --- | --- | --- | --- |
| | K | P | Ca | Mg |
| Wild carrot | 2.77 (0.15) | 0.19 (0.00) | 3.14 (0.07) | 0.36 (0.01) |
| Bitterweed | 3.33 (0.06) | 0.31 (0.01) | 2.71 (0.06) | 0.51 (0.01) |

*Note*: Standard errors are given in parentheses.

concentrations in plots nearest the landfill. The statistical relationship was described adequately by a quadratic polynomial model (Figure 3). Although there was a statistically significant relationship between soil and plant Se, the amount of plant Se found in the controls was small and they contained amounts similar to other control tissues in these experiments (e.g., Tables 1, 3, and 5). This unexpected contamination of the control site could have been due to underground transport of Se, since it was located below the landfill, although the distance was greater than 100 m. Selenium-containing runoff from the surface of the landfill is another possibility. The poor correlation between the concentrations of Se in the soil and in plant tissues might be due to the occurrence of different forms of Se in the dry soils of the landfill and the much wetter soils of the control area. The most oxidized species of Se, selenate ($SeO_4^{2-}$), is also the most mobile and readily taken up by plants, and would be expected to be the most dominant form of Se in the dry soils of the landfill. Selenite ($SeO_3^{2-}$), a more reduced form of Se (as well as elemental Se and selenide), is much less mobile and available to plants, and could be more predominant in the wetter soils of the control area.[19,20] The accumulation of elements in indigenous species at the landfill site was similar to that reported earlier,[16] with minor differences due to the selection of species that were harvested. In bromegrass, milkweed, birdsfoot trefoil, and bitterweed, Mo was elevated in plants grown on the landfill. However, this was not consistent for B content, although B accumulation was a common feature of different indigenous species in the earlier study.

The accumulation ratio (landfill/control) was generally less than 2:1 for all elements (except Se) and for all species of plants studied. The ecological consequences of the slightly elevated concentrations are probably not of great potential significance. One possible exception may be As, for which no accurate analyses were made.

Significant amounts of Se in all crops were found on the landfill compared with the control, even where statistical significance could not be established because of plot-to-plot variability in Se uptake in plants grown on the landfill plots. Accumulation ratios exceeding 50:1 in several species were not unusual. The amount of Se accumulated by the several species of plants was related only in part to depth of rooting. Bromegrass, a shallow-rooted plant, did not accumulate Se, as also reported earlier.[16] Alfalfa, a deep-rooted legume, accumulated relatively high concentrations of Se, which increased in sequential harvests that were made at different times over the growing season. The nominal 0.7 m soil cap would not provide a significant barrier to prevent deep-rooted plants from penetrating to the fly ash layer. Results for rutabaga and carrot, however, indicate that depth of the root system cannot be the only mechanism for uptake and accumulation of Se. Hurd-Karrer[17] reported that members of the Cruciferae family were Se accumulators, while members of the Umbelliferae family were not. In our studies, rutabaga, a crucifer, accumulated much more Se than carrot, an umbellifer. Since neither species is deep-rooted, a special mechanism must exist that imbues certain species or families with this capability. Soluble elements could move upward from the fly ash by capillarity. Shallow roots could also contact fly ash particles mixed into the soil cap by earthworms or other arthropods, or by animal burrowing (a common feature at our landfill). The soil cap was applied by large earth-moving equipment, which could easily result in significant variations in soil thickness. Roots of shallow-rooted plants may also grow deeper during drought years. Also, given sufficient time, Se and other elements would be transported to the upper layers of soil by root decay, and deposition and decay of leaves or other plant litter.[16]

Three corn varieties contained different amounts of Se in various tissues. Agway 135 accumulated more Se than Halsey, the other field corn variety tested. Golden Cross Bantam, a sweet corn variety, gave intermediate values. In all varieties, however, leaves and kernels were the tissues of highest Se content.

We did not detect symptoms of Se toxicity in any species of plant. In fact, the Se content rarely exceeded 10 mg/kg (dry weight) in any of our studies. The absolute amounts of Se found were not exceptionally high, especially if one considers those reported for areas where Se toxicity has occurred.[22] The possible long-term consequences of these relatively high levels of Se in biota cannot be evaluated.

Cattails growing in the drainage channels leading from the landfill to the sedimentation pond, and in the pond itself, exhibited a considerable amount of leaf tip necrosis.[16] The cause of this injury is not known, but chemical analyses have shown that several potential toxic elements were elevated in the tissues. The amelioration of Se uptake from soils by application of gypsum has been known for more than 50 years.[18] Competition between S and Se in root uptake and in metabolic processes reduces Se absorption. Application of gypsum provides a possible means of reducing Se uptake by plants growing on fly ash landfills, and has the added advantage of being a by-product of coal desulfurization. Our preliminary studies at one application rate have confirmed the efficacy of gypsum treatment on Se accumulation in plants. Clearly, dose-response and time-course studies are needed under various edaphic conditions to determine the appropriate application rates and the value of this treatment to plants in general.

## ACKNOWLEDGMENTS

The authors thank the New York State Electric and Gas Corp. and the Empire State Electric Energy Research Corp. for financial support and technical assistance. We are also deeply indebted to Charles R. Tutton for his agricultural expertise and his many helpful suggestions.

## REFERENCES

1. Borisoff, E., testimony, April 26, 1988, American Coal Ash Association, Denver, 1988.
2. U.S. Environmental Protection Agency, Wastes from the Combustion of Coal, EPA Report 530-SW-88-002, 1988.
3. U.S. Department of Transportation, Fly Ash Facts for Highway Engineers, Technology Transfer, USDOT, FHWA-DP-59-8, 1986.
4. Page, A. L., Elseewi, A. A., and Straughan, I. R., Physical and chemical properties of fly ash from coal-fired power plants with reference to environmental impacts, *Residue Rev.*, 71, 83, 1978.

5.  Furr, A. K., Kelly, W. C., Bache, C. A., Gutenmann, W. H., and Lisk, D. J., Multielement uptake by vegetables and millet grown in pots on fly ash amended soil, *J. Agric. Food, Chem.*, 24, 885, 1976.

6.  Furr, A. K., Parkinson, T. F., Hinrichs, R. A., van Campden, D. R., Bache, C. A., Gutenmann, W. H., St. John, L. E., Pakkala, I. S., and Lisk, D. J., National survey of elements and radioactivity in fly ashes: absorption of elements by cabbage grown in fly ash-soil mixtures, *Environ. Sci. Technol.*, 11, 1195, 1977.

7.  Furr, A. K., Parkinson, T. F., Gutenmann, W. H., Pakkala, I. S., and Lisk, D. J., Elemental content of vegetables, grains, and forages field-grown on fly ash-amended soil, *J. Agric. Food Chem.*, 26, 357, 1978.

8.  Churey, D. J., Gutenmann, W. H., Kabata-Pendras, A., and Lisk, D. J., Elemental concentrations in aqueous equilibrates of coal and lignite fly ashes, *J. Agric. Food Chem.*, 27, 910, 1979.

9.  Doran, J. W. and Martens, D. C., Molybdenum availability as influenced by application of fly ash to soil, *J. Environ. Qual.*, 1, 186, 1972.

10. Elseewi, A. A., Straughan, I. R., and Page, A. L., Sequential cropping of fly ash-amended soils: effect on soil chemical properties and yield and elemental composition of plants, *Sci. Total Environ.*, 15, 247, 1980.

11. Elseewi, A. A. and Page, A. L., Molybdenum enrichment of plants grown on fly ash-treated soils, *J. Environ. Qual.*, 13, 394, 1984.

12. Tolle, D. A., Arthur, M. F., and Van Voris, P., Microcosm/field comparison of trace element uptake in crops grown in fly ash-amended soil, *Sci. Total Environ.*, 31, 43, 1983.

13. El Mogazi, D., Lisk, D. J., and Weinstein, L. H., A review of physical, chemical, and biological properties of fly ash and effects on agricultural ecosystems, *Sci. Total Environ.*, 74, 1, 1988.

14. Richmond, M. E., Wischusen, E. W., Weinstein, L. H., Schneider, R. S., and Lisk, D. J., Movement and accumulation of potential toxic elements into the terrestrial and aquatic fauna associated with a soil-capped fly ash landfill, Paper No. 202 (Abstr.), Int. Conf. Metals in Soils, Waters, Plants and Animals, 1990.

15. Weinstein, L. H., Schneider, R. S., Osmeloski, J., Wischusen, E. W., Richmond, M., Beers, A. O., and Lisk, D. J., Elemental cycling of selenium and other elements through the biotic and abiotic systems of a capped coal fly ash landfill, Paper No. 244 (Abst.), Int. Conf. Metals, in Soils, Waters, Plants and Animals, 1990.

16. Weinstein, L. H., Osmeloski, J. F., Rutzke, M., Beers, A. O., McCahan, J. B., Bache, C. A., and Lisk, D. J., Elemental analysis of grasses and legumes growing on soil covering coal fly ash landfill sites, *J. Food Safety*, 9, 291, 1989.

17. Hurd-Karrer, A. M., Factors affecting the absorption of selenium from soils by plants, *J. Agric. Res.*, 50, 413, 1935.

18. Hurd-Karrer, A. M., Relation of sulphate to selenium absorption by plants, *Am. J. Bot.*, 25, 666, 1938.

19. Mikkelsen, R. L., Page, A. L., and Bingham, F. T., Factors affecting selenium accumulation by agricultural crops, in *Selenium in Agriculture and the Environment*, SSSA Special Pub. No. 23, 65, 1989.

20. Elrashidi, M. A., Adriano, D. C., and Lindsay, W. L., Solubility, speciation, and transformation of selenium in soils, in *Selenium in Agriculture and the Environment*, SSSA Special Pub. No. 23, 51, 1989.

21. Olson, O. E., Fluorometric analysis of selenium in plants, *Anal. Chem.*, 52, 627, 1969.

22. Ohlendorf, H. M., Bioaccumulation and effects of selenium on wildlife, in *Selenium in Agriculture and the Environment*, SSSA Special Pub. No. 23, 133, 1989.

# 11

# Accumulation of Mo in Wheat and Alfalfa Grown on Fly Ash-Amended Acid Mine Spoils

R. F. Keefer, D. K. Bhumbla, and R. N. Singh

West Virginia University, Morgantown, WV

## ABSTRACT

Field experiments were conducted to monitor the Mo concentrations in wheat *(Triticum vulgare)* and alfalfa *(Medicago sativa L.*) grown on two acidic mine spoils amended with fly ash. Changes in plant-available Mo were determined by three extraction methods: (1) Grigg's reagent (acidified ammonium oxalate), (2) Dowex 1 × 8 anion exchange resin, and (3) 0.1 $M$ NaOH. Molybdenum concentrations of wheat and alfalfa plants grown during the first year on fly ash-amended mine spoils were > 10 mg/kg and Cu to Mo ratios were < 2. These levels of Mo and Cu to Mo ratios in feed for cattle or sheep would be considered hazardous and require Cu supplementation. Alfalfa grown on fly ash-amended plots were higher than those of the control plots. Soil Mo concentrations determined by an anion exchange resin and 0.1 $M$ NaOH were related to Mo concentrations of the crops; however, Mo extracted by Grigg's reagent was not significantly related to concentrations of Mo in plants.

## INTRODUCTION

### Literature Review

Molybdenum is a multivalent element that can exist in more than one oxidation state, i.e., +3, +4, +5, or +6. In aqueous solution, the most stable oxidation state of Mo is +6. Molybdate ($MoO_4^{2-}$) is stable over a wide range of pH and Eh values.

Coal combustion has been implicated as the largest source of Mo contribution to the environment.[1] About 610 metric tons of Mo were estimated to be added annually to the environment of the U.S. as a coal combustion pollutant compared

0-87371-890-9/93/$0.00+$.50
© 1993 by Lewis Publishers

to 990 metric tons of Mo as a pollutant from all other sources.[2] The mean Mo concentrations in coal is about 3 mg/kg[3] ranging from 0 to 73 mg/kg.[4] Following combustion of coal most of the Mo is retained by fly ash particles, causing Mo concentrations in fly ash to be considerably higher than in the original coal. In power plants burning coal, over 85% of the Mo in the coal appeared in the ashes.[1,5] Furthermore, during combustion all of the Mo is volatilized as $MoO_3$ in the hottest zone of the furnace (temperatures of $> 1400°C$), then condenses evenly and completely. The finer particles, due to their higher surface area, are higher in Mo concentrations than the coarser particles. The Mo concentration in fly ash ranges from 7 to 160 mg/kg,[6] with increasing concentration as fly ash size decreases. These higher concentrations of Mo associated with surfaces of fly ash particles can be taken up by plants and thus cause toxicity to animals, particularly ruminants.

The mean Mo concentrations of 44 and 15 ppm for fly ash and bottom ash, respectively, are many times the mean Mo concentration of soils (2 mg/kg).[7,8] Leaching experiments showed that Mo was released from fly ash more readily than from bottom ash.[7] Thus, Mo associated with fly ash poses an environmental threat due to its high leachability and consequent bioavailability.

The availability of Mo added through fly ash was investigated in a greenhouse experiment by Doran and Martens;[9] Mo was concluded to be equally available from fly ash and sodium molybdate. Elseewi and Page[10] reported that Mo availability from fly ash was not controlled by soils, plants, time of harvest, or alkalinity. In both of these studies, fly ash was used at rates less than 80 g/kg soil. From greenhouse experiments, Elseewi and Page concluded that addition of fly ash at rates exceeding 40 g/kg may produce forages that would be toxic to animals.

Although the pH of fly ash varies from acidic (pH 4.2) to alkaline (pH 12.4),[6] Phung and co-workers[11] have demonstrated that some types of fly ash can be used as liming agents on acid soils. Most of their work suggested that the rate of addition should be less than 80 g/ha of soil. However, when fly ash was used as a topsoil substitute in reclamation of pyritic acid mine spoils, much higher rates were added.[12,13] Fly ash has many potential beneficial properties, such as high water-holding capacity, high neutralization potential, and the presence of plant nutrients, that make it a suitable candidate for use as a topsoil substitute in revegetation of pyritic mine spoils. However, addition of fly ash to soil often results in excessive levels of Mo and low levels of P in plant tissue.[3,14] Much of the P found in fly ash is unavailable for plant uptake[15] because under the alkaline conditions in the ash, insoluble calcium phosphates and other complexes are formed.[16] Phosphorus deficiency in fly ash is difficult to overcome with just the addition of fertilizer. If the pH of the system remains very high, the P will continue to be fixed as calcium phosphates.

Addition of soluble P to fly ash-amended soils may result in high levels of Mo in plant tissue because soluble P releases chemi-sorbed Mo to the soil solution.[17–21] In solution culture, Stout et al.[17] showed that high P levels in culture solutions increased Mo uptake as much as tenfold. Adding P increases the concentration of Mo in plants due to the formation of a complex phosphomolybdate anion which is readily absorbed by plants.[22] In pyritic mine spoils, the plant uptake of Mo may be controlled by either Mo × P or Mo × S interactions. These pyritic mine spoils are high in sulfate due to oxidation of sulfide in pyrite. Huising and Matrone[23] suggested an "oxyanion chemical parameter" concept for $SO_4^{2-}$-induced reduction of Mo uptake. Although Mo is in group VIB and can exist in several oxidation states, it is most stable in the +6 oxidation state where it is usually bound to four oxygen atoms and exists as the oxyanion ($MoO_4^{2-}$). Molybdate and other oxyanions in the same group, e.g., chromate ($CrO_4^{2-}$) and tungstate ($WO_4^{2-}$), can be competitive in soil and plant reactions. Oxyanions of group VIB are very similar to those in group VIA. Since sulfate is the most abundant member of group VIA and is present in soil at much higher concentration than molybdate, it can suppress the uptake of molybdate.

Use of fly ash on mine spoils also provides another mitigating factor, i.e., generation of new oxides/hydroxides of Fe by reaction of oxidation products of pyrite with alkaline materials in the fly ash.[24] The resulting Fe oxides provide surface for adsorption of $MoO_4^{2-}$ ions. The availability of Mo to plants was reduced more by soil reactions than by leaching.[25] Molybdenum was probably slowly converted into discrete secondary compounds, such as Al- or Fe-molybdates, or was adsorbed by positively charged oxides of Al or Fe. The reduced availability of Mo with time was also related to the adsorption of Mo on the surfaces of Al and Fe oxides, which were positively charged at a pH level below their zero point of charge. The principal positively charged sites in spoil for anionic adsorption are hydrous oxides of Al and Fe, if the pH is below their zero point of charge.[26] The importance of Fe oxides for retention of Mo was demonstrated by Norrish,[27] who reported that Mo in soils consistently correlated with Fe concentrations in the soil. Molybdenum retained in chemi-sorbing sites was not likely to convert to a discrete Fe compound, such as $Fe_2(MoO_4)_3$, as this compound is too soluble to persist in soils as an inert phase.[28]

Large indurated Fe concretions could tie up large amounts of Mo in unavailable forms. Soils high in free Fe oxides adsorbed large quantities of $MoO_4^{2-}$ from aqueous solutions.[29] Aluminum oxides were also capable of removing Mo from the aqueous solutions, but their effectiveness was less than that of Fe oxides.

## Objectives

The present field investigation was undertaken to (1) evaluate the effect of addition of large amounts of fly ashes to pyritic mine spoils on Mo concentrations of wheat and alfalfa, (2) monitor the changes in Mo concentrations of alfalfa and Mo in mine spoils over four growing seasons in an acid mine soil amended with fly ash-rock phosphate mixtures, and (3) examine if use of slightly soluble P, such as rock phosphate, affects the Mo concentrations of wheat and alfalfa.

## MATERIALS AND METHODS

### Fly Ash and Rock Phosphate

Fly ash used in this study was obtained from a coal-fired power plant located in northern West Virginia (Fort Martin). This fly ash (Class F; noncementatious) was alkaline (pH 10.7) with a neutralization potential of 120 cmol$_c$/kg. Total

**Table 1**
**Some Physicochemical Properties of Two Mine Spoils**

| Property | Lenox, WV | Westover, WV | Method Ref. |
|---|---|---|---|
| Texture | | | 30 |
| Sand (%) | 26 | 42 | |
| Silt (%) | 42 | 40 | |
| Clay (%) | 32 | 18 | |
| pH | 3.2 | 3.4 | 32 |
| Organic C (%) | 1.05 | 1.95 | 32 |
| Pyritic S (%) | 0.65 | 0.35 | 31 |
| Amorphous oxides of Fe (%) | 1.15 | 0.85 | 32 |
| Total Mo (mg/kg) | 5.15 | 3.75 | 32 |

chemical analyses of the fly ash showed 40 mg of Mo/kg and $SiO_2$ of 38%, $Al_2O_3$ of 20%, $Fe_2O_3$ of 29%, CaO of 6%, MgO of 1%, $SO_4^{2-}$ of 2%, $K_2O$ of 1%, and $P_2O_5$ of 0.5%. The rock phosphate obtained from North Carolina contained 10.5% P and 22.7% Ca.

## Mine Spoils, Experimental Design, and Treatments

The experiments were conducted on two abandoned West Virginia mine sites, at Lenox (Preston County), and Westover (Monongalia County), that were characterized physicochemically (Table 1). Texture was determined by the pipette method,[30] pyritic sulfur by the method of Sobek et al.,[31] and organic carbon, amorphous oxides of Fe, and total Mo by methods of Page et al.[32] Field experiments were established in October 1982 with a factorial, randomized block experimental design with treatments of three rates of fly ash (0, 400, and 800 metric tons/ha), and three rates of rock phosphate (0, 15, and 30 metric tons/ha) replicated five times. The fly ash added at 400 or 800 metric tons/ha contributed 16 or 32 kg of Mo/ha, respectively. Without lime and P, plants will not grow in these acid mine spoils (pH 3.2 to 3.4). Therefore, the 0 fly ash/0 rock P check was amended with lime at 22.5 metric tons/ha and triple superphosphate at 300 kg/ha. The amendments were incorporated by disking.

## Crops Grown, Harvesting Schedule, and Sample Preparation

Plots were seeded to wheat *(Triticum vulgare)* as a nurse crop and alfalfa *(Medicago sativa* L.) inoculated with *Rhizobium meliloti.* Wheat was sampled at jointing and at maturity (separated into grain and straw). Two cuttings of alfalfa were harvested each year. Alfalfa was sampled over a 4-year period at the Lenox site, but only for three cuttings at the Westover site because it was remined before the experiment was completed. Plant samples were washed with distilled water, dried at 65°C in a forced air oven, and ground to < 1 mm in a Wiley mill. Total dry matter yields will be reported elsewhere.

## Plant Sample Analyses

For Mo analysis, plant samples were digested by the method of Ganje and Page.[33] Molybdenum in the acid solutions was analyzed by electrothermal atomic absorption spectrometry. The samples were analyzed by wall atomization and analysis was compensated for reduced sensitivity with a number of firings of the pyro-coated tube comparing with a running series of standards.

## Soil Sampling and Analyses

Soil samples were collected at the end of each growing season and analyzed by standard methods of Page et al.[32] Molybdenum in soils was analyzed after extraction with Grigg's reagent.[34] Chemi-sorbed Mo was determined by extracting soils with 0.1 $M$ NaOH. An ion exchange resin was used to determine plant-available Mo since ion exchange resins are thought to simulate the desorbing effects of plant roots better than chemical extractants.[35]

Anion exchange resin (Dowex $1 \times 8$, $> 60$ mesh) extraction involved three grams of the resin enclosed in a $3 \times 5$ cm 60-mesh polyethylene net bag. The resin was saturated with chloride using 2 $M$ HCl. Excess salt was removed by washing the resin with distilled deionized water. Each resin bag was placed in a 100 mL wide-mouth glass bottle containing 10 g of soil and 40 mL of distilled deionized water. The flasks were shaken for 24 hr horizontally on a reciprocating shaker at 2 cycles/sec. The resin bags were washed free of soil particles with distilled deionized water and molybdate ions were removed form the resin with 2 $M$ HCl and Mo was determined by electrothermal atomic absorption spectrometry.

## Statistical Analysis

All data were analyzed for statistical significance by analysis of variance using the 0.05 level of significance. Least significant difference (LSD) values were calculated and used for comparison among the treatments.

## RESULTS AND DISCUSSION

Generally, 50 to 100 g Mo/ha treatments are needed for growth of most agronomic crops on Mo-deficient soils and as much as 400 g Mo/ha may be needed on vegetable crops, such as cauliflower *(Brassica oleracea* var. *botrytis).*[36] The total concentration of Mo in fly ash used in this study, i.e., 40 mg/kg, falls within the Mo concentration rage of 7 to 160 mg/kg reported by Page et al.,[6] but was higher than the reported mean Mo concentration of fly ash from bituminous coals of 3 mg/kg.[3] Addition of Mo at much lower rates, i.e., $> 1$ kg Mo/ha have been reported to produce plants containing Mo at levels toxic to animals.[34] Thus, the additions of Mo through fly ash in the present investigation were high enough to produce plants with hazardous levels of Mo. However, several investigators [28,37–40] have shown that Mo uptake depends on a variety of factors, including soil pH, soil type, redox potential, Al and Fe concentrations in soil, and the supply of other compounds, such as sulfates and phosphates.

## Molybdenum Concentrations in Alfalfa

Molybdenum concentration in alfalfa tissue increased significantly with an increase in amount of fly ash applied in both the Lenox (Table 2) and the Westover (Table 3) mine spoils. In both spoils, the lowest Mo concentrations occurred in plants on plots that did not receive fly ash, and the highest amounts of Mo (9.84 and 10.5 mg/kg in Lenox and Westover soil, respectively) were found in plants growing in plots which were treated with the highest rate of fly ash. The concentration of Mo for the first cutting of alfalfa ranged between 0.6 and 9.8 mg/kg and 0.6 and 10.5 mg/kg for the Lenox (Table 2) and Westover (Table 3) spoils, respectively. In both spoils, the Mo concentrations in alfalfa grown on plots that did not receive fly ash increased with lime application (i.e., the 0 fly ash/0 rock phosphate treatment), e.g., in Lenox mine spoil, application of lime increased Mo concentrations in alfalfa from 0.6 to 1.3 mg/kg. Thus, an increase in pH from 3.2 to 5.6 at Lenox and from 3.4 to 6.2 at Westover (Table 4) due to liming increased the availability of indigenous soil Mo.

Application of both rock phosphate and fly ash had no effect on Mo concentration of alfalfa in fly ash-treated mine spoil (Tables 2 and 3) presumably due to the relatively slow release of P from the rock phosphate. Thus, P concentrations were not high enough to form complex phosphomolybdate ions which could be responsible for phosphate-induced enhancement in Mo uptake by plants.[22]

## Molybdenum Concentrations in Alfalfa with Cropping

Molybdenum concentrations in alfalfa tissues from the Lenox spoil decreased with cropping (Table 2). A similar trend was observed for the Westover spoil, where data were available for only three cuts of alfalfa (Table 3). Very small differences in Mo concentrations in alfalfa were observed between the two cuts of alfalfa that were harvested during the first year of crop production. However, drastic reduction of Mo concentrations were observed after the first growing season, i. e., between the second and the third cuttings of alfalfa (Tables 2 and 3). The Mo concentrations of alfalfa after the third cutting showed a steady decline with cropping (Table 2) through eight cuttings at Lenox.

Alfalfa plants grown on plots treated with fly ash had higher Mo concentration than those grown on plots that did not receive fly ash (Tables 2 and 3). This trend persisted for the duration of the experiment (4 years; Table 2), but there was much less spread from cut 1 of the third year on. Similar results from greenhouse experiments were reported by Doran and Martens[9] and Elseewi and Page.[10] However, Mo in the fly ash in the present study was four times more concentrated than that reported by Elseewi and Page,[10] but results still showed much lower Mo concentrations in alfalfa tissue grown on fly ash-amended spoil. The differences

**Table 2**
**Molybdenum Concentrations of Alfalfa Grown on Lenox Mine Spoil Treated with Fly Ash/Rock Phosphate Mixtures**

| Rate Added (metric tons/ha) | | Concentration (mg Mo/kg) | | | | | | | |
| --- | --- | --- | --- | --- | --- | --- | --- | --- | --- |
| | | Year 1 | | Year 2 | | Year 3 | Year 4 | Year 5 | Year 6 |
| Fly Ash | Rock Phosphate | Cut 1 | Cut 2 | Cut 1 | Cut 2 | Cut 1 | Cut 1 | Cut 1 | Cut 1 |
| 0[a] | 0 | 1.29 | 1.24 | 1.24 | 1.08 | 1.27 | 1.18 | 1.19 | 1.19 |
| 0 | 15 | 0.60 | 0.61 | 0.50 | 0.71 | 0.58 | 0.50 | 0.58 | 0.58 |
| 0 | 30 | 0.60 | 0.70 | 0.60 | 0.77 | 0.77 | 0.74 | 0.74 | 0.76 |
| 400 | 0 | 7.77 | 7.27 | 2.64 | 2.18 | 1.37 | 1.27 | 1.50 | 1.49 |
| 400 | 15 | 7.54 | 7.18 | 2.49 | 2.49 | 1.47 | 1.58 | 1.36 | 1.47 |
| 400 | 30 | 7.81 | 6.59 | 2.80 | 2.50 | 1.60 | 1.50 | 1.41 | 1.38 |
| 800 | 0 | 9.84 | 9.26 | 3.01 | 2.99 | 2.15 | 1.79 | 1.69 | 1.68 |
| 800 | 15 | 9.33 | 8.76 | 2.97 | 2.96 | 1.60 | 1.70 | 1.28 | 1.29 |
| 800 | 30 | 9.56 | 9.14 | 2.96 | 3.06 | 1.85 | 1.83 | 1.73 | 1.73 |
| LSD[b] (0.05) | | 0.65 | 0.83 | 0.79 | 0.40 | 0.45 | 0.67 | 0.65 | 0.45 |

[a] 0 fly ash 0 rock phosphate treatment received 22.4 metric tons lime and 300 kg/ha triple superphosphate.
[b] LSD = least significant difference.

**Table 3**
**Molybdenum Concentrations of Alfalfa Grown on Westover Mine Spoil Treated with Fly Ash/Rock Phosphate Mixtures**

| Rate Added (metric tons/ha) | | Concentration (mg Mo/kg) | | |
| --- | --- | --- | --- | --- |
| | | Year 1 | | Year 2 |
| Fly Ash | Rock Phosphate | Cut 1 | Cut 2 | Cut 1 |
| 0[a] | 0 | 1.35 | 1.24 | 1.07 |
| 0 | 15 | 0.65 | 0.70 | 0.80 |
| 0 | 30 | 0.60 | 0.75 | 0.70 |
| 400 | 0 | 8.00 | 8.15 | 3.20 |
| 400 | 15 | 8.15 | 8.00 | 3.15 |
| 400 | 30 | 8.25 | 8.19 | 3.25 |
| 800 | 0 | 9.97 | 10.00 | 3.45 |
| 800 | 15 | 10.15 | 9.69 | 3.65 |
| 800 | 30 | 10.50 | 9.89 | 3.90 |
| LSD[b] | (0.05) | 0.54 | 0.45 | 0.37 |

[a] 0 fly ash 0 rock phosphate treatment received 22.4 metric tons lime and 300 kg/ha triple superphosphate.
[b] LSD = least significant difference.

in Mo release for plant uptake may be due to differences in Mo-fixing capacity of the two types of fly ash. The fly ash used by Elseewi and Page had only 2.0% $Fe_2O_3$ while the fly ash used in the present investigation had 28.9% $Fe_2O_3$. Fly ash formed during burning of coal consists of spherical micron-sized particles enclosed in a two-phase Si glass matrix[39] within which elements can become entrapped.[12] The speed with which the external glass matrix is dissolved will govern the availability of Mo to plants. Molybdenum in fly ash used in the present study could be associated with the internal glass matrix and/or be adsorbed on associated oxides or hydroxides of Al and Fe, and thus may be less available to crops.

## Molybdenum Concentrations in Wheat

Molybdenum concentration of wheat grown on fly ash-treated mine spoils increased with level of ash application in the two spoils under investigation (Table 5). The Mo concentrations were highest in plots that received 800 metric tons of fly ash/ha. Molybdenum concentrations in the two spoils at all fly ash levels were higher in wheat grain and young plants at jointing than in the straw at maturity. Thus, plant sampling at jointing can provide a reasonable estimate of Mo concentration in wheat grain at maturity.

Molybdenum concentrations were higher in wheat than alfalfa (Tables 2, 3, and 5). Usually legumes contain more Mo than grasses.[40] On the other hand, Gupta[40] found no differences in the Mo concentrations of red clover, alfalfa, and grasses grown on podzol soils of eastern Canada, but base levels were probably low. The differences in Mo concentrations in wheat and alfalfa in the present study may be related to the time sequence in which these two crops were grown: wheat

**Table 4**
**pH of Two Mine Spoils (0 to 15 cm) Treated with Fly Ash/Rock Phosphate Mixtures and Sampled After Cropping**

| Rate Added (metric tons/ha) | | Years of Cropping | | | | | | | | |
|---|---|---|---|---|---|---|---|---|---|---|
| | | Westover, WV | | Lenox, WV | | | | | | |
| Fly Ash | Rock Phosphate | Year 1 | Year 2 | 1 | 2 | 3 | 4 | 5 | 6 | |
| 0[a] | 0 | 6.2 | 3.7 | 5.6 | 4.0 | 3.4 | 3.2 | 3.1 | 2.9 | |
| 0 | 15 | 4.4 | 3.8 | 3.6 | 3.4 | 3.3 | 3.2 | 3.1 | 3.0 | |
| 0 | 30 | 4.2 | 3.7 | 3.6 | 3.4 | 3.2 | 3.0 | 3.1 | 3.1 | |
| 400 | 0 | 7.6 | 7.0 | 7.2 | 6.5 | 6.5 | 6.5 | 6.5 | 6.6 | |
| 400 | 15 | 7.6 | 7.1 | 7.1 | 6.7 | 6.6 | 6.5 | 6.5 | 6.5 | |
| 400 | 30 | 7.6 | 7.1 | 7.0 | 6.6 | 6.5 | 6.6 | 6.5 | 6.5 | |
| 800 | 0 | 7.7 | 7.2 | 7.6 | 6.7 | 6.6 | 6.5 | 6.7 | 6.7 | |
| 800 | 15 | 7.8 | 7.2 | 7.4 | 6.8 | 6.7 | 6.7 | 6.7 | 6.6 | |
| 800 | 30 | 7.9 | 7.2 | 7.5 | 6.8 | 6.6 | 6.5 | 6.5 | 6.7 | |

[a] 0 fly ash, 0 rock phosphate treatment received 22.4 metric tons lime and 300 kg/ha triple superphosphate.

**Table 5**
**Molybdenum Concentrations (mg/kg) of Wheat Tissue (Year 1) Grown on Two Mine Spoils Treated with Fly Ash/Rock Phosphate Mixtures**

| Rate Added (metric tons/ha) | | Lenox, WV | | | Westover, WV | | |
|---|---|---|---|---|---|---|---|
| Fly Ash | Rock Phosphate | Jointing | Straw | Grain | Jointing | Straw | Grain |
| 0[a] | 0 | 0.22 | 0.38 | 0.20 | 0.20 | 0.36 | 0.20 |
| 0 | 15 | 0.27 | 0.32 | 0.22 | 0.25 | 0.37 | 0.17 |
| 0 | 30 | 0.24 | 0.36 | 0.27 | 0.27 | 0.36 | 0.24 |
| 400 | 0 | 15.85 | 12.70 | 15.50 | 18.00 | 13.50 | 18.65 |
| 400 | 15 | 17.76 | 15.00 | 16.50 | 18.72 | 15.00 | 17.95 |
| 400 | 30 | 19.25 | 14.40 | 16.00 | 18.69 | 15.40 | 19.45 |
| 800 | 0 | 23.90 | 18.90 | 22.17 | 23.25 | 18.70 | 21.35 |
| 800 | 15 | 26.07 | 19.30 | 26.50 | 25.75 | 22.20 | 24.75 |
| 800 | 30 | 27.15 | 20.25 | 26.84 | 28.00 | 23.00 | 26.45 |
| LSD[b] | (0.05) | 4.75 | 1.72 | 3.51 | 4.95 | 6.00 | 5.52 |

[a] 0 fly ash, 0 rock phosphate treatment received 22.4 metric tons lime and 300 kg/ha triple superphosphate.
[b] LSD = least significant difference.

was planted as a nurse crop for the alfalfa, and matured before the alfalfa became established. Thus, wheat plants probably absorbed much of the available Mo before that in the fly ash could be transformed to a plant-unavailable form. This is supported by Mo concentrations in tissues of wheat and alfalfa.

## Molybdenum Reactions in Soil

An increase in soil pH due to liming has been shown to increase the availability of soil Mo due to decreased retention of molybdate by oxides and hydroxides of Al and Fe.[41,42] Molybdenum in fly ash is present as a layer of $MoO_3$ on the surface of the fly ash particles.[1] Availability of this Mo to plants depends on the hydrolysis of the $MoO_3$ on the ash surface to $MoO_4^{2-}$ under alkaline conditions. However, with neutral or acidic fly ash, Mo will likely be associated with oxides of Al and Fe. Adsorbed molybdate ions can also undergo a ligand exchange with phosphate and thereby increase the availability of Mo to plants. However, no increase in Mo concentrations of alfalfa tissue was observed due to the application of rock phosphate in plots that did not receive fly ash (Tables 2 and 3). Generally, the increase in Mo availability with liming or phosphate additions has been considered to be an ion exchange reaction of simple monomeric anions ($MoO_4^{2-}$ and $HMoO_4^{1-}$) for hydroxyls supplied through liming or for $H_2PO_4^{1-}$ supplied through rock phosphate.[42] In the acid mine spoils of the present study, molybdate in soil solution may not be in the monomeric form. In an acid environment, tetrahedral $MoO_4^{2-}$ undergoes the following reaction:[42,43] First, the molybdate ions become protonated, forming monomeric $HMoO_4^{1-}$ and $H_2MoO_4$. Second, condensation begins after the initial protonation, and a series of iso- or hetero-polymolybdates are formed. These polymolybdates are not easily translocated across the plant cell membrane, so that plants growing at low pH do not absorb sufficient Mo even when total Mo is high.

## pH Reactions in Soils Relative to Mo Availability

Plots receiving no fly ash, no rock phosphate, but limestone resulted in a pH of 5.6 at Lenox and 6.2 at Westover (Table 4). These differences in pH could have contributed to the variation in Mo concentrations in alfalfa from these sites (Tables 2 and 3). Molybdate exists as an anion and its adsorption increases as positive charge on the soil constituents increases. Positive charge on the soil clays and free oxides of Fe increases with decrease in pH. Thus, in rock phosphate-treated soils, the low pH and high anion exchange capacity probably prevented release of Mo uptake by plants.

## Molybdenum Fixation in Soils

Molybdenum can become less available with time by fixation reactions with Al and Fe oxides[25] similar to that of P.[44,45] In the present study, fly ash reacted with spoils in the field for 4 years, which should allow sufficient time for molybdates to undergo the slow Mo-fixation reactions.[44] The anions — arsenate, phosphate, and molybdate — are similar in that they have high adsorption affinity, but may require long periods of equilibration before concentrations in the soil solution are substantially decreased. Greenhouse experiments by Doran and Martens[9] for 94 days and Elseewi and Page[10] for 365 days were short enough to preclude large amounts of Mo-fixation.

## Molybdenum Concentrations in Mine Spoils

Application of fly ash to mine spoils increased Mo concentration — as extracted by Grigg's method, i.e., ammonium oxalate (Table 6), anion exchange resin (Table 7), and NaOH (Table 8) — of the two mine spoils, but application of rock phosphate had no effect on the extractable Mo in the spoils (Tables 6–8). This inability of rock phosphate to extract Mo explains the absence of phosphate-induced enhancement of Mo uptake. The solubility of rock phosphate in the spoil was insufficient to increase P concentrations of the soil solution substantially for replacement of chemi-sorbed Mo.

With time the anion exchange resin extractable-Mo (Table 7) and $0.1\ M$ NaOH extractable-Mo (Table 8) were significantly reduced. Thus, plant available Mo decreased with time due to transformation of Mo into a less soluble, i.e., less plant-available form, probably by fixation reaction on the surface of oxides of Al and Fe added to the spoils through fly ash additions. The Mo extracted by ion exchange resin and NaOH results from ligand exchange and should approximate the Mo fraction of the spoil which is potentially available to plants. The decrease in the amount of this Mo fraction with cropping shows that Mo was being transformed into a form in which Mo was not released by ligand exchange, and

Table 6
Grigg's[a] Extractable Mo (mg/kg) with Time in Two Mine Spoils Treated with Fly Ash/Rock Phosphate Mixtures

| Rate Added (metric tons/ha) | | Years of Cropping | | | | | | | |
|---|---|---|---|---|---|---|---|---|---|
| | | Westover, WV | | | Lenox, WV | | | | |
| Fly Ash | Rock Phosphate | 0 | 1 | 2 | 0 | 1 | 2 | 3 | 4 |
| 0[b] | 0 | 112 | 121 | 115 | 130 | 145 | 137 | 127 | 120 |
| 0 | 15 | 104 | 110 | 121 | 120 | 127 | 126 | 120 | 135 |
| 0 | 30 | 107 | 109 | 112 | 140 | 140 | 147 | 130 | 130 |
| 400 | 0 | 2210 | 2270 | 2200 | 2110 | 2240 | 2200 | 2270 | 2170 |
| 400 | 15 | 2250 | 2200 | 2160 | 2090 | 2090 | 2280 | 2190 | 2200 |
| 400 | 30 | 2200 | 2230 | 2260 | 2110 | 2190 | 2210 | 2270 | 2180 |
| 800 | 0 | 3450 | 3550 | 3510 | 3240 | 3200 | 3230 | 3250 | 3200 |
| 800 | 15 | 3510 | 3550 | 3520 | 3320 | 2200 | 2210 | 2260 | 3330 |
| 800 | 30 | 3550 | 3550 | 3540 | 3420 | 3420 | 3320 | 3340 | 3410 |
| LSD[c] | (0.05) | 140 | 170 | 170 | 350 | 310 | 280 | 330 | 400 |

[a] Grigg's reagent is acidified ammonium oxalate.
[b] 0 fly ash, 0 rock phosphate treatment received 22.4 metric tons lime and 300 kg/ha triple superphosphate.
[c] LSD = least significant difference.

Table 7
Extractable Mo (mg/kg) with Time in Two Mine Spoils Treated with Fly Ash/Rock Phosphate Mixtures

| Rate Added (metric tons/ha) | | Years of Cropping | | | | | | | |
|---|---|---|---|---|---|---|---|---|---|
| | | Westover, WV | | | Lenox, WV | | | | |
| Fly Ash | Rock Phosphate | 0 | 1 | 2 | 0 | 1 | 2 | 3 | 4 |
| 0[a] | 0 | 12 | 10 | 11 | 16 | 17 | 17 | 15 | 16 |
| 0 | 15 | 8 | 11 | 9 | 11 | 10 | 10 | 8 | 8 |
| 0 | 30 | 14 | 10 | 10 | 12 | 10 | 11 | 10 | 10 |
| 400 | 0 | 1000 | 280 | 150 | 950 | 310 | 170 | 140 | 120 |
| 400 | 15 | 980 | 320 | 160 | 940 | 320 | 160 | 140 | 140 |
| 400 | 30 | 980 | 310 | 170 | 960 | 300 | 170 | 140 | 140 |
| 800 | 0 | 1230 | 420 | 220 | 1150 | 400 | 180 | 160 | 150 |
| 800 | 15 | 1270 | 380 | 220 | 1210 | 380 | 170 | 150 | 140 |
| 800 | 30 | 1300 | 440 | 210 | 1230 | 390 | 180 | 160 | 150 |
| LSD[b] | (0.05) | 130 | 50 | 40 | 120 | 70 | 50 | 40 | 30 |

[a] 0 fly ash, 0 rock phosphate treatment received 22.4 metric tons lime and 300 kg/ha triple superphosphate.
[b] LSD = least significant difference.

could be chemi-sorbed on the surface of Al and Fe oxides or precipitated as $Fe_2(MoO_4)_3$.

On the other hand, there was no change in ammonium oxalate extractable Mo (Grigg's reagent) with cropping (Table 6). This extractant dissolves noncrystalline Al and Fe oxides.[25] Thus, any Mo complexed by oxides of Al and Fe should be

Table 8
Sodium Hydroxide Extractable Mo (mg/kg) with Time in Two Mine Spoils Treated with Fly Ash/Rock Phosphate Mixtures

| Rate Added (metric tons/ha) | | Years of Cropping | | | | | | | |
| --- | --- | --- | --- | --- | --- | --- | --- | --- | --- |
| | | Westover, WV | | | Lenox, WV | | | | |
| Fly Ash | Rock Phosphate | 0 | 1 | 2 | 0 | 1 | 2 | 3 | 4 |
| 0[a] | 0 | 19 | 17 | 16 | 22 | 21 | 18 | 14 | 15 |
| 0 | 15 | 21 | 18 | 19 | 15 | 13 | 15 | 17 | 15 |
| 0 | 30 | 17 | 19 | 18 | 16 | 14 | 19 | 16 | 19 |
| 400 | 0 | 1300 | 420 | 250 | 1210 | 290 | 160 | 120 | 110 |
| 400 | 15 | 1340 | 440 | 260 | 1170 | 280 | 170 | 130 | 120 |
| 400 | 30 | 1280 | 400 | 260 | 1310 | 300 | 160 | 140 | 120 |
| 800 | 0 | 1490 | 440 | 280 | 1470 | 400 | 180 | 150 | 130 |
| 800 | 15 | 1580 | 470 | 240 | 1580 | 420 | 190 | 150 | 140 |
| 800 | 30 | 1600 | 500 | 280 | 1540 | 430 | 190 | 160 | 140 |
| LSD[b] | (0.05) | 150 | 80 | 50 | 160 | 40 | 30 | 40 | 30 |

[a] 0 fly ash, 0 rock phosphate treatment received 22.4 metric tons lime and 300 kg/ha triple superphosphate.
[b] LSD = least significant difference.

solubilized. However, Smith and Leeper[25] argued that ammonium oxalate does not extract Mo by direct complex formation between a Mo cation and oxalate at pH 3.3 since the isoelectric point of hexavalent Mo was < 2.0. Oxalate ions can not extract Mo by an anion exchange mechanism since Grigg's reagent extracts more Mo than that extracted by NaOH, but in an anion exchange process hydroxyls should predominate over oxalate radicals. Thus, Grigg's method was not suitable for monitoring changes in plant available Mo. However, when used along with the other two methods, it provides information that Mo was becoming less available to plants by being complexed with Al and Fe oxide components of fly ash.

## Molybdenum Toxicity in Animals

Development of Mo toxicity in animals arises from a complex interaction among Mo, Cu, and S.[46] Molybdenum concentrations of > 10 mg/kg in forages are known to induce Cu deficiency in animals.[47,48] In the present study, Mo concentrations exceeded this critical level for alfalfa grown on Westover mine spoil treated with 800 metric tons/ha of fly ash (Table 3) in the first year and for all wheat tissues in the first year at both locations (Table 5) at both fly ash application rates. The alfalfa grown on Lenox spoil (Table 2) treated with 800 metric tons/ha of fly ash for the first year only approached this 10 mg/kg level. Thus, these crops had potentially toxic levels of Mo during the first year. After the first year, Mo levels dropped considerably and were much below the critical level of 10 mg/kg.

The toxicity of any level of dietary Mo is affected by the ratio of dietary Cu to dietary Mo. The critical Cu to Mo ratio in animal feeds with respect to incidence of molybdenosis was reported to be 2.0.[49] Lower ratios than these are reported to result in Mo-induced Cu deficiency.[50] The Cu to Mo ratios in fodder (wheat and

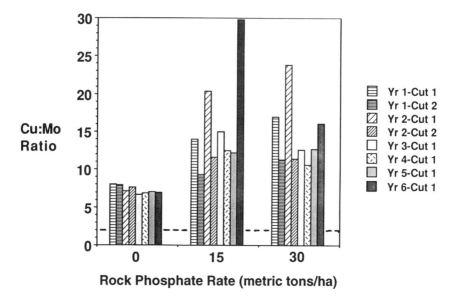

**FIGURE 1.** Cu:Mo ratios in alfalfa grown on Lenox mine spoil treated with rock phosphate but no fly ash (ratios below the dashed line could result in Mo induced Cu deficiency if fed to animals).

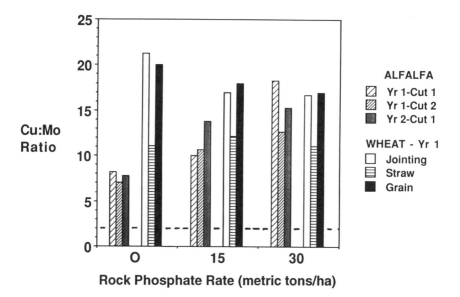

**FIGURE 2.** Cu:Mo ratios in alfalfa and wheat grown on Westover mine spoil treated with rock phosphate but no fly ash (ratios below the dashed line could result in Mo induced Cu deficiency if fed to animals).

alfalfa) harvested in the first year of reclamation (alfalfa cuttings 1 and 2) from both mine spoils not receiving fly ash were satisfactory for animal feed (Figures 1 and 2). The Cu to Mo ratios for wheat from Lenox were similar to those from Westover. On the other hand, the Cu to Mo ratios in both wheat and alfalfa in the first year from fly ash-treated spoils contained hazardous levels of Mo as indicated by the Cu to Mo ratio of < 2.0 when the Cu to Mo ratio was taken as a parameter for risk assessment (Figures 3–5). Data from Westover at rates of 800 and 400 metric tons/ha of fly ash were similar. However, after the first year Cu to Mo ratios in the forage were at acceptable levels, i.e., > 2.0. High levels of S are also known to enhance the excretion of Mo in cattle, reducing the ability of Mo in fodder to cause molybdenosis.[49,50] In the present study, wheat plants contained over two times the S concentration of alfalfa (data not shown) and thus may have less potential Mo toxicity than the Cu to Mo ratios suggest.

## CONCLUSIONS

Crops grown on two fly ash-amended mine spoils contained elevated concentrations of Mo. Molybdenum concentrations of >10 mg/kg, known to induce Cu deficiency in animals, were found in alfalfa grown on the Westover mine spoil treated with fly ash at 800 metric tons/ha in the first year and all wheat tissue in the first year at both Westover and Lenox. Molybdenum concentrations in fly ash-treated mine spoils decreased over successive cuttings. Thus, crops grown after the first year contained nonhazardous levels of Mo and could probably be used as forage for livestock without any adverse effect on animal health. This was substantiated by Cu to Mo ratios in the crops grown. Addition of rock phosphate to fly ash-amended mine spoils had no effect on Mo concentrations of the crops. Thus, rock phosphate can be used to supply the P needs of fly ash-amended mine spoils without adversely impacting the Mo concentrations in the plants. The reduction of Mo availability could have been due to (1) adsorption of Mo on indigenous oxides of Fe, (2) adsorption of Mo on the newly formed Fe oxides from neutralization of reaction products of pyrite oxidation by alkaline materials from fly ash, and/or (3) occlusion of Mo within newly formed oxides of Fe. Molybdenum extracted with Grigg's reagent (acidified ammonium oxalate) did not change with cropping; however, Mo extracted with anion exchange resin or NaOH decreased with time. This relationship demonstrated that Mo availability decreased with cropping due to either chemi-sorption or occlusion of Mo within iron oxides.

## ACKNOWLEDGMENTS

Published with the approval of the Director of the West Virginia Agricultural and Forestry Experiment Station as Scientific Paper No. 2271. This work was supported partially by funds from the WVU Energy and Water Research Center, National Mine Land Reclamation Center, Allegheny Power Service Corporation (via James D. Burnell), and the Hatch Act (NE-159).

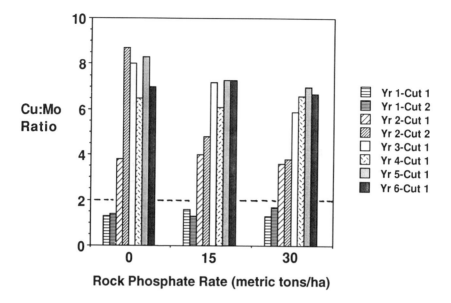

**FIGURE 3.**   Cu:Mo ratios in alfalfa grown on Lenox mine spoil treated with rock phosphate
and 400 metric tons/ha of fly ash (ratios below the dashed line could result in
Mo induced Cu deficiency if fed to animals).

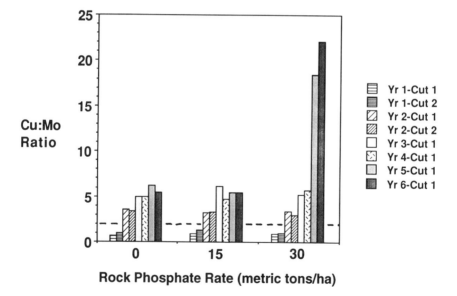

**FIGURE 4.**   Cu:Mo ratios in alfalfa grown on Lenox mine spoil treated with rock phosphate
and 800 metric tons/ha of fly ash (ratios below the dashed line could result in
Mo induced Cu deficiency if fed to animals).

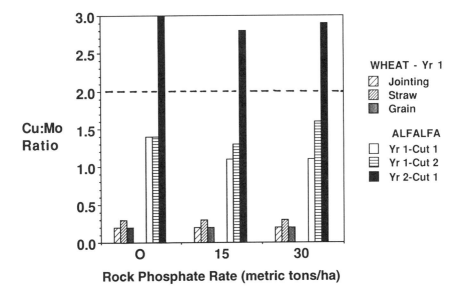

**FIGURE 5.** Cu:Mo ratios in alfalfa and wheat grown on Westover mine spoil treated with rock phosphate and 400 metric tons/ha of fly ash (ratios below the dashed line could result in Mo induced Cu deficiency if fed to animals).

## REFERENCES

1. Kaakinen, J. W., Estimating the potential for Mo enrichment in flora due to fallout from a nearby coal-fired power plant, in *Molybdenum in the Environment*, Chappell, W. R. and Petersen, K. K., Eds., Marcel Dekker, New York, 1977.

2. Anon., National emission inventory of sources and emissions of molybdenum, GCA Corporation, Environmental Protection Agency, NTIS, EPA 450/3-74-009.

3. Adriano, D. C., Page, A. L., Elseewi, A. A., Chang, A. C., and Straughan, I., Utilization and disposal of fly ash and other coal residues in terrestrial ecosystems: a review, *J. Environ. Qual.*, 9, 333, 1980.

4. Valkovic, V., *Trace Elements in Coal,* CRC Press, Boca Raton, FL, 1983.

5. Klein, D. H., Andren, A. W., Carter, J. A., Emery, J. F., Feldman, C., Fulkerson, W., Lyon, W. S., Ogle, J. C., Talmi, Y., Van Hook, R. I., and Bolton, H., Pathways of thirty-seven trace elements through coal-fired power plants, *Environ. Sci. Technol.*, 9, 973, 1975.

6. Page, A. L., Elseewi, A. A., and Straughan, I., Physical and chemical properties of fly ash from coal-fired power plants with reference to environmental impact, *Residue Rev.*, 71, 83, 1979.

7. Ainsworth, C. C. and Rai, D., Selected chemical characterization of fossil fuel wastes, Report EA-5321, Electric Power Research Institute, Palo Alto, CA, 1987.

8. Lindsay, W. L., Inorganic phase equilibria of micronutrients in soils, in *Micronutrients in Agriculture,* Mortvedt, J. J., Giordano, P. M., and Lindsay, W. L., Eds., Soil Science Society of America, Madison, WI, 1972.

9. Doran, J. W. and Martens, D. C., Molybdenum availability as influenced by application of fly ash to soil, *J. Environ. Qual.,* 1, 186, 1972.

10. Elseewi, A. A. and Page, A. L., Molybdenum enrichment of plants grown on fly ash-treated soils, *J. Environ. Qual.,* 13, 394, 1984.

11. Phung, H. T., Lund, L. J., and Page, A. L., Potential use of fly ash as a liming material, in *Environmental Chemistry and Cycling Processes,* Adriano, D. C. and Brisbin, L., Eds., Technical Information Center, U. S. Department of Energy, Augusta, GA, 1978, 504.

12. Capp, J. P. and Engle, C. F., Fly Ash in Agriculture, Bureau of Mines Inf. Circ. 8384, U.S. Department of Interior, Washington, D.C., 1967, 210.

13. Ghazi, H. E., Evaluation of rock phosphate and fly ash amendments to ameliorate toxic conditions in acid mine soils for crop production, Ph.D. dissertation, West Virginia University, Morgantown, 1984.

14. Martens, D. C., Availability of plant nutrients in fly ash, *Compost Sci.,* 12(6), 15, 1971.

15. Macak, J. J., III, and Knight, M. J., Establishment of vegetation on mixtures of bottom ash and fly ash, in *Proc. 7th Int. Ash Util. Symp.,* American Coal Ash Association, 1985, 106.

16. Townsend, W. N. and Gillham, E. W., Pulverized fuel ash as a medium for plant growth, in *The Ecology of Resource Degradation and Renewal,* Chadwick, M. J. and Goodman, G. T., Eds., Blackwell Scientific, Oxford, 1975, 287.

17. Stout, P. R., Meagher, W. R., Pearson, G. A., and Johnson, C. M., Molybdenum nutrition of crop plants. I. The influence of phosphate and sulfate on the absorption of molybdenum from soils and solution cultures, *Plant Soil,* 3, 51, 1951.

18. McLachlan, K. D., Phosphorus, sulfur, and molybdenum deficiencies in soils from Eastern Australia in relation to nutrient supply and some characteristics of soil and climate, *Austr. J. Agric. Res.,* 6, 673, 1955.

19. William, C. and Thornton, I., The effect of soil additives on the uptake of molybdenum and selenium from soils from different environments, *Plant Soil,* 36, 395, 1972.

20. Gupta, U. C. and Monro, D. C., Influence of sulfur, molybdenum, and phosphorus on chemical composition and yields of Brussel Sprouts and of molybdenum on sulfur content of several plant species grown in the greenhouse, *Soil Sci.,* 107, 114, 1969.

21. Singh, M. and Kumar, V., Sulfur, phosphorus, and molybdenum interactions on the concentration and uptake of molybdenum in soybean plants, *Soil Sci.,* 127, 307, 1979.

22. Barshad, I., Factors affecting the Mo content of pasture plants. II. Effect of soluble phosphates, available nitrogen, and soluble sulfates, *Soil Sci.,* 71, 387, 1951.

23. Huisingh, J. and Matrone, G., Interaction of molybdenum in animal nutrition, in *Molybdenum in the Environment,* Chappell, W. R. and Petersen, K. K., Eds., Marcel Dekker, New York, 1977, 125.

24. Bhumbla, D. K., Singh, R. N., and Keefer, R. F., Water quality from surface mined land reclaimed with fly ash, in *Proc. 9th Int. Ash Util. Symp.,* American Coal Ash Association and Electric Power Research Institute, Palo Alto, CA, 1991, 57-1.

25. Smith, B. H. and Leeper, G. W., The fate of the applied molybdate in acidic soils, *J. Soil Sci.,* 20, 246, 1969.

26. Schwertmann, U. and Taylor, R. M., Iron oxides, in *Minerals in Soil Environment,* Dixon, J. B. and Weed, S. B., Eds., Soil Science Society of America, Madison WI, 1989, 408.

27. Norrish, K., Geochemistry and mineralogy of trace elements, in *Trace Elements in Soil-Plant-Animal Systems*, Proc. Jubilee Symp., Waite Agric. Res. Inst., Nicholas, D. J. D. and Egan, E. R., Eds., Academic Press, New York, 1975, 55.

28. Vlek, P. L. G. and Lindsay, W. L., Molybdenum solubility relationships in soils irrigated with high-Mo water, American Society of Agronomy, Madison, WI, *Agron. Abstr.,* 1974, 126.

29. Karimian, N. and Cox, F. R., Adsorption and extractability of molybdenum in relation to some chemical properties, *Soil Sci. Soc. Am. J.,* 42, 757, 1978.

30. Page, A. L., Miller, R. H., Kenney, D. R., Eds., *Methods of Soil Analysis, Part 1,* 2nd ed., Soil Science Society of America, Madison, WI, 1982.

31. Sobek, A. A., Schuller, W. A., Freeman, J. R., and Smith, R. M., Eds., Field and Laboratory Methods Applicable to Overburdens in Mine Soils, U. S. Environmental Protection Agency, Cincinnati, EPA-600/2-78-054, 1978.

32. Page, A. L., Miller, R. H., and Keeney, D. R., Eds., *Methods of Soil Analysis, Part 2. Chemical and Microbiological Properties,* 2nd ed., Soil Science Society of America, Madison, WI, 1982.

33. Ganje, T. J. and Page, A. L., Rapid acid dissolution of plant tissue for cadmium determination by atomic absorption spectrophotometry. *Atomic Absorp. Newslett.,* 13, 131, 1974.

34. Reisenauer, H. M., Molybdenum, in *Methods of Soil Analysis*, Black, C. A., Evans, D. D., Ensminger, L. E., White, J. L., Clark, F. E., and Dinauer, R. C., Eds., American Society of Agronomy, Madison, WI, 1965, 637.

35. Wright, R. J. and Hosner, L. R., Molybdenum release from three Texas soils, *Soil Sci.,* 138, 375, 1984.

36. Gupta, U. C. and MacLeod, L. B., Effects of sulfur and molybdenum on the Mo, Cu, and S concentrations of forage crops, *Soil Sci.,* 119, 441, 1975.

37. Lindsay, W. L., Inorganic phase equilibria of micronutrients in soils, in *Micronutrients in Agriculture*, Mortvedt, J. J., Giordano, P. M., and Lindsay, W. L., Eds., Soil Science Society of America, Madison, WI, 1972, 41.

38. Berger, K. C. and Pratt, P. F., Advances in secondary and micronutrient fertilization, in *Fertilizer Technology and Usage*, McVicar, M. K., Bridger, G. L., and Nelson, L. B., Eds., Soil Science Society of America, Madison, WI, 1965, 313.

39. Pal, U. R., Gossett, D. R., Sims, J. L., and Leggett, J. E., Molybdenum and sulfur nutrition effects on nitrate reduction in burley tobacco, *Can. J. Bot.,* 54, 2013, 1976.

40. Sims, J. L., Leggett, J. E., and Pal, U. R., Molybdenum and sulfur interaction effects on growth, yield, and selected chemical constituents of burley tobacco, *Agron. J.,* 71, 75, 1979.

41. Warren, C. J. and Dudas, M. J., Weathering processes in relation to leachate properties of alkaline fly ash, *J. Environ. Qual.,* 14, 405, 1984.

42. Gupta, U. C., Effect of method of application and residual effects of molybdenum on the molybdenum concentrations and yield of forages on podzol soils, *Can. J. Soil Sci.,* 59, 183, 1979.

43. Reisenauer, H. M., Tabikh, A. A., and Stout, P. R., Molybdenum reactions with soils and the hydrous oxides of iron, aluminum, and titanium, *Soil Sci. Soc. Am. Proc.,* 31, 637, 1962.

44. Reyes, E. D. and Jurinak, J. J., A mechanism of molybdate adsorption on $\alpha$-$Fe_2O_3$, *Soil Sci. Soc. Am. Proc.,* 31, 637, 1967.

45. Ferriero, E. A., Hemly, A. K., and Bussetti, S. G., Molybdate sorption by oxides of aluminum and iron, in *Soils Fert.,* 49, 108, 1985.

46. Barrow, N. J. and Shaw, T. C., The slow reaction between soil and anions. II. Effect of time and temperature on the decrease in phosphate concentration in the soil solution, *Soil Sci.,* 119, 167, 1975.

47. Barrow, N. J. and Shaw, T. C., Effects of solution:soil ratio and vigor of shaking on the rate of phosphate adsorption by soil, *J. Soil Sci.,* 30, 67, 1979.

48. Evans, J. L. and Davis, G. K., Copper-sulfur-molybdenum interrelationships in the growing rat evaluated via the factorial arrangement of diet treatments, in *Molybdenum in the Environment,* Chappell, W. L. and Petersen, K. K., Eds., Marcel Dekker, New York, 1977, 149.

49. Kubota, J., The geochemistry and mineralogy of trace elements, *J. Range Manage.,* 28, 252, 1975.

50. Vlek, P. L. G. and Lindsay, W. L., Molybdenum concentrations in Colorado pasture soils, in *Molybdenum in the Environment,* Chappell, W. L. and Petersen, K. K., Eds., Marcel Dekker, New York, 1977, 619.

51. Multimore, J. E. and Mason, J. L., Copper to molybdenum ratio and molybdenum and copper concentration in ruminant feeds, *Can. J. Anim. Sci.,* 51, 193, 1971.

52. Gupta, U. C. and Lipsett, J., Molybdenum in soils, plants, and animals, *Adv. Agron.,* 34, 73, 1981.

# 12

# Elements in Coal and Coal Ash Residues and Their Potential for Agricultural Crops

M. P. Menon, K. S. Sajwan, G. S. Ghuman, J. James, and K. Chandra

Savannah State College, Savannah, GA

## ABSTRACT

Fly ash samples were collected from six coal-fired power plants in and near the U.S. Department of Energy (DOE) Savannah River Site to study the effects of the application of fly ash/organic compost mixture to soils on the availability and uptake of elements by various plant species. Power plants with emission control devices were efficient in removing atmospheric pollutants from the precipitated ash. The availability of elements to plants varied with the texture of the various types of fly ash. Fly ash-amended homemade compost generally improved plant growth and enhanced yield. Yields of collard greens *(Brassica oleracea* L.) and mustard greens *(Brassica juncea* L.) increased from 400 to 500% for organic compost-treated soil and soil treated with fly ash-amended compost over the bare soil control. Soils treated with fly ash-amended compost often gave higher concentrations than the control for K, Ca, Mg, S, Zn, and B in the Brassica crops.

## INTRODUCTION

Coal is one of the most abundant fossil fuels and provides a cheap source of electrical power throughout the world. The national energy plan indicates that coal production in the U.S. alone will reach an all-time high of 1.91 billion metric tons by the end of 2000.[1] To achieve the national goal of energy independence, coal consumption is expected to increase rapidly.

Following burning of coal for electrical power, large amounts of ash are produced with an estimated annual production of 67 million metric tons, of which about 84% is fly ash.[2] During combustion, products of the burned coal are usually partitioned among the slag (bottom ash), fly ash, and flue gases (vapor and particulate emissions). In plants equipped with particulate emission control devices, i.e., electrostatic precipitators, most (>99%) of the fly ash is efficiently removed.[3,4] Fly

0-87371-890-9/93/$0.00+$.50

ash is collected dry, but bottom ash is usually slurried with water and pumped into ash basins or lagoons.

Less than 15% of the fly ash is used — as a filler in asphalt mix and a stabilizer in road bases, for brick-making, thermal insulation, deodorization of animal wastes, etc.[5] Therefore about 85% of the ash is disposed of, primarily in lagoons or landfills. Large amounts of fly ash need to be disposed of with negligible environmental impact. As the stockpiles of fly ash continue to grow from the increased demand for electricity, there is more concern about the environmental impact of this ash, especially the leaching of the inorganic constituents of ash into domestic water supplies.

Fly ash typically contains a wide range of elements — some essential and others toxic to both plants and animals[6] — which exist in the form of silicates, oxides, sulfates, borates, or borosilicates, phosphates, and carbonates.[7,8] The elemental concentrations of fly ash will depend on a number of factors, such as coal used for combustion, boiler conditions during burning, and trapping efficiency of air emission-control devices. Fly ash usually contains high concentrations of some plant elements, such as Ca, Mg, Fe, K, S, B, and small amounts of Cu, Mn, Mo, P, and Zn. Bituminous and sub-bituminous coals are mostly located in the eastern and midwestern regions of the U.S., whereas the lignites are predominant in the western U.S. The former are characteristically high in S and sometimes produce acidic ash as low as pH 4.5. On the other hand, lignites have lower S and produce ash up to pH 12.[6,9–11] Fly ash from western states is distinctly higher in B but generally lower in other trace elements (i.e., As, Cd, Co, Cr, Pb, Sb, and Zn) than fly ash from eastern and midwestern coals.[10–14]

The chemical composition of fly ash samples collected from six coal-fired power plants in and near the DOE Savannah River Site was investigated in the present study. The objectives were to determine: (1) the difference in the physicochemical characteristics of ash collected with or without electrostatic precipitators, (2) the vertical and horizontal distribution of selected elements in ashes collected and stored at a power plant for elemental enrichment studies, (3) the extent of vertical transport of elements in soil by column experiments, and (4) the feasibility of using fly ash as a mixture with organic compost to grow selected agricultural crops. The main emphasis will be on the behavior of elements that have beneficial or detrimental effects on the growth of plants in soils modified by fly ash or fly ash-amended composts.

## LITERATURE REVIEW

Coal and coal combustion residues have been characterized for elemental analysis by a number of laboratories.[15–20] Swaine and co-workers[15,16] examined Australian coals and compared the elemental composition with coals from several parts of the world, and concluded that the ranges for most trace elements are similar among coals irrespective of their origin. They also observed an inverse relationship between B and ash content, suggesting that B is organically bound in the coal.

The chemical composition and physical properties of fly ash under different experimental conditions have been studied by many researchers to determine the environmental impact of the inorganic constituents at the disposal sites and also to exploit their beneficial effects on agricultural crops.[1,4,11,17–20] Page et al.[11] discussed the particle size distribution, mineralogy, and chemical composition of fly ash and fly ash extracts. Block et al.[19] used neutron activation analysis to determine 43 elements simultaneously in coal ash, bottom ash, and fly ash, and suggested that most elements are associated with the mineral phase, but some may have been organically bound in the coal. They also observed significant differences in the chemical compositions of coal and ash products. Klein et al.[21] reported concentrations and ratios of 37 trace elements in outlet and inlet fly ash and slag from a large cyclone-fed power plant. Elseewi et al.[17] demonstrated that some ionic concentrations in fly ash aqueous extracts increased with dilution, although this system does not represent a true equilibrium. Menon et al.[20] found that at least 5 days were required for equilibration of fly ash-water systems. Davison et al.[22] reported marked increases in the concentrations of some of the trace elements and S with decreasing particle size in fly ash emitted from coal-fired power plants. They suggested a mechanism of high temperature volatilization with subsequent condensation of some trace elements onto small ash particles.

Several investigators[23–30] have used fly ash to amend soil properties for raising agricultural crops. Except for N, fly ash contains most elements required for plant growth. Stone[26] and Page et al.[11] have reported that soil amended with fly ash can retain more available water for plant growth. Hodgson and Holiday[31] recommended the use of surface soil to cover fly ash disposal sites before planting a crop. Salter et al.[25] did not observe any increase in crop yields of carrots, lettuce, radishes, and beets grown on soil amended with fly ash at rates up to 753 metric tons/ha. However, Stone[26] found yield increases of 30% or more for onions and 14% more for cabbages, but no improvement in sugar beets and broad-bean yields when they were grown in surface soil treated with fly ash up to 502 metric tons/ha. Ciravolo and Adriano[28] reported that the yields of corn, bush beans, and white clover in a greenhouse did not increase with ash application — probably from excess salinity, B toxicity, and/or P and N deficiency. Furr et al.[29] also observed that crop yields of seven different vegetables grown on soils amended with 10% fly ash showed inconsistent results. Other workers[32–35] have examined multielement uptake by crops grown on fly ash-amended soils. Fly ash added to acid strip mine soils (pH from 4.0 to 6.0) enhanced the growth and development of soybeans.[36]

Fly ash can also be used with sewage sludge to grow plants.[37] To overcome the deficiencies of N and P in fly ash amended soils, Adriano et al.[2] suggested mixing high-alkaline fly ash with highly carbonaceous material to make composts for soil treatment for agricultural crops. Organic materials composted with fly ash should provide N and P deficient in the ash.[38]

## Essential Elements for Plant Growth

Analytical technique for elemental analyses of coal and coal ashes have included atomic absorption (AA) spectrophotometer in aqueous extracts using different extractants, including DTPA,[17,18] and nondestructive neutron activation analysis (NAA).[19,21]

The metals present in coal or fly ash may be classified as: (1) detrital or occluded in the crystalline state, (2) organically bound, (3) ion-exchangable, and (4) water soluble.[39] Nonmetallic elements such as P, S, and B may exist in the anionic state within the coal or residue as phosphate, sulfate, and borate. Soluble salts can be extracted with water and are usually readily available for plants. On the other hand, organically bound species may need extraction with DTPA or acids. Elseewi et al.[17] have shown that extraction of Ca, Mg, and S with an acetate buffer removes more than extraction by water. This is probably due to the ion-exchange properties of the adsorbed species. Since the primary objective of our work was to determine the usefulness of fly ash and its elements for agricultural crops in the natural environment, the analysis of all elements in fly ash and fly ash-amended composts was performed on only water extracts.

## Vertical and Horizontal Distribution of Elements in Ash Residues

Limited information[11,21] based on the NAA of the residues exists on the dispersion of elements in ash residues derived from coal combustion. Vertical distribution of elements refers to their partition among bottom ash, precipitator ash, dust collector (fly) ash, and flue gas. Flue gas from power plants equipped with electrostatic precipitators consists mostly of volatile elements that amount to only 0.7% of all elements in the coal. Horizontal distribution refers to the concentration profile of elements present in crude coal, weathered ash, slag, and ash from lagoons. The latter samples, except slag, will be depleted in water-soluble constituents due to weathering and leaching processes. Relative enrichment or depletion factors for elements in all of these samples can be calculated if water extracts of the samples are analyzed.

### Physical and Chemical Properties

Elements available in fly ash for agricultural crops originate in coal regardless of the form present in the fly ash. Although the color, nature, and mineralogy of coals from different parts of the world vary, the ranges of elemental content for most of the trace elements are similar.[16] However, the S content and the fly ash generated from the coal combustion are significantly different, depending on the type and origin of the coal,[6] ranging from 0.4 to 4% and 4 to 26%, respectively. An important physical property of fly ash is its capacity to retain water for agricultural systems.[25,34] This aids plants to assimilate elements from fly ash-amended compost applied to soil. Another physical property of coal ash that determines the incorporation of macroelements and other trace elements to the residue is the size of its particles. Since surface area is inversely related to particle size and larger surface area favors metal bonding with the particle, fly ash with smaller particles seems to concentrate the elements more than the larger-sized particles.

Most of the chemical analyses reported for macroelements and other trace elements in fly ash are based on the NAA technique. Some analyses have also been performed in other aqueous systems mixed with fly ash.[17] Extractions of ionic species in an acetate buffer were found to be much higher than those in water. Only water-soluble elements are considered here as these elements are available to plants.

## EXPERIMENTAL METHODS

### Materials Used in the Study

Fly ash samples from five coal-fired power plants, i.e., SRS 484-D, SRS 784-A, SRS 184-K, SRS 105-P, and SRS 200-H at the DOE Savannah River Site were taken by plant operators. Only the SRS 484-D plant was equipped with an electrostatic precipitator which yielded very fine fly ash. Fly ash from other plants contained larger-sized particles. A fine fly ash sample for comparison was collected from the South Carolina Electric and Gas (SCE&G) power plant equipped with an electrostatic precipitator. Crude coal and other ash residues — including bottom ash and boiler ash from vertical hoppers of SRS 484-D plant, and weathered ash and ash from a lagoon — were also furnished by the Savannah River Operation Office. Only fresh fly ash obtained from SRS 484-D was used to make the composts. Composting used three different types of organic manure: "Gotta Grow" (a rich manure) and "Compost Toast" (a low grade manure) were purchased from Bricker's farm in Augusta, GA, and another, "homemade," was composted at our site.[38]

### Extraction of Elements from Coal and Coal Residues

Elements and other ionic species have, in the past, mostly been extracted from fly ash using different extractants such as deionized water, NaCl acidified to pH 3, NaCl alone, or acetic acid-acetate buffer solutions at various fly ash/solution ratios.[17] Except for coal, elements were extracted with water to release elements, as elements in this form will be available to plants grown in fly ash-amended composts. Fly ash samples were extracted by hand-shaking 10 g of the dried ash with 100 mL of deionized distilled water (1:10 dilution) for 1 min, agitating for 1 hr, and allowing them to stand at room temperature for at least 5 days.[20] The pH and electrical conductivity of all water extracts were measured, and elements were analyzed in the aqueous extracts by AA emission. One gram of dried coal was ashed in an electric furnace at 600°C, digested with 1 $M$ HCl, diluted with pure water, filtered, and made up to 100 mL to prepare the stock solution for coal analysis. Two fly ash samples, SRS 484-D and SCE&G, were extracted with water similarly to determine the extent of extractability of water-soluble salts.

The fly ash sample collected from SRS 184-K plant that was not equipped with an electrostatic precipitator was fractionated using standard U.S. testing sieves with different openings. The separated fractions were extracted with water as described and pH, electrical conductivity, and concentration of elements were measured.

## Column Studies to Measure Downward Movement of Elements in Soil

The downward transport of elements ensures that elements are within the reach of plant roots. This can be examined by a profile of their concentrations with depth in the soil. Two identical PVC columns (110 cm ht × 15 cm diam.) filled with a sandy loam soil were brought to our greenhouse from the Savannah River Ecology Laboratory. The columns were provided with five outlets on the side (labeled A-E from top to bottom) for water flow. The columns were extended 45 cm by gluing similar PVC tubes at their tops to introduce fly ash-amended composts and saturated with deionized, distilled water. A blank or control run was made by collecting 100 mL of water from the bottom port E of each column, always keeping the top surface wet. Two types of fly ash — SRS 484-D (fine fly ash) and SRS 184-K (coarse fly ash) — were used to make fly ash-amended composts (20% fly ash) with "homemade" compost. Three kilograms of fly ash-amended composts were placed in each column marked as 484-D and 184-K. A 500 mL aliquot of megapure water was added to each column and mixed with the compost for a few minutes. A 100 mL fraction of the eluant was collected from three ports of each column. These effluents and the controls were analyzed for elements. After 4 weeks, both columns were cut with a power saw and eight horizontal sections of soil, including the fly ash-compost section, were removed for analysis. The soil samples were dried in an oven at 60°C and extracted with deionized, distilled water for analysis.

## Preparation of Fly Ash-Amended Composts

The fly ash to compost ratio was established by mixing fly ash with "homemade" compost in various proportions (10 to 60% fly ash) and testing the manure for plant growth. Plant yields and element concentrations in plants from fly ash-amended composts were the indicators of effective composition. The fly ash-amended compost to soil ratio was established using a fixed amount of "Compost Toast" mixed with 20% fly ash, and varying amounts of soil in pots before sowing the seeds.

## Greenhouse Study

Corn *(Zea mays* L.) and sorghum *(Sorghum vulgare* L.) as spring crops and collard greens *(Brassica oleracea* L.) and mustard greens *(Brassica juncea* L.) as fall crops were seeded in pots lined with plastic bags containing 1.75 kg of fly ash-amended compost of an assigned composition and 5.25 kg of sifted sandy loam soil previously mixed in a twin-shell blender. There were eight sets of pots (four in each set) for each plant including the control with no compost. The plants were thinned to three when they reached a height of about 7 cm, watered regularly with deionized distilled water, and harvested after 8 weeks. The plant tops were dried in the oven for 3 to 5 days, depending on the type of plant, and the oven-dry weights were measured. Three other plants — string beans *(Phaseolus vulgaris*

L.), bell pepper *(Capsicum frutescens* L.), and eggplant *(Solanum melongena* L.) — were also grown in the greenhouse using fly ash-amended compost containing 20% fly ash. Control plants in soil and soil treated with organic compost alone were also grown, harvested, and dried as specified. The dry plant tissues were ground in a stainless steel blender. A subsample of 1 g of each treatment was dry ashed in a quartz crucible at 500°C in a furnance. The ash was digested with 10 mL of 1 *M* HCl, diluted with water, filtered, and brought to 100 mL volume for analyses.

## Chemical Analysis

The concentrations of six elements (K, Ca, Mg, Zn, Mn, and Cu) in all water extracts and in plant samples reported in this work were measured by flame AA. Boron was determined in all samples spectrophotometrically using the improved azomethine-H method.[40] Dissolved orthophosphate was also measured by a spectrometric method using a single reagent consisting of 5 *M* $H_2SO_4$, antimony potassium tartrate, ammonium molybdate, and ascorbic acid.[41] Sulfate was measured by a turbidimetric method using a conditioning agent that consisted of ethanol, NaCl, and glycerol with $BaCl_2$ as precipitating agent.[41] Nitrate was determined potentiometrically using the digital ion analyzer and double junction (Orion) electrode with 2 *M* $(NH_4)_2SO_4$ solution for filling the outer chamber. Nitrate determination by this method was used throughout as a way for measuring N in plants. Although this method does not indicate total N in the plant material, it does provide a reasonable estimate of N present in Brassica plants of the cruciferae family only[42,43] and N by this method was not determined for the other plants grown.

## RESULTS AND DISCUSSION

### Physical and Chemical Properties of Coal and Coal Ash

There were significant differences in the concentration of elements among the three particle sizes (Table 1). Water extracts from fly ash collected from two power plants (SRS 484-D and SCE&G) equipped with electrostatic precipitators were either neutral or basic in water solutions (Table 2). The other fly ash samples were acidic with considerably lower pH values.

There were considerable differences in the chemical composition of samples collected from power plants with and without the emission control devices (Table 2). Calcium was enriched and the microelements Zn, Mn, and Cu were depleted in fly ash from 484-D and SCE&G, as opposed to the samples from those with no precipitators.

The electrical conductivity of the extracts and concentrations of Ca and K in the solution from the two fly ash samples increased with time after mixing and usually reached a plateau after about 100 hr (Figure 1). The conductivity of

**Table 1**
**Some Physical Properties of and Concentrations (mg/kg) of Essential Elements[a] in Fractionated SRS 184-K Fly Ash Samples**

|  | Fraction 1 | Fraction 2 | Fraction 3 |
|---|---|---|---|
| Property |  |  |  |
| Size | <180 μm | <500 μm | >500 μm |
| Percentage | 70.2 | 20.4 | 9.4 |
| EC (dS/m)[b] | 1609 ± 2.8 | 1142 ± 83 | 1438 ± 35 |
| pH | 3.98 | 3.98 | 3.86 |
| K | 733.4 ± 42 | 542.1 ± 22 | 904.8 ± 16.4 |
| Ca | 1715.0 ± 212 | 517.0 ± 58 | 411.0 ± 100 |
| Mg | 291.0 ± 29 | 244.0 ± 6 | 352.0 ± 8 |
| Zn | 28.4 ± 0.4 | 25.2 ± 0.7 | 36.2 ± 1.7 |
| Mn | 46.6 ± 0.7 | 44.2 ± 1.6 | 67.6 ± 3.9 |
| Cu | 0.6 ± 0.3 | 3.1 ± 0.5 | 4.5 ± 2.6 |
| B | 66.2 ± 2.8 | 76.8 ± 5.0 | 76.9 ± 4.3 |
| S | 3562.0 ± 309 | 2960.0 ± 5.0 | 3260.0 ± 227 |

*Note:* ± figures are standard deviation from mean of three replications.

[a] N, P, Fe, and Mo were not measured.
[b] EC = electrical conductivity.

equilibrated water extracts of fly ash was plotted as a function of number of extractions (Figure 2). About 90% of the soluble salts were removed with the first extraction and the percent of extraction dropped to about 10% before leveling off at the third extraction. This suggests that water-soluble elements from fly ash are only briefly available to plants. Elseewi et al.[17] have also observed the effect of aging of fly ash aqueous systems on the basic characteristics of the suspension.

Concentrations of water-soluble elements in three different organic composts and the corresponding fly ash-amended composts (20% fly ash) were compared (Table 3). Data from the 20% fly ash composition were used for comparison because earlier it was found that this composition enhanced the yield of plants when grown in soil mixed with the amended compost.

## Vertical Deposition of Elements in Coal Ashes

There were considerable differences in the deposition of water-extractable elements present in the various residues. The elemental content of these residues is related to the percent of the ash produced from the coal and the efficiency of removal of the element from the pathway and may be represented by the equation:

$$C_X(ash) = F_{ash} \cdot C_X(coal) \cdot REF_X \qquad (1)$$

where $C_X$ (ash) in ash and $C_X$ (coal) in coal is the concentration of an element X, $F_{ash}$ is the fraction that produced the ash, and $REF_X$ is the removal efficiency factor for the element. In the absence of correct estimate of F values, one can define a relative removal efficiency factor (RREF) for each element, choosing a reference element to estimate the relative efficiencies for the deposition of elements in the

**Table 2**
**Electrical Conductivity, pH, and Concentrations (mg/kg) of Essential Elements in Water Extracts of Fly Ash from Different Power Plants**

| | Power Plant Sampled | | | | | | Literature Values |
| | SRS 484-D[a] | SRS 784-A | SRS 184-K | SRS 105-P | SRS 200-H | SCE&G[a] | Conc. Range[b] |
|---|---|---|---|---|---|---|---|
| EC[c] (dS/m) | 927 | 1199 | 1816 | 973 | 999 | 1095 | — |
| pH | 6.96 | 4.25 | 3.94 | 4.47 | 4.58 | 8.37 | — |
| Element | | | | | | | |
| K | 184 | 510 | 124 | 150 | 461 | 449 | 1,900–30,000 |
| Ca | 1895 | 875 | 1517 | 1867 | 1443 | 1942 | 1,100–50,000 |
| Mg | 41 | 97 | 240 | 107 | 86 | 109 | 400–10,000 |
| Zn | 0.2 | 39.3 | 47.2 | 5.3 | 29.8 | 0.3 | 70–1,000 |
| Mn | 4.6 | 34.6 | 724 | 63 | 53 | 7.5 | 100–1,000 |
| Cu | 0.7 | 5.9 | 4.9 | 1.0 | 8.9 | 0.6 | 33–1,000 |
| B | 29.3 | 35.3 | 65 | 33 | 27 | 48 | — |
| P | ND | 0.2 | 2.2 | ND | ND | 1.8 | — |
| S | 683 | 1833 | 3667 | 1117 | 1217 | 1517 | — |
| N[d] | 7.9 | 6.6 | 5.1 | 7.5 | 6.3 | 6.9 | — |

a  Equipped with electrostatic precipitators.
b  Davidson et al., Ref. 22; analysis was done mostly by NAA.
c  EC = electrical conductivity.
d  Nitrate method.

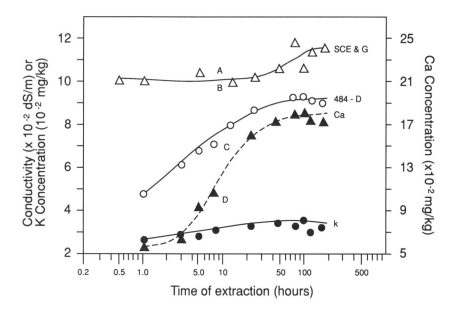

**FIGURE 1.**    Measured conductivity and the concentration of K and Ca vs. extraction time (A: conductivity of SCE&G sample; B: conductivity of SRS 484-D sample; C: concentration of Ca of SRS 484-D sample; D: concentration of K of SRS 484-D sample).

residues. Using Mg as a reference element, the RREF can be expressed by the equation:

$$RREF_X = REF_X \,/\, REF_{Mg} = \left( C_X(ash) \,/\, C_X(coal) \,/\, C_{Mg}(ash) \,/\, C_{Mg}(coal) \right) \quad (2)$$

when one uses a value of 1.0 for $REF_{Mg}$. The values for RREF compared to Mg for each of the elements studied were calculated (Table 4). The relative efficiency for the removal of Ca from the flue gas and deposition on the residues was very high. Bottom ash accumulated S and N in the form of sulfate and nitrate very efficiently. Potassium was also removed efficiently by the precipitators. The basic character of ash residues collected from power plants equipped with electrostatic precipitators may be attributed to the efficient removal of base metals from the gas stream onto the residues. Fly ash samples collected from all power plants that do not have electrostatic precipitators were acidic (Table 2).

## Distribution of Elements in Coal and Coal Ashes

Although NAA data on the distribution of a large number of elements among coal, slag (bottom ash), inlet fly ash (precipitator ash) and outlet fly ash have been reported,[21] elemental content of weathered ash, and lagoon ash were seldom reported; probably because these residues may exhibit significant

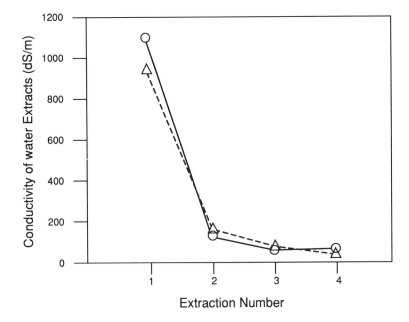

**FIGURE 2.**  Conductivity of fly ash extracts vs. extraction number (0: water extracts from SCE&G samples; Δ: water extracts from SRS-D samples).

variation in the elemental content due to leaching of the inorganic constituents by rainfall and other natural processes. Periodic chemical analysis of these ash residues may reveal the extent of leaching of the chemicals from these systems if one constructs a model using the average daily input of ash residues to the system, chemical state of the elements present, and the natural processes occurring at the site.

The present study examined the element content of coal combustion residues at the time of their collection (Table 5). Potassium was high in bottom ash and Ca was high in ash from the lagoon. The large amount of water-extractable Ca in the crude coal is difficult to explain. The enrichment of Ca in ash from the lagoon may be due to the accumulation of the insoluble $CaCO_3$ and/or slightly soluble CaO. The high solubility of K in water precludes the rapid leaching of this element. All of the three transition metals — Zn, Mn, and Cu — are much more concentrated in weathered ash than in others. Although bottom ash removes a significant fraction of S, the weathered ash accumulated more S as sulfate than the rest. Weathered ash and ash from lagoons were richer in B than in the bottom ash.

## Downward Transport of Elements in Soil

The first 0 to 15 cm section represents the unused fly ash-amended compost, the second 15 to 20 cm section records the concentrations in the interphase between the compost and soil and the remaining sections are soil only (Table 6).

**Table 3**
**Comparison of the Analysis of Essential Elements (µg/g) of Three Organic Composts Amended with 20% SRS 484-D Fly Ash[a]**

| Element | "Gotta Grow" | | "Compost Toast" | | "Homemade Compost" | |
|---|---|---|---|---|---|---|
| | 0% FA | 20% FA | 0% FA | 20% FA | 0% FA | 20% FA |
| K | 23,026[a] | 22,709 | 783 | 657 | 3197 | 2952 |
| Ca | 1,352 | 1,915 | 411 | 149 | 294 | 544 |
| Mg | 436 | 228 | 67 | 42 | 175 | 385 |
| Zn | 69 | 53 | 3.7 | 0.5 | 2.6 | 1.3 |
| Mn | 6.2 | 8.2 | 6.1 | 2.3 | ND[b] | ND |
| Cu | 17 | 14 | 2.9 | 3.1 | ND | ND |
| B | 80 | 33 | 9.8 | 29.7 | 23.3 | 18.1 |
| P | 316 | 324 | 2409 | 2041 | 560 | 367 |
| S | 2550 | 2185 | NM[c] | NM | 4951 | 4308 |
| N[d] | 939 | 692 | NM | NM | 893 | 1320 |
| pH | 8.5 | 8.1 | 7.0 | 7.2 | 7.6 | 7.2 |
| E.C. (dS/m) | 9050 | 5350 | 573 | 402 | 419 | 591 |

[a] Each value is mean of three samples.
[b] Not detected.
[c] NM = not measured.
[d] Nitrate method.

Concentrations of most of the elements decreased with depth and probably leveled off at a depth of about 80 cm. More Mn was evident in sections below 80 cm depth in both columns, indicating that Mn probably existed as soluble $Mn.^{2+}$ If it existed as an anionic species, its rate of movement would probably have been smaller.

The elution pattern of the elements is somewhat different from the soil deposition or binding characteristics of the metal ions. The concentration of all the metals except Zn were highest in the eluate from the bottom port (Port E) in SRS 484-D column (Table 7). The eluate collected from Port E of this column, before the addition of fly ash-amended compost (control), also had significant amounts of K, Mg, and Ca. On the other hand, in column SRS 184-K eluate from the uppermost port (Port A) shows the maximum amounts of all metals. Control eluate from this column had low amounts of these metals. This difference in the elution pattern may be attributed to the coarse texture of the fly ash in column 184-K used for the amendment of the compost. Elution in 484-D probably followed the ion-exchange mechanism involving sorption and desorption, while in column 184-K water-soluble elements moved down with the flow of water so fast that ion exchange did not occur.

The flow rate in column 184-K was much higher than that in column 484-D, indicating that the water retention capacity of fine fly ash is much higher than that of the "coarse" fly ash. This may be applicable for the utilization of coarse fly ash in landfills where fast drainage is required. Outlet ash (i.e., "coarse fly ash") from power plants without an ash precipitator, such as 184-K, drained quickly.

**Table 4**
**Vertical Distribution (Concentration in mg/kg) and Removal Efficiency of Coal Ashes for Essential Elements During Coal Combustion in SRS 484-D Plant[a]**

| Element | Crude Coal Conc. | Bottom Ash Conc. | Bottom Ash RREF[b] | Boiler Ash 1 Conc. | Boiler Ash 1 RREF | Boiler Ash 2 Conc. | Boiler Ash 2 RREF | Boiler Ash 3 Conc. | Boiler Ash 3 RREF | Fresh Fly Ash Conc. | Fresh Fly Ash RREF |
|---|---|---|---|---|---|---|---|---|---|---|---|
| K | 1883 ± 126 | 15.8 | 0.88 | 261.4 | 2.7 | 1152.5 | 3.54 | 1164.6 | 3.54 | 138.3 | 1.44 |
| Ca | 405 ± 26 | 51.2 | 13.2 | 1245.8 | 59.9 | 1574.5 | 22.5 | 1581.3 | 22.1 | 779.3 | 37.8 |
| Mg | 725 ± 21 | 6.9 | 1.0 | 37.2 | 1.0 | 125.5 | 1.0 | 128.3 | 1.0 | 36.9 | 1.0 |
| Zn | 26.3 ± 2 | 0.34 | 1.34 | 0.45 | 0.33 | 3.9 | 0.84 | 5.2 | 1.1 | 0.90 | 0.67 |
| Mn | 27.6 ± 4 | 0.30 | 1.1 | 1.26 | 0.88 | 4.7 | 0.96 | 4.7 | 0.96 | 0.82 | 0.58 |
| Cu | 35.8 ± 4 | 0.71 | 2.1 | 0.50 | 0.27 | 5.5 | 0.88 | 5.9 | 0.93 | 0.43 | 0.23 |
| B | 62.3 ± 6 | 8.3 | 13.9 | 298.5 | 93.4 | 823.5 | 76.4 | 842.5 | 76.5 | 69.4 | 21.9 |
| S | 3445 ± 751 | 895.8 | 37.6 | 0.0 | 0.33 | 3854.3 | 0.33 | 3906.5 | 0.33 | 999.8 | 0.29 |
| N[c] | 93 ± 22 | 296.8 | 333.8 | 80.4 | 16.9 | 14.2 | 0.88 | 5.4 | 0.33 | 108.3 | 22.9 |

*Note*: ± figures are standard deviation from mean of three replications.

[a] Each value is mean of three samples.
[b] RREF = relative removal of efficiency factor.
[c] Nitrate method.

**Table 5**
**Distribution of Essential Elements (Concentration in mg/kg) in Coal and Coal Ashes Accumulated in the Vicinity of the Power Plants[a]**

| Element | Crude Coal | Weathered Ash | Bottom Ash | Ash from Lagoon |
|---|---|---|---|---|
| K | 14.4 | 5.0 | 15.8 | 9.6 |
| Ca | 1039.0 | 114.0 | 51.2 | 346.4 |
| Mg | 117.4 | 109.4 | 6.9 | 16.2 |
| Zn | 6.2 | 4.5 | 0.34 | 0.39 |
| Mn | 11.2 | 16.1 | 0.30 | 0.30 |
| Cu | 2.8 | 3.7 | 0.71 | 0.43 |
| B | 23.9 | 31.9 | 8.3 | 40.3 |
| S | 3448.0 | 4166.8 | 895.8 | 958.5 |
| N[b] | 84.0 | 33.4 | 296.8 | 85.3 |

[a] Water-soluble elements; P was not detected, Fe and Mo were not measured.
[b] Nitrate method.

**Table 6**
**Movement of Essential Elements (Concentration in mg/kg) in Two Soil Columns After Elution with Water[a]**

| Column Section | K | Ca | Mg | Zn | Mn | Cu | B | P | S | N[b] |
|---|---|---|---|---|---|---|---|---|---|---|
| Column1[c] | | | | | | | | | | |
| 0–15 cm | 402 | 332 | 121 | 0.75 | 3.63 | 0.38 | 33.7 | 12.1 | 708 | 140 |
| 15–20 cm | 102 | 128 | 58 | 0.40 | 0.42 | 0.25 | 6.5 | 1.8 | 250 | 136 |
| 20–40 cm | 29 | 62 | 32 | 0.47 | 1.56 | 0.17 | 2.4 | 0.3 | 167 | 72 |
| 40–60 cm | 24 | 24 | 12 | 0.31 | 3.97 | 0.21 | 5.5 | 1.5 | 242 | 40 |
| 60–80 cm | 12 | 27 | 10 | 0.35 | 1.28 | 0.25 | 2.4 | 2.7 | 275 | 38 |
| 80–100 cm | 19 | 37 | 16 | 0.53 | 10.70 | 0.03 | 14.4 | 0.9 | 267 | 47 |
| 100–115 cm | 14 | 38 | 6 | 0.23 | 2.75 | 0.04 | 7.4 | 1.5 | 275 | 45 |
| 115–125 cm | 5 | 11 | 7 | 0.23 | 5.97 | 0.08 | 1.7 | 0.0 | 242 | 42 |
| Column 2[d] | | | | | | | | | | |
| 0–15 cm | 286 | 92 | 58 | 0.85 | 2.00 | 0.28 | 11.4 | 18.7 | 258 | 92 |
| 15–20 cm | 57 | 24 | 16 | 0.30 | 0.22 | 0.17 | 2.4 | 8.9 | 133 | 149 |
| 20–40 cm | 35 | 36 | 12 | 0.91 | 2.22 | 0.25 | 12.2 | 0.0 | 200 | 157 |
| 40–60 cm | 12 | 40 | 10 | 0.52 | 1.75 | 0.38 | 12.0 | 2.4 | 300 | 29 |
| 60–80 cm | 16 | 37 | 10 | 0.56 | 2.89 | 0.04 | 0.0 | 1.5 | 184 | 138 |
| 80–100 cm | 17 | 33 | 10 | 0.46 | 4.11 | 0.17 | 2.0 | 0.0 | 133 | 174 |
| 100–115 cm | 18 | 32 | 2 | 0.19 | 3.31 | 0.10 | 1.7 | 1.2 | 92 | 127 |
| 115–125 cm | 8 | 8 | 1 | 0.53 | 2.91 | 0.03 | 1.7 | 0.6 | 133 | 87 |

[a] Average of three samples; analytical error is less than 10%.
[b] Nitrate method.
[c] Filled with sandy loam soil covered with 3 kg 484-D and fly ash-amended compost.
[d] Filled with sandy loam soil covered with 3 kg 184-K fly ash-amended compost.

## Analysis of "Homemade" Compost and Fly Ash-Amended Composts

Some of the elements from fly ash were released to the soil during composting (Table 8). Mass balance calculations using Equation 3 and data from Table 2, column 2 and Table 8, column 4 revealed that fly ash-amended compost (20% fly ash) was enriched by 12% in K and 62% in Mg, whereas Ca, Zn, and B were depleted by 13, 63 and 37%, respectively; N and S were also enriched by 46 and 4.5%, respectively.

**Table 7**
**Concentrations (mg/L) of Five Metallic Elements in 100-ml Eluates in Two Columns Using Surface-Applied Fly Ash-Amended "Homemade" Composts[a]**

| Element | Column 1[b] | | | | Column 2[c] | | | |
|---|---|---|---|---|---|---|---|---|
| | Control | Port B | Port D | Port E | Control | Port A | Port C | Port E |
| K | 2.18 | 1.15 | 1.48 | 2.25 | 1.03 | 10.51 | 1.98 | 2.18 |
| Mg | 6.19 | 1.44 | 1.37 | 2.73 | 0.47 | 10.66 | 1.50 | 2.30 |
| Ca | 27.40 | 3.67 | 6.76 | 25.10 | 1.93 | 32.40 | 4.25 | 21.00 |
| Zn | 0.04 | 0.05 | 0.12 | 0.04 | 0.01 | 0.32 | 0.24 | 0.05 |
| Mn | 0.08 | 0.12 | 0.39 | 0.40 | 0.02 | 4.24 | 0.12 | 0.44 |

*Note:* Ports A, B, C, D, and E are outlets in the column starting from the top for leachate collection.

[a] Average of three samples, analytical error is less than 10%.
[b] Covered with 484-D fly ash-amended compost (very fine and slightly basic fly ash).
[c] Covered with 184-K fly ash-amended compost (coarse and acidic fly ash).

$$\%E = 100\left(\left\{C_X(FA-Comp)_{obs.} - C_X(FA).a_{FA} + C_X(Comp).b_{Comp}\right\}\right)$$
$$/C_X(FA-Comp)_{obs}$$

(3)

where $\%E$ is the percent enrichment, $C_{Xs}$ are concentrations of X in fly ash, compost and fly ash-amended composts and $a_{FA}$ and $b_{COMP}$ are the proportions of fly ash and compost in the fly ash-amended compost.

Similar enrichment and depletion, but of different magnitude, were also observed in fly ash-amended composts of other compositions, suggesting that chemical interaction between the fly ash and the organic compost occurred, changing the solubility of the elements within the growth media. Fly ashes amended with composts resulted in additional elements such as Mg, Zn, P, S, and N from the compost that may not be available in fly ash for plant nutrition. Fly ash and compost soil ratios also affected crop growth, with the optimum ratio being 1:3 for maximum yields.

Aging of the organic compost before mixing fly ash also affected the release of elements in fly ash-amended composts (Table 9). Soil usually had the least amount of elements while coal, fly ash, and leaf ash (ash obtained after burning leaves) differed greatly in their chemical composition with respect to water-soluble elements. Great differences were also observed in the element content of fly ash-amended composts (20% fly ash) compared with compost 1 or compost 2.

## Effect of Fly Ash-Amended Compost on Agricultural Crops

Yields of corn and sorghum have been reported elsewhere.[38] Dry matter yields of both mustard greens and collard greens were higher when treated with fly ash-amended composts than with organic compost alone (Figure 3). Yields were even much higher than those when the plants were grown on soil only (control) with no compost. Yields and element concentrations of the four plants grown on soil (control) and soil mixed with fly ash-amended composts of

**Table 8**
**Elemental Analysis (Concentration in mg/kg) of "Homemade" Fly Ash-Amended Composts[a]**

| Element | Compost 1 (0% Fly Ash) | Compost 2 (15% Fly Ash) | Compost 3 (20% Fly Ash) | Compost 4 (30% Fly Ash) | Compost 5 (40% Fly Ash) | Compost 6 (50% Fly Ash) | Compost 7 (60% Fly Ash) | F[b] Variance |
|---|---|---|---|---|---|---|---|---|
| K | 3197[c] | 2575 | 2952 | 2059 | 1501 | 1501 | 1189 | 25.8 |
| Ca | 294 | 274 | 544 | 343 | 347 | 363 | 528 | 19.5 |
| Mg | 175 | 159 | 385 | 187 | 155 | 163 | 157 | 79.3 |
| Zn | 2.6 | 1.9 | 1.3 | 0.7 | 0.6 | 0.5 | 0.2 | 15.4 |
| B | 23.3 | 29.1 | 18.1 | 32.4 | 30.9 | 40.3 | 41.4 | 11.1 |
| P | 560 | 269 | 367 | 133 | 69 | 73 | 24 | 69.7 |
| S | 4951 | 3986 | 4308 | 4044 | 3469 | 3760 | 3292 | — |
| N[c] | 893 | 913 | 1320 | 761 | 1238 | 1624 | 1077 | 16.7 |

[a] SRS 484-D fly ash was used for amendment.
[b] Two-way variance analysis using results from replicate analysis, significant at $p = 0.01$.
[c] Nitrate methods.

**Table 9**
**Essential Element Concentrations (mg/kg) in Soil, Coal, Ashes, Composts, and Fly Ash-Amended Composts**

| Element | Soil | Coal | Fly Ash | Leaf Ash[a] | Compost 1[b] | Compost 2[c] | Fly Ash-amended Compost[d] |
|---|---|---|---|---|---|---|---|
| K | 261 | 1,883 | 184 | 12,630 | 7,557 | 19,357 | 16,322 |
| Ca | 222 | 405 | 1,895 | 1,750 | 697 | 8,864 | 4,522 |
| Mg | 22 | 725 | 41 | 2,188 | 415 | 3,210 | 1,668 |
| Zn | 0.4 | 26.3 | 0.2 | 0.1 | 6.2 | 1.6 | 0.7 |
| Mn | 0.2 | 27.6 | 4.6 | 0.2 | ND[e] | 1.1 | 0.3 |
| Cu | 0.0 | 35.8 | 0.7 | 0.1 | ND | 0.4 | 0.1 |
| B | NM[f] | 23.9 | 29.3 | 10.2 | 23.3 | 2.3 | 9.9 |
| P | 75.6 | ND | ND | 28.1 | 560.0 | 222 | 195 |
| S | 875 | 3445 | 683 | 725 | 4,951 | 3,186 | 3,372 |
| N[g] | 444 | 92.9 | 7.9 | NM | 2,116 | 4,673 | 2,009 |

a   Leaf ash = ash obtained after burning leaves.
b   Six week-old organic compost.
c   Eight month-old organic compost.
d   Compost 2 was amended with 20% 484-D fly ash.
e   ND = not detected.
f   NM = not measured.
g   Nitrate method.

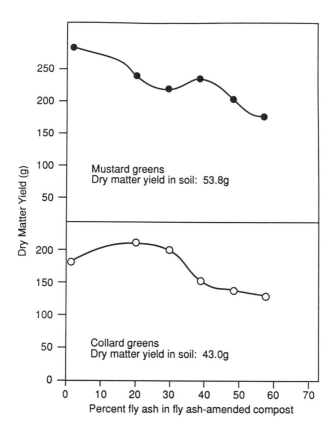

**FIGURE 3.** A plot of plant yields as a function of percent fly ash in fly ash-amended compost (o: collard green; •: mustard green).

optimum composition were calculated (Table 10). Maximum yields were obtained for corn and sorghum from the 30% fly ash/compost mixture, whereas for collard and mustard greens they were 20 and 40% fly ash in the mix, respectively. At these levels of fly ash, the yield increases over the control were 114% for corn, 106% for sorghum, 378% for collard greens, and 348% for mustard greens . Although an increase in crop yields can be expected for these plants on fly ash-amended compost compared to soil, the magnitude of the increase will depend on the nature of the plant and the proportion of fly ash mixed with the compost.

The concentrations of certain elements, especially Ca, were higher in plants grown on bare soil than in plants grown on fly ash-amended compost. Mustard greens, green beans, eggplants, and bell peppers grown in organic compost had the highest yields (Table 11). String beans, bell peppers, and eggplants grew poorly on base soil or soil treated with 20% fly ash-compost, but collard greens had highest yield on soil treated with 20% fly ash-amended compost. Potassium concentrations were higher in all plants treated with compost than with base soil, whereas Zn concentrations were high for fly ash-treated pots (Table 12). Magne-

**Table 10**
**Dry Matter Yields (g) and Essential Element Concentrations (mg/kg) in Four Plants Using Fly Ash-Amended Composts[a]**

| | Corn | | Sorghum | | Collard Greens | | Mustard Greens | |
|---|---|---|---|---|---|---|---|---|
| | Control (Soil) | Compost 1 (30% Fly Ash) | Control (Soil) | Compost 2 (30% Fly Ash) | Control (Soil) | Compost 3 (20% Fly Ash) | Control (Soil) | Compost 4 (40% Fly Ash) |
| Dry Plant Yield (g) | 21 | 45 | 16 | 33 | 43 | 206 | 54 | 241 |
| Element | | | | | | | | |
| K | 11,000 | 31,000 | 12,000 | 38,000 | 28,500 | 86,400 | 20,700 | 70,400 |
| Ca | 12,000 | 9,200 | 11,100 | 9,500 | 43,300 | 40,300 | 23,700 | 18,000 |
| Mg | 11,000 | 3,300 | 7,400 | 7,300 | 2,330 | 6,930 | 1,400 | 3,010 |
| Zn | 33 | 23 | 14 | 19 | 17 | 66 | 25 | 42 |
| Mn | 51 | 18 | 38 | 28 | 13 | 14 | 11 | 9 |
| Cu | 12 | 8 | 10 | 11 | 20 | 14 | 19 | 15 |
| B | 26 | 75 | 24 | 21 | 26 | 60 | 74 | 58 |
| P | 1,240 | 2,320 | 1,780 | 3,160 | 5,480 | 5,660 | 3,430 | 3,290 |
| S | 5,300 | 9,300 | 6,200 | 6,100 | 8,470 | 19,030 | 8,330 | 7,080 |
| N[b] | ND[c] | ND | ND | ND | 2,750 | 2,240 | 2,950 | 2,370 |

a  Maximum yields with most effective fly ash/compost ratios were used for comparison.
b  Nitrate method.
c  ND = not determined.

**Table 11**
**Dry Matter Yield (g) of Plants Grown in Bare Soil, Soil Treated with Organic Compost, and Soil Treated with 20% Fly Ash-Amended Compost[a]**

| Plant Species | Bare Soil | Soil + Organic Compost | Soil + 20% Fly Ash Amended Compost |
|---|---|---|---|
| Collard greens (Brassica oleracea L.) | 43.0 | 172.3 | 205.7 |
| Mustard greens (Brassica juncea L.) | 53.8 | 284.6 | 237.8 |
| String Beans (Phaseolus vulgaris L.) | 12.2 | 25.8 | 5.3 |
| Eggplants (Solanum melonguna L.) | 7.0 | 22.7 | 11.9 |
| Bell peppers (Capsicum frutescens L.) | 6.0 | 17.6 | 9.9 |

[a] Average of three samples; standard deviation is less than 10%.

sium concentrations were higher than the control for most treated plants (Table 13).

In order to evaluate the relative plant capacity to absorb elements under different nurturing conditions, an enrichment factor (EF) or a deficiency factor (DF) was defined as follows:

$$EF \ or \ DF = \frac{C_{FA-Comp}(X) - C_{soil}(X) \ or \ C_{org-Comp}(X)}{C_{Sb(ave)}}$$

where $C_{FA-Comp}(X)$ is the concentration of element X in plant nourished by fly ash-amended compost, $C_{soil}(X)$ is its concentration in plants raised in soil, $C_{org-Comp}(X)$ is its concentration in plant treated with organic compost and $C_{Sb(ave)}$ is the average content of Sb in these plants. The concentration of Sb was nearly constant in all treatments, so it was used as a reference element for normalization to account for the differences in moisture content, uniformity of medium, and the biomass production of plant. Numbers that are positive indicate an enrichment and those that are negative show deficiency of that element in plants nourished by fly ash-amended compost relative to those raised in soil or soil treated with organic compost.

Except for eggplants grown on fly ash-amended compost, there were no noticeable N deficiencies in plants (Table 14). There were deficiencies for Ca in beans, peppers, and eggplants that corresponded to the relatively low Ca concentrations of these plants (Table 14). The enrichment factor for B was also much higher for all plants. Thus, some phytotoxic effect of B along with relatively low Ca concentrations in beans, bell peppers, and eggplants may have been responsible for their low yield (Table 12).

## Correlation Studies

Plant element concentrations were correlated with composition of fly ash-amended "Homemade" composts for corn and sorghum (Table 15) to determine

**Table 12**
**Concentration of Essential Elements (mg/kg dry matter) in Collard Greens and Mustard Greens Grown in Bare Soil, Organic Compost, or 20% Fly Ash-Amended Compost**

| Element | Collard Greens | | | Mustard Greens | | |
|---|---|---|---|---|---|---|
| | Bare Soil | Organic Compost | 20% Fly Ash-Amended Compost | Bare Soil | Organic Compost | 20% Fly Ash-Amended Compost |
| N[b] | 2750 | 3500 | 2240 | 2950 | 3440 | 2430 |
| P | 4476 | 4834 | 4824 | 3493 | 4213 | **4382**[c] |
| K | 19171 | 39978 | **48563**[c] | 21669 | 63497 | **70836** |
| Ca | 36300 | 34536 | 37985 | 27055 | 19417 | 23970 |
| Mg | 1964 | 3381 | 4079 | 1107 | 2037 | **2172** |
| S[d] | 8470 | 10280 | 19037 | 8330 | 8060 | 5830 |
| Zn[e] | 17.4 | 51.0 | **65.9** | 25.3 | 40.9 | **52.4** |
| Mn | 13.3 | 22.7 | 13.5 | 10.7 | 12.7 | 8.7 |
| Cu | 6.3 | 6.8 | 1.0 | 18.5 | 15.8 | 13.3 |
| B | 19.3 | 8.0 | **27.8** | 17.3 | 6.3 | 18.6 |
| Fe | 55.2 | 72.5 | 67.5 | 62.4 | 63.1 | 64.7 |
| Mo | 8.8 | 1.3 | 4.3 | 5.2 | 0.0 | 1.6 |

a Each value is mean of three samples.
b Nitrate method.
c Boldface means elevated level.
d Turbidimetric.
e Atomic absorption; all the rest by ICP.

**Table 13**
Concentration of Essential Elements (mg/kg dry matter) in Beans, Pepper and Eggplant Grown in Potted Soil Under Different Conditions[a]

| Element | String Beans | | | Bell Pepper | | | Eggplant | | |
|---|---|---|---|---|---|---|---|---|---|
| | Bare Soil | Organic Compost | Fly Ash-Amended Compost | Bare Soil | Organic Compost | Fly Ash-Amended Compost | Bare Soil | Organic Compost | Fly Ash Amended Compost |
| P | 3468 | 3911 | 4101 | 3592 | 3907 | **5006**[b] | 3232 | 3634 | **5566** |
| K | 9179 | 28928 | **45144**[b] | 38142 | 63120 | **71795** | 16707 | 44009 | **46325** |
| Ca | 23340 | 30084 | 12316 | 12507 | 12520 | 10938 | 17482 | 22019 | 16292 |
| Mg | 3809 | 4584 | 2747 | 4286 | 4600 | 4487 | 3548 | 3914 | 4076 |
| S[c] | 9100 | 7920 | 9430 | 8470 | 7010 | 7990 | 8190 | 7990 | **12080** |
| Zn[d] | 68.7 | 71.3 | 56.3 | 81.1 | 103.0 | 60.6 | 67.5 | 93.0 | 81.9 |
| Mn | 123.2 | 42.2 | 64.9 | 140.8 | 45.7 | 68.5 | 71.8 | 54.8 | 83.9 |
| Cu | 10.8 | 11.1 | **20.3** | 13.3 | 13.1 | 14.6 | 11.6 | 15.5 | 13.0 |
| B | 65.1 | 75.8 | **177.3** | 60.8 | 60.5 | **252.8** | 50.9 | 51.9 | **158.1** |
| Fe | 137.8 | 167.7 | 126.8 | 245.6 | 151.0 | 132.4 | 169.5 | 185.4 | 190.8 |

[a] Each value is mean of three samples.
[b] Boldface means elevated level.
[c] Turbidometric.
[d] Atomic absorption; all the rest by ICP.

**Table 14**
Enrichment (+) or Deficiency (−) Factors for Several Essential Elements in Plants Grown in Bare Soil or 20% Fly Ash-Amended Compost

| | Collard Greens | | Mustard Greens | | String Beans | | Bell Pepper | | Eggplant | |
|---|---|---|---|---|---|---|---|---|---|---|
| | Soil | Compost | Soil | Compost | Soil | Compost | Soil | Compost | Soil | Compost |
| Ave. Sb Conc. (mg/kg) | 12.6 ± 0.6 | | 13.2 ± 2.3 | | 16.2 ± 0.4 | | 14.4 ± 0.8 | | 12.5 ± 0.1 | |
| Element | | | | | | | | | | |
| Na[a] | −40.0 | −100 | −39.0 | −76.5 | ND[b] | ND | ND | ND | ND | ND |
| P | 27.6 | 0.8 | 67.3 | 12.8 | 39.0 | 11.7 | 98.0 | 76.3 | 187 | 155 |
| K | 2333 | 681 | 3725 | 556 | 2220 | 1001 | 2337 | 602 | 2369 | 185 |
| Ca | 134 | 274 | −234 | 345 | −680 | −1097 | −109 | −110 | −95 | −458 |
| Mg | 168 | 55 | 81 | 10.2 | −66 | −113 | 14.0 | −7.8 | 42.4 | 13.0 |
| S | 638 | 694 | −189 | −184 | 20.4 | 93.2 | −33.0 | 68.0 | 311 | 327 |
| B | 0.67 | 1.6 | 0.1 | 0.93 | 6.9 | 6.3 | 13.3 | 13.3 | 8.6 | 8.5 |

*Note:* ± figures are standard deviation from mean of three replications.

[a] Nitrate method.
[b] ND = not determined.

**Table 15**
**Correlation Coefficients ("r" values) for the Dependence of Essential Element Concentrations of the Plants on the Composition of Fly Ash-Amended "Homemade" Composts**

| Element | Corn (Zea mays L.) | Sorghum (Sorghum vulgare L.) |
|---------|--------------------|-------------------------------|
| K | -0.005 | 0.726 |
| Ca | 0.702 | 0.209 |
| Mg | 0.266 | -0.086 |
| Zn | 0.182 | NM[a] |
| B | 0.608 | 0.607 |
| P | 0.507 | 0.792 |
| S | 0.700 | 0.909 |

[a] NM = Not measured

(1) if plant element contents of the fly ash-amended composts were related to plant yield and (2) if there was any relation between the elemental content of the plant and its concentration in the fly ash-amended compost. The linear equations corresponding to these plots and the "r" values were calculated (Figures 4 and 5). There were positive correlations in all cases except B, where yield decreased with an increase in B content of the compost.

Correlation coefficients for the dependence of the element concentrations in corn and sorghum plants on the percent of fly ash in fly ash-amended composts ("Homemade") were calculated (Table 15). Positive and significant correlations existed for K in sorghum, Ca in corn, and B, P, and S in corn and sorghum. It appears that positive and meaningful correlations depend on the nature of the plant. Some elements in fly ash were directly or indirectly assimilated by plants.

## SUMMARY AND CONCLUSIONS

Outlet fly ash recovered in the dust chamber of a power plant equipped with electrostatic precipitator(s) was usually very fine. On the other hand, fly ash collected from power plants not equipped with this device was "coarse" with particle size ranging from 183 μm to 500 μm. Concentration of water-extractable elements in ashes with different particle sizes varied. Fly ash samples collected from plants with a precipitator, as is the case with SRS 484-D and SCE&G power plants, were generally neutral or basic, while others seem to be acidic depending on the type of coal burned in the power plant.

Release and availability of elements in fly ash-amended composts depended on the type of organic compost used for amendment. A chemical interaction between fly ash and organic compost may have occurred during composting. Aging of the organic compost probably released elements to the system, creating high salinity that may be toxic for plants. "Homemade" compost, aged for 1 or 2 months, was beneficial when amended with fly ash because it released elements needed for plant growth.

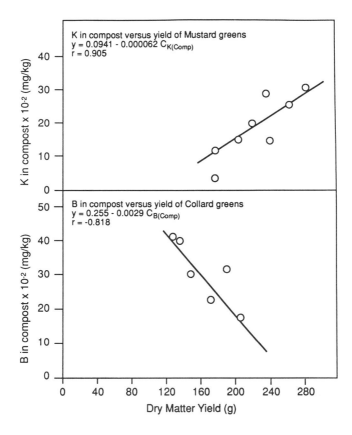

**FIGURE 4.**    Dry matter yield of collard and mustard green vs. the element content of fly ash-amended compost.

Water-extractable elements present in coal ashes removed at various stages at the SRS 484-D power plant exhibited marked differences in concentrations, indicating the relative efficiencies of the precipitator for the removal of elements from crude coal during combustion. The emission control devices installed in this plant were efficient in removing elements in the precipitated ash. Weathered ash and ash from the lagoon had a relatively high concentration of macroelements, such as Ca, Mg, and S, probably from the presence of carbonate and/or metal oxides and $CaSO_4$. Potassium was leached more readily than any other elements from weathered ash and ash from the lagoon. Studies on the downward transport of elements through column experiments showed that the availability of elements in soil for plant growth was limited to a depth of about 80 cm from the top and the fine ash from SRS 484-D had larger water retention capacity than the "coarse" ash from SRS 184-K power plant.

Fly ash-amended "Homemade" compost generally improved plant growth and enhanced yield. Collard greens and mustard greens (Brassica crops) benefited

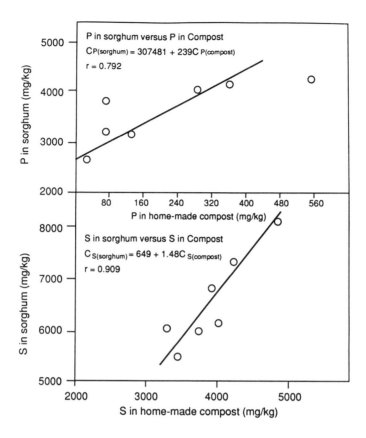

**FIGURE 5.** Correlation between elements in plant and elements in fly ash-amended compost.

more than the other plants by treatment with fly ash-amended compost. Yields and element uptake by these plants were higher than control when compared with corn and sorghum. The fly ash-amended compost gave poor yields for string beans, bell peppers, and eggplants. There may be some potential in using fly ash-amended composts for certain agricultural crops.

## ACKNOWLEDGMENTS

This work was made possible by a research grant from the Department of Energy, No. DE-FG09-88SR18047. The manuscript preparation was supported by contract DE-AC09-765R00-819 between the U.S. Department of Energy and the University of Georgia's Savannah River Ecology Laboratory. The authors express their gratitude to Dr. Robert F. Keefer for his helpful suggestions, review and comments in the preparation of this manuscript. The authors are also grateful to Dr. Domy C. Adriano for his constant encouragement, helpful suggestions, and help to carry out this research. Laboratory assistance rendered by several undergraduate students is also greatly appreciated.

## REFERENCES

1. Gordon, R. L., The hobgoblin of coal: policy and regulatory uncertainties, *Science,* 200, 153, 1978.
2. Adriano, D. C., Page, A. L., Elseewi, A. A., Chang, A. C., and Straughan, I. R., Utilization and disposal of fly ash and other coal residues in terrestrial ecosystems, *J. Environ. Qual.,* 9, 333, 1980.
3. Anon., Savannah River Laboratory Report, Part II, E. I. duPont de Nemours and Company, Wilmington, DE, 1986, 59.
4. Klein, D. H., Andren, A. W., and Bolton, N. E., Trace element discharges from coal combustion for power generation, *Water Air Soil Poll.,* 5, 71, 1975.
5. Bolch, W. E., Jr., Solid waste and trace element impact, in *Coal burning issues,* Green, A. E., Ed., University of Florida Press, Gainesville, 1980, 231.
6. Furr, A. K., Parkinson, T. F., Hinrichs, R. A., van Campen, D. R., Bache, C. A., Gutenmann, W. H., St. John, L. E., Jr., Pakkala, I. S., and Lisk, D. J., National survey of elements and radiochemistry in fly ashes and absorption of elements by cabbage grown in fly ash-soil mixtures, *Environ. Sci. Technol.,* 11, 1194, 1977.
7. Capp, J. P. and Spencer, J. D., A Summary of Applications and Technology, Bureau of Mines Inf. Circ. 8483, U.S. Department of Interior, Washington, D.C., 1970, 1.
8. Cope, F., The development of soil from an industrial waste ash, in Trans. Int. *Soc. Soil Sci., Comm. IV and V,* Palmerston North, New Zealand, 1962, 859.
9. Bern, J., Residue from power generation: processing, recycling, and dispersal, in *Land Application of Waste Materials,* Soil Conservation Society of America, Ankeny, IA, 1976, 226.
10. Natusch, D. F. S., Barer, C. F., Matusiewicz, H., Evans, C. A., Baker, J., Loh, A., Linton, R. W., and Hopke, P. K., Characterization of trace elements in fly ash, in *Proc. Int. Conf. Heavy Metals in the Environment,* 1975, 553.
11. Page, A. L., Elseewi, A. A., and Straughan I. R., Physical and chemical properties of fly ash from coal-fired power plants with reference to environmental impact, *Residue Rev.,* 71, 103, 1979.
12. Plank, C. O. and Martens, D. C., Boron availability as influenced by application of fly ash to soil, *Soil Sci. Soc. Am. Proc.,* 38, 974, 1974.
13. Page, A. L., Bingham, F. T., Lund, L. J., Bradford, G. R., and Elseewi, A. A., *Consequences of Trace Elements Enrichment of Soils and Vegetation from the Combustion of Fuels Used in Power Generation,* Research and Development Series 77-RD-39, Southern California Edison Company, Rosemead, CA, 1977, 225.
14. Abernathy, R. F., Spectrochemical Analysis of Coal Ash for Trace Elements, Investigations 7281, U.S. Department of Interior, Washington, D.C., 1969.
15. Brown, H. R. and Swaine, D. J., Inorganic constituents of Australian coals. I. Nature and mode of occurrence, *J. Inst. Fuel,* 37, 422, 1964.
16. Clark, M. C. and Swaine, D. J., *Trace Elements in Coal. I. New South Wales coals,* Commonwealth Scientific and Industrial Research Organization, Adelaide, Australia, 1962.

17. Elseewi, A. A., Page, A. L., and Grimm, S. R., Chemical characterization of fly ash aqueous systems, *J. Environ. Qual.,* 9, 424, 1980.
18. Phung, H. T., Lund, L. J., Page, A. L., and Bradford, G. R., Trace elements in fly ash and their release in water and treated soils, *J. Environ. Qual.,* 8, 171, 1979.
19. Block, C. and Dams, R., Study of fly ash emission during combustion of coal, *Environ. Sci. Technol.,* 10, 1011, 1976.
20. Menon, M. P., Ghuman, G. S., James, J., Chandra, K., and Adriano, D. C., Physico-chemical characterization of water extracts of different coal fly ashes and fly ash-amended composts, *Water Air Soil Poll.,* 50, 343, 1990.
21. Klein, D. H., Anders, A. W., Caster, J. A., Emery, J. F., Feldman, C., Fulkerson, W., Lyon, W. A., Ogle, J. C., Talmi, Y., Van Hook, R. I., and Bolton, N., Pathways of thirty-seven trace elements through coal-fired power plants, *Environ. Sci. Technol.,* 9, 973, 1975.
22. Davison, R. L., Natusch, D. F., Wallace, J. F., and Evance, Jr., C. A., Trace elements in fly ash. Dependence of concentration on particle size, *Environ. Sci. Technol.,* 8, 1107, 1974.
23. Martens, D. C., Availability of plant nutrients in fly ash, *Compost Sci.,* 12, 15, 1971.
24. Plank, C. O. and Martens, D. C., Boron availability as influenced by application of fly ash to soil, *Soil Sci. Soc. Am. Proc.,* 38, 974, 1974.
25. Salter, P. J., Webb, D. S., and Williams, J. B., Effects of pulverized fuel ash on the moisture characteristics of coarse-textured soils and on crop yields, *J. Agric. Sci.,* 77, 53, 1971.
26. Stone, D. A., Effects of fine-particle amendments on the soil moisture characteristics of a gravelly sand loam and on crop yields, *J. Agric. Sci.,* 81, 303, 1973.
27. Martens, D. C. and Beahm, B. R., Chemical effects on plant growth of fly ash incorporation into soil, in *Environmental Chemistry and Cycling Process,* Adriano, D. C. and Brisbin, I. L., Eds., U.S. Department of Commerce, Springfield, VA, 1978, 637.
28. Ciravolo, T. G. and Adriano, D. C., Utilization of coal ash by crops under green-house conditions, in *Ecology and Coal Resource Development,* Wali, M., Ed., Pergamon Press, Elmsford, NY, 1980, 958.
29. Furr, A. K., Kelly, W. C., Bache, C. A., Gutenman, W. H., and Lisk, D. J., Multielement uptake by vegetables and millet grown in pots on fly ash amended soil, *J. Agric. Food Chem.,* 24, 885, 1976.
30. Martens, D. C., Schnappinger, M. G., Jr., Doran, J. W., and Mulford, F. R., Fly Ash as a Fertilizer, Inf. Circ. 8488, U.S. Bureau of Mines, Washington, D.C., 1970, 310.
31. Hodgson, D. R. and Holiday, R., The agronomic properties of pulverized fuel ash, *Chem. Ind. (London),* 785, 1966.
32. Elseewi, A. A., Bingham, F. T., and Page, A. L., Availability of sulfur in fly ash to plants, *J. Environ. Qual.,* 7, 69, 1978.
33. Doran, J. W. and Martens, D. C., Molybdenum availability as influenced by application of fly ash to soil, *J. Environ. Qual.,* 1, 186, 1972.
34. Townsend, W. N. and Gillham, E. W. F., Pulverized fuel ash as a medium for plant growth, in *The Ecology and Resource Degradation and Renewal,* Chadwick, M. J. and Goodman, G. T., Eds., Blackwell Scientific, Oxford, 1975, 287.
35. Cherry, D. and Guthrie, R. K., The uptake of chemical elements from coal ash and settling basin effluent by primary producers. Relation between concentrations in ash deposits and tissues of grasses growing on the ash, *Sci. Total Env.,* 13, 27, 1979.

36. Fail, J. L., Jr. and Wochok, Z. S., Soybean growth on fly ash-amended strip mine spoils, *Plant Soil,* 48, 472, 1977.

37. Adriano, D. C., Page, A. L., Elseewi, A. A., and Chang, A. C., Cadmium availability to sudan grass grown on soils amended with sewage sludge and fly ash, *J. Environ. Qual.,* 11, 197, 1982.

38. Menon, M. P., Ghuman, G. S., James, J., and Chandra, K., Effects of coal fly ash-amended composts on the yield and elemental uptake by plants, *J. Environ. Sci. Health,* 27, 1127, 1992.

39. Menon, M. P., Ghuman, G. S., and Obie Emeh, C., Trace element release from estuarine sediments of south mosquito lagoon near Kennedy Space Center, *Water Air Soil Poll.,* 12, 295, 1979.

40. John, M. K., Chuah, H. H., and Neufeld, J. H., Application of improved Azomethine-H method to the determination of boron in soils, *Anal. Lett.,* 8, 559, 1975.

41. Environmental Monitoring Support Laboratory, Methods for Chemical Analysis of Water and Wastes. Methods, 365.2 and 375.4, U.S. Environmental Protection Agency, Cincinnati, 1979.

42. Goodwin, T. W. and Mercer, E. I., *Introduction to Plant Biochemistry,* Pergamon Press, Oxford, 1983, 328.

43. Mengal, K. and Kirkby, E. A., *Principles of Plant Nutrition,* International Potash Institute, Bern, Switzerland, 1978.

# Index